BUILDING PRODUCT MODELS:

Computer Environments Supporting Design and Construction

CHARLES M. EASTMAN

CRC Press

Boca Raton London New York Washington, D.C.

Library of Congress Cataloging-in-Publication Data

Eastman, Charles M.
    Building product models : computer environments supporting design
and construction / by Charles M. Eastman.
        p. cm.
    Includes bibliographical references and index.
    ISBN 0-8493-0259-5 (alk. paper)
    1. Computer–aided design. 2. Building materials—Computer
simulation. 3. Architectural design—Computer simulation.
I. Title.
TA643.E27 1999
720′.22—dc21
                                                        99-29602
                                                          CIP

No claim to original U.S. Government works
International Standard Book Number 0-8493-0259-5
Library of Congress Card Number 99-29602
Printed in the United States of America 1 2 3 4 5 6 7 8 9 0
Printed on acid-free paper

PREFACE

We are coming to the end of an epoch. Architecture, civil engineering and building construction have relied upon paper-based drawings as the primary working representation for over two millennia. However, within the next few decades, these practices will move to a new electronic, digital representation. The new digital building representation is called a *building product model*. A building product model is a digital information structure of the objects making up a building, capturing the form, behavior and relations of the parts and assemblies within the building. A building product model is potentially a richer representation than any set of drawings and can be implemented in multiple ways, including as an ASCII file or as a database. The data in the model will be created, manipulated, evaluated, reviewed and presented using computer applications, some of which are extensions of the current computer-based design and engineering tools. Others will be new capabilities enabled by this new information base. Traditional drawings will be one of the many reports that this new representation will generate.

Major efforts are being made throughout the world to develop such a representation. Among the industry groups involved in making this conversion are the US-based PDES (Product Data Exchange using STEP) organization and its international counterpart, ISO-STEP (International Standards Organization - Standard for The Exchange of Product model data), which have AEC (Architecture Engineering and Construction) subgroups. Other efforts are being undertaken by numerous research groups sponsored, for example, by the European Union in Europe and by the National Science Foundation in the US. Another significant effort is the IAI (International Alliance for Interoperability), an industry consortium initiated in the US and spread worldwide with over five hundred members and eight national chapters. At the same time, several of the leading CAD (computer-aided design) vendors are developing new product lines that endeavor to provide many of the same capabilities.

Building product models eventually will be used by most of the people associated with the building and real estate businesses: architects and engineers, contractors, owners and facility managers. Making the transition to a new representation that cuts across such a broad and informally organized sector of the economy will be a slow and potentially arduous undertaking. Later, this change is likely to open up important new possibilities for doing design, engineering and construction in more effective ways than are now practical. Possible examples include detail planning of construction while still doing design, or the use of CAD/CAM in construction. Other possibilities include analysis of a buildings operation and physical plant, undertaken remotely over the Web. In the longer run, building product models is likely to affect how most building organizations do business, how the industry is regulated and how its professionals are trained.

While there have been many prototypes, only a few building product models now exist. And the current technology is not be suitable for many needs. Those building models that have been developed thus far have been specified by consortia of users and vendors. The range of technical expertise—both regarding computer and information technology, and also the expertise in various domains of building—are beyond the capabilities of any one organization. That is why it is generally agreed that a standard is the needed solution, to which all organizations, large and small, can direct their efforts.

It is fairly certain that the transition to building product models will occur. Currently, the building industry is the only major industry that produces 3D products that is NOT using some form of product model. While the information technologies used in the aerospace, automotive, manufacturing and electronics industries that have made the transition ahead of the building industry offer insights, those particular solutions do not fit the particular needs of the building industry, and some new strategies and methods are necessary.

This book was written to provide the first thorough presentation of concepts, technology, and methods now being used to develop building product models. The book has two main goals: (1) to provide the field of building product modeling with a firmer foundation, based on a standard reference. Thus it is expected to be useful to industry and research people working in the area; (2) to provide a single comprehensive text for university courses in building product models. The book is organized into three sections: history; current tools, concepts and results; and new issues and research. It reviews the concepts, supporting technologies, and the resulting standards and demonstration projects in a form allowing full review of the work in building product models up to the current time. It brings together many diverse efforts and puts them in a common, easily read and understood framework. It incorporates numerous examples supported by over 150 figures.

Up to now, all the efforts in building modeling have been reported in research journals and conference proceedings, or recorded in minutes of standards meetings, or made available as draft standards on the Internet. The only books available in this area are published conference proceedings, most of which are cited here and listed in the bibliography at the end of the book. The book brings together the essential aspects of this work, from its beginnings in the early 1970s up through the present time.

The book is written to be appropriate for use in graduate level courses in the US, Europe and Asia, for classes in Civil Engineering, Architecture and Building Construction that deal with building product models or information technology in construction. To address this broad target, past and current work in both the US and Europe are reviewed and analyzed. Examples are drawn from a variety of domains covering the building lifecycle.

Students taking the course and other readers are assumed to have some knowledge of current computer-aided design systems and some familiarity with computer software development. The book reviews concepts of programming and software engineering, as needed to deal with the technical issues regarding building product models, but it is not a book about programming or software development techniques. Although the material covered has been truncated in many places, it is still too large to address well within an academic quarter. Semester-length courses will have to pick and choose what to address. The materials can be covered well as a two-quarter course. A useful augmentation to the materials covered here is the availability of a suite of EXPRESS tools, allowing students to develop and parse models, or to develop simple interfaces. A list of vendors of such products is offered in an Appendix.

Building product modeling is an active research and development area. New developments emerge frequently, like most areas of technological innovation, which will date some of the work included here. At the same time, I have attempted to cover the foundation of the area in such a way that readers, faculty and students will be able to add to the materials here easily, without rejecting what has been covered. This volume surveys much technical work carried out by many individuals. There are bound to be errors. If you find any, I invite you to let me know. Another edition may eventually be produced, allowing me another chance to improve and extend what is written.

I could organize and present the breadth of material covered here only with the help of many of the researchers and developers who have made fundamental contributions to the field. I am particularly indebted to people who have taken busy time out to review parts of the book describing the work they manage or with which they are deeply involved. They have helped to make the material much clearer and more accurate, though the errors that remain are this author's. People who have reviewed parts of the book include Godfried Augenbroe, Bo-Christer Bjork, Martin Fischer, Ed Hoskins, Paul Richens, Richard See, and Alastair Watson.

I also had much help in preparing the manuscript. Samia Rab and David Craig added greatly to its clarity and helped produce many of the figures. Saif El-Haq helped with some figures. I am also deeply indebted to my wife Mary Claire, who has provided crucial help in resolving issues of writing style as they arose.

Last, I am deeply indebted to my colleagues at Georgia Tech, especially those who are working on some of the issues addressed here. I am indebted to Fried Augenboe, whose broad experience and insight into the field will certainly contribute to future advances. I also thank Tay Shang Jeng, a daily collaborator. While university faculty assume to teach most of their students (or think they do) they also learn with and from their best Ph.D. students. The creative insights offered in discussions with Tay Sheng, both in implementation issues and recently in the processes of design and design management, have changed my view of the role and subtlety of the processes we daily engage in when we collaborate together to produce any design or product.

The impetus for doing this book grew out of the 1995 CIB W-78 conference at Stanford University. During this event, which brought together from around the world many of the people involved in building product modeling, there was much discussion about the newest capabilities and concepts. After asking several impertinent questions to a panel of experts, I realized that most of us—myself included—were all addressing slightly different issues and making different assumptions about the field, what was important, and what should be considered "progress". We all had only a partial perspective of the field. Partly as an act of penitence for my own lack of knowledge, and partly as a needed base for future work, I began making notes for a book on building modeling. However, it was only in the winter of 1995-96 that I began in earnest. I planned to get the job done in a year. However, the efforts in the field were expanding, both through the IAI, CIMsteel and new EU efforts. In the end, it has taken almost three years.

A book such as this one will never "cover" a field. Rather it can only serve as—at best—a strong introduction that is balanced and that provides foundations, allowing readers to extend the work presented. These were my goals for what follows.

Charles M. Eastman
Atlanta, Georgia, USA
December, 1998

ABOUT THE AUTHOR

Charles M. Eastman

Charles Eastman is Professor in the Colleges of Architecture and Computing at Georgia Institute of Technology, Atlanta, where he leads a program in Design Computing. Prof. Eastman was one of the early researchers in architectural computer-aided design, developing base concepts as well as production systems for over thirty years. He was the lead designer of five CAD systems, two of which were commercial efforts. He has consulted for many industry groups, in construction and in CAD, aerospace and manufacturing. He has written more than 80 papers in the areas of CAD, engineering databases and geometric modeling.

After graduating from the University of California at Berkeley in Architecture and a short period teaching at the University of Wisconsin, he joined the faculty at Carnegie-Mellon University with joint appointments in Architecture and Computer Science. There, he rose from Assistant Professor to Professor. Later, he also joined the faculty of the School of Public Policy there, now known as the Heinz School. At Carnegie-Mellon, he directed several laboratories, including the Institute of Physical Planning, the Center for Building Sciences and CAD-Graphics Lab.

In 1982, he co-founded Formtek, a CAD and engineering data management startup, where he served as President and in other positions. In 1988, Formtek was sold to Lockheed Martin.

He went to the University of California at Los Angeles in the Graduate School of Architecture and Urban Planning and he became the Director of the Center for Design and Computation at UCLA. In 1996, he joined the faculty at Georgia Tech.

Professor Eastman was founder and first President of ACADIA, the Association for Computer-Aided Design in Architecture, and is a representative in the International Alliance for Interoperability and ISO-STEP. He currently holds editorial positions on five journals: *Research in Engineering Design*, *Automation in Construction*, *Computer-Aided Design*, *Design Studies* and the electronic journal *Information Technology in Construction*.

TABLE OF CONTENTS

PART ONE:

The Context and History of Building Models

INTRODUCTION

Part One surveys the general context and background of processes and technologies now applied to building representations throughout a building's lifecycle. It includes a survey of computer technology, both software and hardware, primarily reviewing the evolution of CAD and similar applications. It does so to provide a history of the previous efforts to those reviewed later. The purpose of Part One is to identify for the reader a shared background and set of issues that can serve as a framework for the later chapters.

In Chapter One, we review the current procedures and information used in the five lifecycle phases of a building: Feasibility, Design, Construction Planning, Construction, Facility Management and Operation. The term "phase" is used consistently to refer to these large units within the lifecycle. Within each phase, there are a variety of stages, activities and tasks, each of which have information and communication needs. These processes define many issues regarding information flows and content. It is emphasized repeatedly throughout this volume that information content and processes are interdependent; if processes change, the information the new processes will need will be different than the old processes. The chapter winds up with a discussion of the differences between the building industry and other major industries, such as manufacturing and aerospace. This section identifies some of the economic and structural contexts that make information technology applications in building different from those in other business sectors.

Chapter Two examines the past and current tools used to generate and represent information about buildings. It focuses on computer-aided design systems (CAD) and the impressive technological advances that have supported CAD system development. Alternative conceptualizations and uses of CAD systems are examined. The chapter also examines some early efforts at building modeling that are precursors to the current efforts. Several systems in the UK and the US were developed in the 1970s and early 1980s that had similar ambitions to the goal of current generation intelligent CAD systems—development of an integrated environment to support design and construction. The system developed many of the concepts and some of the structures relied on today.

As early building-oriented CAD and information technology systems were developed, the need for data exchange between applications became apparent. The development of early, proprietary exchange formats such as SIF and DXF, and the formal standards such as IGES and SET, were responses to this need. These early exchange formats and standards are reviewed in Chapter Three. Based on the experience of using these exchange capabilities to their fullest, the need for a new generation of more powerful exchange technology became apparent.

CHAPTER ONE
The Context of Design and Building

Arrangement includes the putting of things in their proper places and the elegance of effect which is due to adjustments appropriate for the character of the work. Its forms of expression are these: groundplan, elevation and perspective... . All three come of reflexion and invention. Reflexion is careful and laborious thought, and watchful attention directed to the agreeable effect of one's plan. Invention, on the other hand, is the solving of intricate problems and the discovery of new principles by means of brilliancy and versatility.

Vitruvius
Book One, Chapter 2
<u>*The Ten Books of Architecture*</u>

1.1 INTRODUCTION

Electronic computation is impacting all fields of human endeavor. The automatic teller machines (ATMs) and electronic commerce practices used in banking, the automatic inventory management used in grocery stores, and the CAT scanners and other medical diagnostic devices used in hospitals and clinics, for example, all rely on computer technology. These developments—which some call a revolution—highlight the role of information technology in most areas of work. This book is about the impact of computers and other information technologies on one area in particular, the building industry.

Although we may think otherwise, the building industry, which includes architecture, civil engineering and construction, has been less affected by computers up to now than many other fields. Even though there have been many computing applications in design and construction, they have had only a minor impact on professional practices and methods of operation. In other areas, digital technologies have led to major re-organizations regarding how people structure and carry out their work and how they communicate. The building industry can be expected to receive similar changes and impacts to those in other fields, resulting from the future availability of almost free computer cycles and instantaneous communication of huge volumes of information. At present, however, only a few of the new structures, methods of operation and practices are readily apparent.

This book addresses a critical aspect of these impending changes—the digital representation of buildings and information about buildings. Information technologies that apply to all commerce and business can be adopted by the building industry—such as email, digitizing and electronic distribution of documents, use of CD-ROM and other new digital media, scheduling and cost estimating software, graphical presentation software, and even software for immersive three-dimensional displays. However, the organization and structuring of building product information and the processes operating on that information cannot be adopted from elsewhere but must be generated by the building industry itself. This book addresses the background, history, concepts and technologies involved in developing such electronic representations. This volume is directed toward organizing the information and concepts from which effective approaches to digital representation of building information can evolve.

This initial chapter reviews the context for computation within the building industry. In any country or region, the building industry is embedded in a complex web of relationships that determine procedures and practices within the industry. Changes to these contextual relations will change the needed information technologies. Four types of context are identified below and shown diagrammatically in Figure 1.1.

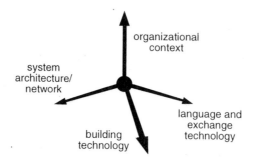

Figure 1.1: Four different contextual dimensions that influence the form of information technology in the building industry.

1. The most obvious type of context is *organizational.* Different organizational structures will have different information exchange and integration needs. A traditional architectural or engineering firm that teams with other consultants to form a virtual organization for a single project will have integration needs different from a large integrated construction firm that has internal design, engineering and construction divisions. As the organizational structure of a firm evolves, or as the structure of the building industry itself changes so, too, will the information technology need to change.

2. Another type of context deals with *software technology.* New languages, function libraries, exchange standards and other tools become available that facilitate new methods of integration and influence what kind of integration may be attempted. Recently, Java has been touted as a radical new software technology for distributing operation and processes. It is an example of how software technologies may alter the possibilities for the general use of information technologies in the building industry.

3. A third kind of context is also technological, dealing with the *system architecture.* Historically, time-shared computers provided an early context for integration, with multiple users sharing a single processor. It allowed the first chat programs and email. A client-server system and network provided a different context. The World Wide Web is perhaps the most recent context affecting information technology in the building industry. New system architectures will likely make different forms of integration easier or more difficult.

4. The last type of context is the kind of *building technology* being modeled. If a firm works with many different construction technologies and also many different building types, its integration needs will take one form. A firm that designs a single building type like auto parking structures will have its own integration needs. A firm that utilizes numerical control machines and prefabrication will have yet still other integration needs.

In this and several following chapters, we will consider each type of context of information technology in the building industry, some more deeply than others. In the rest of this chapter, we consider the current organizational context. Chapter Two reviews the history of hardware and software information technologies leading up to our current context, especially those supporting computer-aided design. Chapters Three, Five and Six survey the current software technologies supporting data exchange and integration. Chapters Seven, Eight and Nine examine current standards efforts. A heavy emphasis throughout this book is given to various software technologies supporting integrated representation of building information. In Chapter Ten, we review alternative system architectures and their implications for integration. In Chapter Eleven, we return to software technology issues and survey current research issues in that area. Although we will not consider construction technologies in detail, the reader is encouraged at various points to relate them to other issues raised. Throughout, we also develop examples from the spectrum of construction technologies.

Current building industry practices are highly varied, depending upon organizational types and other conventions specific to the country of practice. However, in order to introduce some aspects of the organizational context and the information involved, this chapter offers a general description of the processes that operate on building information throughout a building's lifecycle. Each building lifecycle phase is described according to its general information processing activities and the types of data used. The description is drawn from the traditional building industries of the US and Europe. It is not presented as an ideal model (or even as necessarily accurate), but rather as a basis for critique and comparison. Thus readers are encouraged to identify differences between the organizations and processes given and those encountered in different countries or particular business organizations.

The end of this chapter discusses some distinctions between the building industry and other industries, such as manufacturing and aerospace. These differences highlight why information technology developed in other industries cannot be readily adapted to the building industry. Given these two perspectives—of the building lifecycle, and the particular qualities of building that distinguish it from other industries—we begin to identify the current context for information technology within the building industry. We also can begin to identify some of the impediments to realizing the benefits of information technology in architecture and construction.

1.2 CURRENT PRACTICES IN BUILDING

Current practices in architecture and building are the result of centuries of incremental adaptation and change. They reflect the gradual addition of many new construction materials and methods, the evolving recognition of public safety with respect to building design, and changing societal mores and customs. Many practices vary from region to region and country to country, but over the long term, effective practices tend to be adopted widely while inferior practices become obsolete (though often quite slowly).

Here, we consider the building lifecycle generically. There are different ways to name and partition the phases of a building's lifetime. We adapt our classification from one proposed by Gielingh, who structured the lifecycle according to its major transition points. Our interest is not so much the transitions, but the periods of time between them, which we refer to as phases. A phase designates a temporal period before a transition and suggests the primary activities during that period. We use a lifecycle classification consisting of these six phases, as shown in Figure 1.2: *feasibility, design, construction planning, construction, operation* and *demolition planning*.

Each phase involves many activities that are becoming increasingly reliant on computer-based operations. From the 1950s onward, a growing number of applications have been developed that attempt to facilitate the tasks and processes within the building lifecycle. These range from simple drafting or scheduling tools to more complex applications for automatic detailing or fabrication of parts, to automatic monitoring of building plant operations. Other complex applications include material procurement and tracking applications for use during construction and equipment simulation programs for assessing equipment operation strategies during a building's operation.

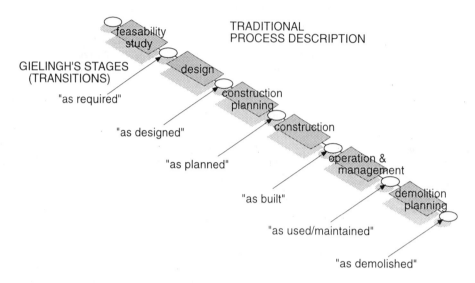

Figure 1.2: One classification of the building lifecycle, addressing both phases and transitions.

A serious impediment to the use of multiple applications has been the information preparation required to apply them. Almost all applications require information that defines the building and/or its context. However, there is often no simple way to generate such information other than manual entry. Even when the information already exists electronically, there may be no way to automatically transfer it to the application. It is not the use of the application itself, in the sense of its user interface or the operations it performs (which are separate issues), but rather the practical integration of the application into a working process. Integration requires access and incorporation of appropriate data, interpretation of results and possibly iterative use and exchange with other members of the building team. The problems of data exchange and integration are significant problems in the building industry and are a major focus of this book.

In order to develop an understanding of the processes that representations of buildings must support, we review those processes currently carried out in the various phases of the building lifecycle. In the following subsections, each phase included in Figure 1.2 is described in some detail. Attention is given to the processes now followed, the support provided by computer-based applications, the representations relied on, the actors and organizations, and the information involved. Each process is diagrammed schematically, using a simple representation showing data enclosed in solid boxes, processes in dashed boxes and inputs and outputs without boxes.

1.2.1 FEASIBILITY

Before a building project is initiated, a feasibility study is usually undertaken. The feasibility study matches the goals of a proposed building project against resources and identifies those special issues requiring response. The feasibility study may be generated by a real estate investment specialist or corporate planner. Alternatively, a family may plan what quantity and quality of housing it can afford. Feasibility typically relies on a quantitative representation of building information. Applications use an accounting package or spreadsheet to model sizes, cash flows and other financial aspects, possibly in conjunction with multiple secondary representations.

The high-level goals of a project involve the quantity or size of different types of space to be built, usually defined as counts or areas. The total space needed is typically derived by estimating major use spaces and functions and then using those to estimate supporting spaces, such as those required for circulation, mechanical systems and utilities. The major resource required for a project is money, expressed either as a total amount or as a cash flow. The desired amount of space compared with the costs of construction identifies the resources required, which is usually constrained. The site for the project often constrains possible building geometries.

The building function may require special study. It may be elaborated to address how the building should respond to special user populations (a different culture, people with special needs) or specialized activities (special manufacturing facilities, needs of a unique organization). Studies of the intended use of space, the activities within them and the interactions between groups may be required. These studies have several names, such as building programming or building operational analysis. Various techniques have been developed to collect such material and organize it to facilitate design.

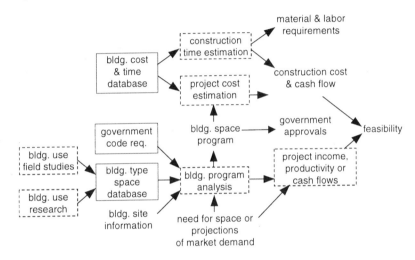

Figure 1.3: A typical feasibility process model of information sources and flows.

The feasibility study involves several processes, presented in Figure 1.3. The central process is to derive the project cost based on space quantities and construction costs. One set of inputs relies on building use studies, government codes and site information to derive a building space program. Another is to adjust standard building cost and time information for the particular context. Major inputs include the typical time and costs associated with the construction of different types of space: needed contracts, licenses and other fees, and costs of capital to cover construction. Construction costs and time are based on a specified level of construction. Construction time may also be an issue, if cash flows are important. If the project involves sales or rentals, the demand and market absorption rate for the types of space planned may be an input. If a commercial venture, it considers the revenues generated after the building is completed, based on market projections, rental or lease prices or the value of activities within the building. It may be necessary to estimate building operating costs: utilities, staffing, maintenance and taxes. This estimate relies on good information of building costs and construction times. It is premised on mortgage and cash flow inputs.

Other aspects of feasibility deal with various approvals from government units: zoning, transportation impacts, and, increasingly, environmental impacts of a proposed project. Projects may be impacted by building code requirements, addressing access, waste requirements, parking and transportation, and safety regarding natural events (fires, storms, wind). If the necessary approvals cannot be obtained (possibly later in the building lifecycle), the project cannot proceed.

THE BUILDING PROJECT FEASIBILITY PHASE: APPLICATIONS	TYPE OF DATA	PERFORMANCE
building program analysis: - total units, rental or usable space in terms of functional service provided - special populations or cultural studies	building quantities and qualities	units of function
project income, productivity or cash flows: - market absorption rate or model	building quantities	functional units per time unit
- rentals, leases and other income	money	income
- operating costs: amortization, utility and other operating costs, taxes	money	building operating costs
- production process simulation	time, equipment	time, costs
project cost estimation: - project costs: design, construction, license and bonds	money	building process costs
- material costs	money	costs
- labor costs	money	costs
- cash flows	money	money per unit time
construction time estimation: - project schedules, from conception to operation	activities and time	building process
- other time-based models of planning, design and construction	activities and time	times
government approvals: - zoning, transportation and environmental impacts analysis	building function, in terms of people, autos, noise, etc.	units of function
material and labor availability	units of labor and material over time	construction units per unit time

Figure 1.4: Potentially useful applications, given current practices, in the building feasibility phase. The type of data relied on and their behavior of interest are also shown.

During the feasibility studies, a number of trial development scenarios are often generated, exploring the relations between the size and scope of a candidate project with times to completion of phases and a cash flow and income analysis. A range of objectives may be involved, including income, depreciation and operating costs. The number of decisions and scale of a feasibility plan can vary greatly, from the economics of building a single family home, to a development involving hundreds of millions of dollars.

Some computer applications that support feasibility processes (the dashed boxes in Figure 1.3) are shown in Figure 1.4, along with the type of data they require and performances of interest. In the US and Europe, various organizations collect construction cost data, which are heavily relied upon for estimating the costs of new projects.

In large scale projects, other factors also must be considered. Feasibility addresses the adequacy of infrastructure for the project: access roads, utilities and power. If these are inadequate, then their expansion becomes part of the project and must be assessed in the costs. Very large projects may tax the ability to produce or acquire particular materials. For such projects, material quantities may be estimated at this phase, not just to determine their cost, but also their availability from regular sources. Computerized planning programs that provide estimates of time, cost, material and labor resources are already available.

The successful output of this phase is a building project specification that has been judged feasible according to the parameters estimated. These parameters include units of construction, the size of the project in terms of floor area, number of seats, or other appropriate unit. They are qualified by the quality of the space and desired special features. It will also include estimates of the cost, in terms of money, other resources, and possibly time. Many of the functional requirements used later in the design, and many of the cost and time figures associated with resources used as targets throughout the building's lifecycle, are derived during this phase.

1.2.2 DESIGN

Based on an agreed upon project scope, generated either intuitively or through a more careful feasibility study, design of the building or civil project is initiated. The fundamental task of design is to further specify the project in sufficient detail so as to allow its construction and to verify that it realizes the intentions of the client (a person or organization).

Current organizational practices for building design are varied. Primary responsibility will often be with an architect, but it can also be with a civil engineer, contractor or project management group. In some states in the US, it is mandated that certain building types be designed by a licensed architect. Regardless of who takes the lead, a range of other professionals quickly gets involved. A project may involve a structural engineer, a soils and foundation engineer for earthwork, a mechanical engineer, an acoustician, a fire safety expert, and other specialists or consultants who address particular functional issues arising from the nature of the project. These specialists may be employed by the design firm or hired as consultants by the architect or construction manager. Big city firms traditionally have had distinct advantages because of direct access to various specialists, but recently this advantage has lessened due to electronic communication that allows design teams from different parts of the world to join for a singular project.

The process of architecture is taught tacitly, through sequences of required submissions and by example. In studio exercises different processes are conveyed in these ways by different faculty with the knowledge that "process determines product". Architects are expected to develop a process that is attuned to their own type of work. At the same time, business practices lead to office conventions in architectural firms that will allow it to be efficient at what it does. Our focus here is on those architectural business practices.

The American Institute of Architects (AIA), similar to most other national professional bodies, outlines seven standard stages of design and offers a corresponding recommendation for billing of fees. Large users of architectural services, such as the US Army Corps of Engineers, have similar lists. All stages include meetings and presentations, reviews of client information, approvals required from agencies and public bodies, and iterative building cost estimates. The stage-specific activities and typical responsibilities and coverage include

1. **predesign:** determination of space requirements, spatial relations, operating functions, future expansion or flexibility objectives; survey of existing facilities; marketing and economic feasibility studies; identification of context, neighborhood issues and environmental requirements; preparation of general project schedules and budget and finance plans. (This phase overlaps the feasibility phase.)

2. **site analysis:** site analysis and selection; site development planning; site utilization studies; circulation and open space requirements; geo-technical studies; planning and zoning assistance; utility and services studies; environmental studies

3. **schematic design:** site plan preparation showing building footprint and general landscape planning; paving and drainage; floorplan layouts that satisfy space requirements in program; equipment and furniture layout; sections and elevations showing vertical dimensions, candidate materials and finishes; structural system types and building sections; solar design and conservation studies; mechanical equipment spaces and duct and chase spaces; fire protection requirements; electrical, fire, security, power and communication system requirements and types

4. **design development:** grading, utility, paving, demolition and landscaping plans; detailing of floorplans; roof plans, typical wall sections, and typical details; 3D models or drawings for review; room finish schedule; furniture and equipment selections, including materials and finishes; interior elevations with material selection; structural system plan including framing and foundation plans with member sizes; outline of structural specifications; mechanical equipment plan schematics and schedules and approximate sizing; acoustical, vibration and visual impact study; approximate plumbing diagrams and plans; electrical and lighting plans; communication equipment plans; fire equipment plans; outline of material specifications

5. **construction documents:** detailed plans for demolition, utility, grading, site, and paving with profiles; final calculations of all drainage, landscape includes plant identification, paths, walls, strips irrigation plan; includes architectural door, window, finish schedules, roof and ceiling plans, all equipment layouts; includes window, door framing, equipment and other details; all interior drawings and final specifications; all structural plans and sections; schedules; final member sizing and joint detailing calculations; foundation plan with test locations and results; all control and expansion joint details; define all clearances; show all notches and penetrations; add control systems, seismic anchorage, all points of connection; locate all pipe routing and controls with final sizes and calculations; equipment schedules with specifications; all power sizing calculations, sizing and installation details; conductor sizing; define circuits, panel boards; preparation of bidding documents

6. **bidding and negotiations:** organization of bidding conferences; review of alternatives/ substitutions; review of bids; bid evaluation

7. **construction contract administration:** coordination with other design/project members; construction observation and representation; preparation of supplemental documents; change order review; shop drawing review; record of payments; project close-out

This list has been amalgamated from several sources, including the AIA Code of Practice and the US Army Corps of Engineers submittal requirements. The Corps of Engineers recommend a payment schedule of 15% for schematic design, 30% for design development, and 55% for construction documents (ignoring the other stages). These fees cover all consultants, unless the scope of the project is such that additional consultants are agreed upon up front and paid by the client.

Because of the complexity of design processes, we consider some of the stages in detail. **Predesign** consists of those design services needed to complete a feasibility study, if one has not been undertaken. These tasks were covered in the previous section.

As the billing schedule indicates, architectural design begins with **schematic design**. The schematic design stage produces the general design concept for the building project that will be refined in the later stages. Design concept means that the definition of the building's general form with interior spaces is laid out and activities allocated to them. The general inputs are shown in Figure 1.5.

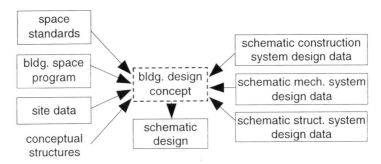

Figure 1.5: Some of the inputs to the conceptual design process.

Within this stage, the designers map into built form the intentions of the project as specified in the feasibility study. The largely quantitative description developed earlier is translated into a geometric form. Spaces are laid out in schematic floorplans, responding to site and orientation issues and relative access between spaces. The floors are composed together into an overall building structure, responding to circulation, structural requirements, and mechanical equipment issues. Preliminary calculations are made to make sure all systems can perform as required. The building mass is enclosed within a shell, giving the building its form. This form and its interior spaces are composed within the site, responding to contextual issues. The character and qualities of all major spaces are partially determined. This stage is the most creative one for architects and is the one emphasized in architectural schools.

Each of these aspects requires both expertise and extensive supporting data. The specific project data includes the building's space program that is drawn from the feasibility study and also site data. A wide range of general information is applied to the schematic design data, including space standards and schematic design information regarding building technologies and systems. Also included are conceptual organizing structures, based on a particular style, building module or other organizing principles.

This stage may be undertaken by a single, very experienced designer, possibly with advice from others, or by a team composed of experts who contribute their various intellectual viewpoints. Special issues are identified, such as those pertaining to site conditions or special functions. The schematic design is fully reviewed and approved before the design proceeds further.

Today, only a single drawing may be required for the schematic design for a house, while twenty-five to fifty drawings may be needed to explain the various aspects of a larger project. Drawings might include the building's program of spaces, as realized by the schematic design (possibly in comparison with the feasibility program), floorplans, elevations and critical building sections.

Design development refines the schematic design by defining all construction materials and the general details associated with them. Design development also defines all mechanical, structural and other systems and services. The sizes of major elements are determined, ranging from those that make up wall and floor systems, to structural, mechanical, communication and other system elements. Special issues—for example, those pertaining to structural, acoustical and vibrational performance—are identified and resolved.

As one of the oldest fields of human endeavor, building construction has available a tremendously rich range of construction technologies, from concrete frame structures to prestressed shells, to stone or rubble infill to curtain walls, to catenary "tent" structures. Panel enclosures may be framed in place to be permanent or constructed as modular prefabricated units to allow easy replacement. A wide range of mechanical systems can be utilized, using some combination of water, steam or air as the heating or cooling medium. This richness allows widely different combinations of technologies to be applied to a building's design. Different technologies have different components, different rules of composition, and often different methods of performance analysis. This palette of technologies is the medium for the architectural art. For growing numbers of such technologies, computer applications exist that support interactive or automatic design. These applications incorporate the special components, design rules and methods of analysis for the associated building technology. The applications mostly target the design development and contract drawing stages of design.

Similar to the richness of construction methods, a broad range of building types and special-purpose spaces exist and must be identified and composed into a spatial layout. Building type knowledge is derived from articulation of the activities enclosed in the type and from analysis of previous instances, for example, of schools, hospitals, and housing. Many books survey particular building types and are the current means for conveying this knowledge to architects and users. They consider the proper design of spaces, in terms of area requirements, environmental conditions, appropriate circulation and access, the quality of spaces, as well as their overall composition in terms of circulation and other frameworks. In the future, computer applications can be expected to convey much of this knowledge. A few examples of such programs already exist today.

The combinatorial possibilities of all the different types of construction, building systems, and building uses are huge. The range is so great that it is not infrequent that a major building project will involve a combination of these issues never encountered in this combination before. A problem of digital building representations is to support this range of combinatorial possibilities.

At this stage, for all but the smallest project, there is a team of specialists working on the project. The design is a team effort, each member with his or her own realm of responsibility. Their responsibilities may be functional (e.g., structural or acoustic) or system-based (e.g., dealing with communications or curtainwalls).

The working out of all the various technical, performance, economic, aesthetic and other issues currently involves many different representations: scale drawings, piping and electrical diagrams, 3D wood or plastic models, various datasets for use in analytic methods, and so forth. Today, these representations are becoming computerized and integrated. Some of the types of computer applications that may be used, given current practices in design, are listed in Figure 1.6. Their use and integration into design processes are highly varied and only approximated in Figure 1.7. The different representations can be considered loosely both coupled and maintained separately, according to the distributed responsibilities of the design team.

Even though they are computerized, applying a computer application for designing or for analyzing the performance of a building is not straightforward. Data describing the building context and its properties must be prepared and presented to the computer in a given format. Defining the load assumptions and performance requirements for an application so as to get meaningful results requires significant expertise, involving, for example, assessment of loads, boundary conditions and simplifying assumptions. Because of these issues, analyses of thermal performance, structures or lighting are now used primarily to check the design for acceptability. Only occasionally are they used iteratively to evaluate alternative designs and to help select those that achieve higher levels of performance.

THE BUILDING DESIGN PHASE: APPLICATIONS	TYPE OF DATA
CAD system, defining geometric layout and materials	geometry, material properties
analyses of design in terms of:	
- structural safety	material performance units
- energy costs for heating	energy units
- vibration, other special performance dimensions	functional units
- construction costs	money
simulation models of building behavior:	
- mechanical system operation	energy units, comfort levels, climatic conditions
- elevators and transport systems	time, traffic flows
- lighting simulation	lighting units
- acoustic simulation	reverberation time, decibels
- people and traffic flows	human densities, speed
automatic and interactive design and detailing for:	
- standard components: stairs, roofs, particular products, special function spaces	geometry, materials, equipment and finishes
- standard detailing conditions	
- particular stylistic intentions	
expert system support, for example, advising on:	
- energy efficient design	geometry and materials,
- material and part selection	knowledge base of technical information in various areas
- operating and maintenance issues	
- construction guidance	
- water and moisture	
building code evaluation, for such issues as:	
- fire, structural, earthquake safety	geometry, use data, material data,
- access for handicapped	energy and movement
- habitability, fresh air and light	
site development, in terms of:	
- grading	site contours
- road, walkways, planters and landscaping	ground cover
- water and drainage systems	soil types
- wind simulation	wind conditions
site investigation:	
- soil and stone boring	location data, soil and geological coding
- geological studies	stability

Figure 1.6: The potential applications useful, given current practices, in the building design phase. The type of data used is also shown.

At the end of design development, all the building technologies and systems comprising the design have been defined and laid out. The spatial layout and dimensions of all spaces and significant building elements have been specified, including those required for various support systems. Materials specifications and critical procedures are defined in a separate document that goes with the drawings. Governmental codes have been reviewed and their requirements incorporated into the design. Testing of materials or assemblies (for wind, fire safety) has been undertaken. All work completed during this stage is reviewed and approved by the client.

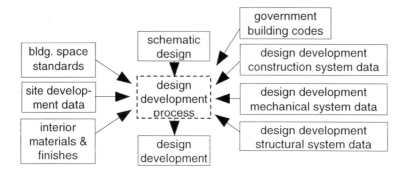

Figure 1.7: General description of the design development process.

The next stage of design is **construction documents**. During this stage, all details of the design are reviewed and refined to support acquisition and assembly of materials on the construction site. Issues of assembly are addressed, including design details of windows, doors and other premade units showing how they are integrated into the building. Expansion joints and other fixtures for weather and special site conditions are detailed. Provisions for demolition of existing site facilities and temporary facilities on the construction site are planned. Other detailed investigations of the construction site are carried out during this stage, including borings and geological investigations. Site clearance, transport of materials and equipment to the site, temporary site services, temporary construction, including public safety shields and scaffolding must be planned.

A recognized feature of design practice is the need to make changes to previous aspects of the design as the process proceeds. Iteration is often required, where earlier actions at one level of detail must be adjusted or changed when more detailed issues are addressed. Only in standardized design situations can effective design be done without backtracking or making changes. Another feature of design practice is the varied membership of design teams, sometimes led by an architect, sometimes by a construction manager, sometimes organized hierarchically, other times horizontally. These features—designing at multiple levels of detail, multidisciplinary teams, iterative design sequences, and varied team organizations—provide a context to which different information flow architectures must respond.

The processes associated with the construction documents stage are similar to those associated with design development, but operate at a finer level of detail. All processes beyond conceptual design involve iterative refinement. One characterization views it as a set of spiraling cones, as design information is created, mostly based on the actions of previous stages but occasionally based on changed strategies, as shown in Figure 1.8. Different domain specialists collaborate at one level of detail until an effective design is reached, then collaborate to develop the design at the next level of detail, and so forth. The level of detail corresponds to the design stage. In the latter stages, a large amount of design information is created, mostly based on the actions of previous stages, but occasionally based on changed strategies.

A complete specification of a building is impossible since a physical product has an infinite number of potential properties. Those parts of the design not specified, whether a paint color, the detailing of some joint, or the finishing of some ductwork, are left to the builder or building part fabricator to execute. Common law requires that a builder should work according to "standard practice". That is, the contractor is expected to follow the conventions of construction practice for all unspecified aspects of the design (a carry-over of previous craft practices). From the designer's viewpoint, quality control requires that the design specification be detailed such that those aspects of the product that are not specified will, if constructed using standard practice, realize the intentions of the design. Judgment is required regarding what standard practice is.

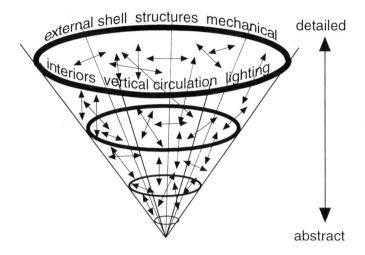

Figure 1.8: One common conceptualization of the design process.

A finished design must satisfy the needs of its different users:
- it must allow the designer to verify that the future product will satisfy his or her intentions, which incorporate and expand those of the client
- it must allow the client to be satisfied that it will realize the client's intentions, as a check upon the designer
- it must provide the builder with adequate information to procure all materials needed for the building and to construct it

The final design representation is currently required by law to be in the form of drawings, called *contract drawings,* plus other written documents specifying the quality of materials and equipment specifications. Today, a set of construction drawings may vary from a dozen or so for a residence to as many as a thousand for a large facility. Since the drawings are used as part of the contract between the client or the design team and a contractor, they have legal status within US courts of law. In the US, these documents are submitted to the regulating government institution, typically a city or county, for a *plan check*, which is a review of the design against the building code requirements. Plans are approved or returned for revision. Upon an acceptable plan check, a building permit is issued. In addition, all the important behaviors expected of the design are assumed to have been predicted, to the degree possible, and have been shown to respond satisfactorily (or modified or removed as an issue).

The other contracting stages of architectural services involve the contracting and construction supervision aspects of service, not design. These stages involve review and coordination with the building contractor and are discussed later.

In summary, design involves many varied issues. For information technology, there are four particular challenges:

- supporting the project-specific organization of design teams that involve both a range of expert knowledge and a varying set of team members
- supporting the widely varying combinatorial set of technical knowledge involved in building design
- coordinating and integrating the multiple levels of detail that are used in developing the design by the different specialists
- and supporting the different processes followed by different designers and the iterative nature of design decision-making

1.2.3 CONSTRUCTION PLANNING

The construction planning phase involves the bidding and tendering processes. It includes the development of a general construction plan, the estimation of construction costs and the submission of a competitive bid to win the construction contract. The bid is based on the contract documents produced during the design phase. This phase is an artifact of the current process of bidding for construction work. It is common practice in most countries and is required by most governments in support of open contracting. It is carried out by each of the various contractors and subcontractors bidding to construct the project.

Tendering involves different forms of contract, each requiring a different type of estimate. Most common is a fixed or *lump sum* bid, although a call for bid sometimes is for a *cost plus a fixed fee* or a *schedule of rates*. In the US, public buildings must be put out for tender to multiple bidders, thus preventing any chance of integrating detailed construction planning with design. Since bidding is competitive, involving multiple bidders with only one or a few winners, the process in this phase is limited to that necessary to generate sufficient information for making cost and time commitments for the project that the contractor is confident can be realized.

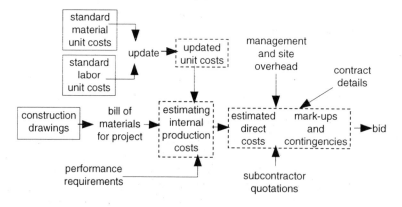

Figure 1.9: The general information flows for cost estimation of a building project, using the bill of materials estimating method.

Construction planning involves adding further detail to the contract documents produced during design so as to extract a detailed list of construction units to which material and labor costs can be associated. One of many possible information flows is characterized in Figure 1.9.

The starting point in cost estimation is the project's cost units and the estimated prices multiplied against them. Typically, a cost unit is a material piece of the construction, such as a wall or floor slab. Average prices per cost unit are supplied by construction industry information suppliers, although large contractors often maintain their own unit cost database. Many material placement operations involve a sequence of preparatory tasks, such as excavation or formwork, which must be included. Costs may be adjusted due to location, expected weather and other contingencies.

They also may be adjusted in response to performance requirements, such as deadlines or reviews, approvals or other conditions associated with particular pieces of work. From these are estimated internal production costs. Subcontractor quotations are added and augmented with the associated management supervision and site work required for the subcontractors. The sums are adjusted for contract details: risk, contingencies and profit to result in the final bid. Most commonly, cost estimates for each contract of work are made by a single person, each subcontractor working separately. On large projects, the project may be partitioned so estimators can work in parallel. Sometimes, company practice is to have two people generate estimates, allowing comparison and reducing risk of omissions or errors.

Some applications and the information used in them are shown in Figure 1.10. Quantities of cost units involve detailed measurements taken from the plans. Current work increasingly relies on digitizing techniques to read coordinates and compute distances, possibly multiplying them with other derived distances. Such data input tools can expedite the derivation of cost units. In the US, plan rooms are a service offered by different organizations, providing a central site for subcontractors to review plans for bidding. They provide some of the services offered by quantity surveyors in the UK.

THE CONSTRUCTION PLANNING PHASE:		
APPLICATIONS	TYPE OF DATA	PERFORMANCE
CAD system description, defining geometric layout and all materials	geometry, material properties	(state description)
construction task planning:		
- extracting in-place units of construction	material cost units	(state description)
- association of units of work and	material costs,	production costs
material costs w/ units of construction	time and work crews	production times
detail product specifications:		(state description)
- for outsourcing		performance specs.
regional resource planning for large projects:		
- production capacities for local materials	material or work units	production capacities
- availability of regional labor pools	over time	
- acquisition plans to deal with shortages		

Figure 1.10: Given current practice, the types of applications that are likely to be useful during the construction planning phase.

During this phase, the general contractor may determine what components of the design are to be outsourced (procured from outside vendors or contractors). Parts of the design are translated into product specifications, which are put out to bid to subcontractors. The tradeoff between outsourcing and local production of units of construction becomes interesting when large scale prefabrication is considered. The general contractor may receive multiple bids for building parts and for various subsystems or rely on a single source. Occasionally the subcontractors bid directly on the project, and the resulting team can be completely new to each other. More generally, the general contractor selects the subcontractors and produces the final integrated construction plan.

Bonding of work that insures the owner against mishap or failures must also be obtained. On certain very large-scale projects, certain materials or labor requirements may exceed local production or resource availability. These capacities must be identified and alternative procurement plans developed.

In summary, this phase involves detailing and augmenting the earlier design representations to incorporate materials, associated cost units, and units of time. Later, this estimated construction plan is used as the target for the actual construction.

Bid selection decisions vary. One is to select the lowest bid, though in some cases the "next-to-lowest" bid has been mandated. In non-public projects, other considerations apply, such as construction time, conditions involving incremental completion of parts of the project and differing strategies of site development. At the end of this phase, a single general contractor and set of subcontractors are selected. Various rules may be imposed in the selection process. It is only at this point that the design team is combined with the construction team, given current practice.

The practice of "blind bidding", where contractors have only a set of plans from which to make their estimates, emphasizes reliance on standard construction practices, so that contractors can reliably estimate costs without undue risks. Extensive non-standard design practices raise costs because of contractor unfamiliarity. Also, the boundary between construction planning, as described here, and construction, the next phase, is a vague one. When unconventional building practices are required, more detailed construction planning is required, which conventionally is only undertaken when the project is secured. The extra cost for preparing an unconventional bid is often finessed by adding a contingency cost to a bid, as a risk factor. The activities of this phase change dramatically if it is merged with design, as when contracts in the US are framed as "design-build".

1.2.4 CONSTRUCTION

The winning contractor(s) must now build the project within the constraints set by their bids. The major planning representation during construction is a construction schedule. The construction schedule is an entirely new representation of the building project, identifying construction sequences and assembly operations for each unit of the building. The schedule may be derived from the cost units determined in the construction planning phase. Alternatively, it may be based on new more detailed units of construction derived from the construction drawings. For each piece of construction and every aspect of the site conditions defined in the construction drawings, the contractor must identify the sequence of actions needed to realize the final conditions. Even the simple job of inserting a prehung door in an opening typically involves inserting and shimming the doorframe, hanging the door, installing hardware and finishing the door and frame. Many units of construction involve many steps and these sequences of operation, along with the manpower, equipment and materials they require, are essential aspects of the expertise of the contractor. Many of these sequences of action are informally but widely known and can be assumed in "standards of practice." For example, the construction of wood or metal stud interior walls, covering them with gypsum board and adding doors is so widely understood that the individual tasks are not usually broken down. Schedules for interior walls may be based on a linear measure of walls and number of doors. In other cases, a unique technology may be used, requiring careful planning and possibly coordination between architect and builder. Often materials are required for some supporting tasks, beyond those finally installed in the building. Examples include formwork for concrete and temporary supports for steel frames before they are braced.

The general information flows for developing a construction schedule are shown in Figure 1.11. As an abstract planning problem, development of a construction schedule requires first identifying the individual units of construction and then the sequences of tasks needed to achieve them. A partial ordering of the tasks exists that forces precedence among them. The ordering must be identified and applied both to tasks supporting a single unit of construction and across the units of construction. Other parts of the construction schedule include the necessary reviews and approvals, based on both client and/or building code requirements. In addition, the schedule must include the operations needed to erect temporary structures, to protect pedestrians or store materials or machinery, and to manage the construction on site. To this abstract task sequence, manpower, equipment, materials and other resources needed to complete each task must be identified. The estimates may have associated statistical distributions, resulting in probabilistic values for different variables, such as the time required for some task.

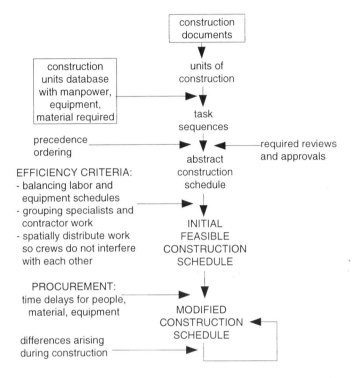

Figure 1.11: A process model for the development of a construction schedule.

A wide range of abstract schedules is technically possible. Different tasks can achieve the same units of construction; for example, a wall can be constructed off-site, transported to the site and installed, or, alternatively, can be constructed in-place. Different equipment may significantly vary the time and labor required to complete a task. Various choices and strategies of construction are embedded in the abstract schedule.

A variety of efficiency criteria are applied to the abstract schedule. These criteria are what distinguish cost-effective construction from those that simply complete the work. Some efficiency criteria that are applied to the schedule are

- balancing labor and equipment usage, so that a small fixed number can do the work
- grouping specialists and subcontractors so their travel time to the site and setup are reduced
- spatially distributing work, so work crews do not interfere with each other
- just-in-time scheduling materials and equipment so that it is available for tasks, but does not accumulate on the site beforehand, allowing damage or taking up space needed for other activities

Scheduling of a large construction project may involve up to a dozen or so managers, each responsible for planning an aspect of the project, based on global agreement of how the project will be partitioned and on general strategies. Some level of coordination is required to make sure no parts are left out, falling "in the gap" between two or more partitioned aspects of the project. After the abstract schedules are prepared, they must be integrated and reconciled, based on efficiency criteria. When separate portions of the overall schedule are generated by different people, merging may be undertaken by a senior person or through a collaborative problem-solving activity. Some construction scheduling software facilitates merging.

Applying the efficiency criteria leads to adjustments of the abstract schedule, resulting in an initial practical schedule. Such a schedule is not the end but rather the starting point for detailed construction planning and management. Either before or after merging, the procurement actions needed to acquire all materials/equipment/components necessary for each unit of construction must be defined to guarantee that the schedule can be realized. These procurement plans are often made in parallel and then merged. Efficiency criteria are applied to them before they are executed.

Each workperson or crew that is not already hired must be hired at this point; each piece of equipment not owned must be rented, leased or purchased; all the materials for each construction task must be acquired. Sometimes these procurements are accomplished easily, while others are complex. Alternative products may have to be evaluated, requiring access to multiple product specifications. Ideally, this information would include availability and delivery times. Some material purchases may require performance testing, review and acceptance by architect and/or client, and may require long lead times for fabrication and transportation to the site. In general, there is a sequence of procurement actions behind every unit of manpower, material or equipment—what are often called *resources*. The procurement actions typically involve ordering, tracking and delivery of the required resources. These sequences of procurement actions need to be identified so they can be closely tracked. Sometimes material acquisition is handled by a materials resource planning (MRP) application that deals with equipment orders, delivery dates and schedules. In other industries, real-time monitoring of equipment orders is becoming available, where fabrication work in a factory can be monitored by a customer, showing the exact status of the work on the factory floor. These systems provide a "virtual supply chain" model.

Construction requires a high degree of coordination among a wide set of fabricators providing materials to the site. The vagaries and delays of construction material deliveries are well known. Companies should demand higher levels of coordination from building component suppliers, including real-time access to the fabrication status of building components. Management systems are available to deal with these issues in other industries.

Most often, construction plans are developed for the complete building project. With growing frequency, however, they are broken into parts and developed separately. For example, the site work, auxiliary buildings, foundation, and different floors of a high-rise building may be planned, code approvals gained, contracted for and built separately. In some cases, this breaking up of the project is for political or economic reasons. Occasionally, it is done so that construction can begin before other aspects of the design are completed. In the US, this is called "fast-track" construction.

In order to fabricate the needed components and install them, the general contractor or subcontractors may generate new detailed drawings called *shop drawings*. Shop drawings provide detailed part, folding, welding, drilling and other data needed to fabricate a building element. Shop drawings are often required for curtainwall systems, sheetmetal ductwork, special cabinet work, and other custom fabricated elements used in the building fabric. These are reviewed and coordinated so that they meet the general specification of the construction drawings and coordination with other construction trades. Further decisions of design detail are imparted in these drawings, which may subtly change the design from that described in the more general contract drawings. In this way, other parties are involved in details of the final design, changing aspects sometimes in significant ways or applying their own interpretation to "standard practice".

Some of the computer applications now used in construction are shown in Figure 1.12. Most applications are coordinated through a scheduling program, which provides a common structure around which all procurement, task planning, equipment procurement and other activities are coordinated.

THE CONSTRUCTION PHASE: APPLICATIONS	TYPE OF DATA	PERFORMANCE
CAD system description, defining geometric layout and all materials, augmented by material quantities and units of labor	geometry, material properties, units of labor, units of material	(state description)
procurement:		
- PO procurement scheduling and tracking	POs, dates, actions	purchasing &
- inventory management		delivery times, money
detail construction task planning:	geometry, material	work rates
- task breakdown and sequencing	units of labor, units of time	
- heavy equipment leasing and/or scheduling	equipment, time	
- job scheduling, tracking and status reporting		
- work crew assignment	people, time	
- custom drawing for production crews	geometry, process plans, materials	work processes
shop drawings:		
- fabrication and assembly drawings	geometry, material properties	(state description)
surveying and geodesy for construction layout:	3D geometry	(state description)
- site surveying	geometry	
- construction layout & alignment	geometry	
- excavation drawings, material storage geometry		
temporary construction:		behavior of unstable
- scaffolding and shoring	geometry, materials	materials
- temporary structures	geometry, materials	
as-built documentation:	geometry, materials	(state description)

Figure 1.12: The possible applications used during the construction phase, given current practice.

The building is constructed by executing the construction schedule, allocating the people, equipment and materials for each task and properly instructing each crew regarding what they will do. Their work is coordinated by a foreman, using the contract or shop drawings as a guide. Conflicts are often encountered that require adjustments and changes. These range from variations in weather, to subterranean conditions that were not anticipated, to crew illness and accidents, to tasks taking different times than were allocated. Such conditions result in significant differences between the construction plan and actual construction activities. Significant changes may require alteration of the intent of the construction drawings and construction schedule.

Some changes to the construction schedule are the result of inadequacies or inconsistencies in the construction drawings, leading to design changes called *change orders*. Change orders can add significant costs to a project. Sometimes the costs are real; in other cases they are used by contractors to raise their profit, by adding charges to the client. The client usually has no recourse but to accept the additional cost of change orders, though they may be passed to the architects or other designers. As work proceeds, there is typically a requirement to summarize the expenses and percentage completion, for example, for partial payments.

As the construction schedule is executed, it is desirable to collect a variety of data. Some clients require the provision of a revised set of construction drawings, noting all changes from those provided by the architect. These are called *as-built drawings*. As-built drawings correspond to an as-delivered description of the building, useful for the later operation phases. Another issue is that the actual costs associated with the units of construction may be significantly different from those used to estimate the project. Collecting such information for later reuse is an important aspect of the corporate knowledge of the contractor. One process for doing this, as an adaptation of the process defined in Figure 1.11, is shown in Figure 1.13.

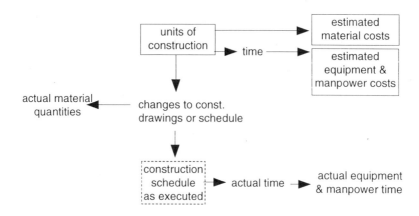

Figure 1.13: A process that can be used to capture the actual time and costs for units of construction.

Throughout construction, US practice has an on-site building inspector scrutinizing the work at critical times to see that it is being carried out according to the requirements of the building code, as approved in the plan check done at the end of design. Approval of the final on-site inspection results in an occupancy permit.

1.2.5 FACILITY OPERATION AND USE

What was a building project for architects and contractors is a *facility* for operations people. Quite often, a building project is planned as an addition to a campus of buildings or as part of some larger facility, such as a hospital, factory or college. In this latter case, a project-level description is merged into a more aggregate-level facility description. For building operations, it is this more aggregate description that is operated on and managed. The accretion of facilities through occasional projects is characterized in Figure 1.14, suggesting that in a large setting, there is a stock of facilities that is added to or modified by various projects over time.

As a building is operated, the issues concerning it fall within three general areas:

(1) *facilities management.* Facilities management involves the use of the building's space, its allocation and utilization as a resource. It deals with space ownership. It manages changes to spaces due to changing wall partitions, adding or blocking doors, etc. It coordinates changes in the allocation of people or activities to spaces. A database that records and tracks all of the entities and relations is required. In addition, space management usually tracks utility and telecommunications distribution and access, the location and inventory of furniture and equipment and other capital investments of the organization. Facility management is especially needed in large organizations, such as universities and hospitals.

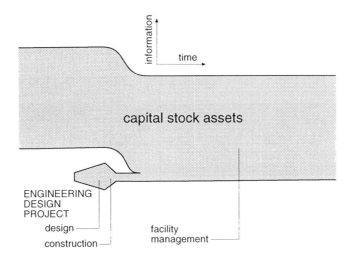

Figure 1.14: The flow of information defining the current building stock and the contribution of a new project.

The information needs for facility management include schematic floorplans for space allocation to units within the organization, electrical and telecommunications diagrams for tracking changes in these systems, and tabular data showing equipment allocations. Facilities management programs thus involve simple graphical support (in comparison to CAD) and fairly rich database support. In order to maintain a facility management database, a well-defined and managed reporting structure is required that captures changes as they occur. Thus, they are typically tied into building maintenance and remodeling procedures.

The base application supporting facility management is a database of building spaces that carries each space's properties and organizational allocation as well as the equipment located within them. Typically, the database is linked to a display or CAD system providing plans of the spaces and graphical querying. A possible information flow is shown in Figure 1.15. In most facilities, allocations are considered static, but in such facilities as school classrooms, and hospital rooms, allocations are made separately for each period of the day. A variety of reports can be generated, including lease data, space allocated per some unit function (such as average square foot per laboratory workstation), and general inventories.

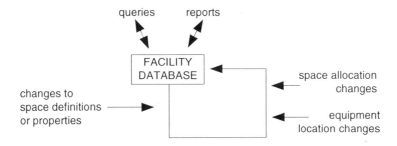

Figure 1.15: Information flows for maintaining and using a facility database.

In addition to the recording of actions taken, space management systems often involve decision support systems that can generate good or optimal space allocation plans, in terms of communication, material transport or other costs. Other applications evaluate these properties of the layout.

(2)   *mechanical equipment operation*. Mechanical equipment typically has operating cycles, such
      as the schedules and algorithms used over the course of the day by elevators, air-conditioning
      control systems, and other mechanical equipment installed within the building. This
      equipment and other parts of the building consume energy and other utilities at some rate that
      is usually metered. These data provide a basis for assessing the performance of the building
      as a complex machine and for assessing the operations of its mechanical equipment in terms
      of efficiency and possibly of quality of service. Such data are often collected and monitored
      as a means to better operate the dynamic behavior of the building.

Several systems in modern buildings have control systems—for fire protection, security,
elevator or vertical circulation operation, mechanical equipment operating cycles, exterior
lighting systems. The number of systems with electronic controls continues to grow and the
recent interest in "smart buildings" suggests that this growth will continue. Certainly for the
public utilities of electrical power, natural gas, fresh and waste water, it is possible to tune a
building's operation to significantly affect the quantities of these services consumed, as well
as the quality of the environment created within the building. Sustainability of natural
resources emphasizes conservation in the use of these services.

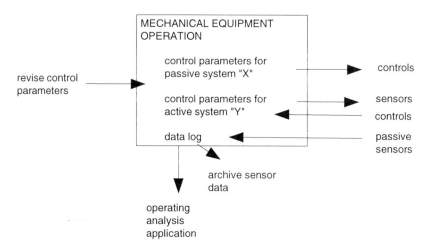

*Figure 1.16: A diagram showing information flows in a possible operation control
computer, managing different types of data collection and control functions.*

In modern buildings, mechanical equipment operation is increasingly controlled by one or
more computer facilities, with associated backup systems to guarantee 24-hour continuous
operation. Some possible functions are shown in Figure 1.16. Passive controls are assumed to
be implemented on a fixed schedule, with the only input being the computer's clock. Active
control systems rely on sensors, monitoring temperature, sunlight, rainfall, the location of
people, or other information useful in controlling the operation of the facility. Active sensor
data may be logged (that is, collected and stored), for possible later use in tuning the systems.
Other sensors may be used as monitors, a common example being the slow scan video
cameras used for security purposes. An extreme example of monitoring is the installation on
a few high-rise buildings of active anti-seismic lateral deflection equipment. Essentially these
consist of heavy weights near the top of the building that are moved in the opposite phase of
any sensed sway of a building. A recent well-known example is the Citicorp Building in New
York City. A computer and a maintenance person continuously monitor this equipment.

Currently, tuning of the system is based on extracting logged data and possibly running simulations of the facility to assess different operating parameters. Remote access through the Internet provides the opportunity for building operations to be monitored remotely, with possibly hundreds of buildings' mechanical systems being monitored, assessed and tuned from a central site.

(3) *building maintenance and repair.* Electrical and mechanical equipment have service cycles requiring scheduled maintenance. These typically include tensioning rubber drive belts, lubricating motors and chain drives and cleaning filters. Such equipment is located in obscure corners of a building and maintenance workers must know where they are. Another use of a facility management system is to support such maintenance operations, with equipment location and technical data encoded in the facility database. At some point, equipment breaks down and must be repaired or replaced. These actions both define a cycle of activities and also provide potentially useful data on serviceability. The information used in such activities are records of two types: a maintenance schedule of dates for servicing different parts of the building, and repair records documenting the frequencies and types of problems with building components.

THE BUILDING OPERATION PHASE: APPLICATIONS	TYPE OF DATA	PERFORMANCE
CAD system, defining geometric layout and all materials	geometry, material properties	(state description)
facilities management:	space IDs, attributes	space utilization
- allocation of space		circulation flows
- assignment of furniture and other movable items		
- assignment of telecommunications, utility services		communication flows to spaces
- changes to interiors and services		activities w/in spaces
maintenance data:		
- maintenance records of surfaces, weather-stripping, other fixed aspects of building	dates and actions taken	behavior of materials
- repair records for equipment		behavior of equipment
- assembly and material failures		behavior of water & air
mechanical equipment operations:	actions, time	
- real-time elect. monitoring of equip.		real-time behavior
- equipment specifications		
- simulating new operating schedules		
- operating instructions		
- recommended maintenance schedule		
layout of electrical, telecommun., other utilities	topology, controls	utility use
- connections and circuits		
archiving for later use		
- remodeling		(state description)
- demolition		

Figure 1.17: Some of the applications used during the building operations phase.

Materials in high wear areas also require maintenance, such as waxing and polishing. Records of such actions are often recorded, so as to better track total building expenses. In the longer cycle, these data provide information about the longevity and lifetime costs of flooring and other materials.

While there are several organizations that provide repair information for automobiles, only a few building operator associations provide maintenance information for such building components as escalators, elevators, bathroom flushing devices and other heavy use components of a building.

It is apparent from the summary of uses presented in Figure 1.17 that building operation involves recording large amounts of tabular textual data, often associated with the physical location of the material, space or equipment, or with operating data. The initial information for this phase comes from the as-built information, which is continuously augmented by building operation and maintenance information.

The last phase of a building model is in support of demolition. Demolition can be viewed as a special remodeling operation, likely to require as-built design information. It often involves recovery of expensive materials and finishes. Slowly, it is being recognized that building demolition is an area of importance to recycling, sustainability and landfill and waste issues. Eventually, designers and builders may have to plan how a building will be disassembled, as the automobile industry is beginning to do.

1.2.6 LARGER CYCLES IN THE LIFECYCLE

For many types of product generation, a single person, small group or single organization is responsible for most of the product's lifecycle, especially design and fabrication. Building is possibly the most extreme case of an opposite organization, one that is highly decentralized with many parties contributing to design and fabrication, and possibly with many different owners and users over the building's lifetime.

Because of a building's long lifetime, some failures only arise after the original designers or builders have moved on or passed away. Also, the long-term performance needs of an organization or society change, modifying or making obsolete the original design intentions. Thus feedback of successes or failures of a building are more difficult than in other product domains. Correspondingly, many aspects of the building industry are resistant to changes because the asserted outcomes have not been demonstrated to withstand "the test of time". An important aspect of building information management is to support and improve the feedback and learning cycle.

Methods for systematically documenting and evaluating building performance are not technologically complex. They are only difficult because of time and cost and the lack of any one party with a continuous construction investment to warrant their use. Most of these methods apply to facility management and involve relating failures to the design decisions that led to them. Sometimes, detective work is required—to determine the source of moisture in a wall or roof, for example. Large corporations, building manager and operator societies are the ones who have the most to gain professionally from such knowledge. One might hope to see such monitoring and reporting in the future, eventually using digital representations of buildings as a base.

Some of the dramatic changes in manufacturing and other areas of product design and realization have been to improve the feedback process by applying control system principles. From this viewpoint, the building lifecycle is largely an open-loop system, with decisions made early based on the best estimates of consequences occurring much later. The move to more closed-loop controls, where corrections are made frequently or continuously have led to spectacular improvements in quality, timeliness, and cost reduction in manufacturing. Closed-loop controls require well-developed measures of performance and frequent monitoring of those measures.

Control system concepts apply to all phases of the building lifecycle. They can apply to the feasibility phase, where the program for a facility can continuously evolve, possibly to the point that a building's space can adapt during use. The design stage can involve many types of prediction. However, traditional architecture practice does not emphasize nor do clients demand strong prediction and control processes in design. The issues of feedback and control in design range from energy efficiency and quality of space to the longevity and maintenance required of particular materials. Feedback and control in construction are already embedded to varying degrees within current practice. The updating of weekly construction schedules and short-term adjustments of crews are examples. Long lead times restrict short-term corrections of actions taken. However, delivery schedules to the site, productivity of equipment, and scheduling of work are all tasks that are highly amenable to closed-loop controls. Mechanical equipment operation is typically a closed-loop control. Other systems, including lighting, security and elevator service are now being considered for closed-loop or adaptive control.

The application of control system concepts to buildings requires regular data collection and analysis. The development of better building product information and record keeping may facilitate improved building planning, design, construction and maintenance. A challenge for all members of the construction industry is to reduce the negative effects of fragmentation now existing in the design, production and use cycle of buildings.

1.2.7 SUMMARY

A review of the above descriptions of the different building lifecycle phases makes it apparent that the information used in each phase has a common thread—each phase uses the general spatial layout of the building. Layout drawings are the main archive used throughout the building lifecycle today. The various elaborations of the building layout vary greatly, ranging from rental/income or other financial data, to design abstractions developed by architects, to detailed shop drawings during construction and building operations data. How to redefine the core building description in a digital format so that it can support the various forms of elaboration over the building's lifetime is a basic question we will explore throughout this volume.

1.3 RELATION OF BUILDING TO OTHER ENGINEERING FIELDS

Architecture and construction have many similarities with other forms of engineering and manufacturing. Until the middle nineteenth century, architecture was the most complex, most resource and technology intensive and most time consuming of all design and fabrication activities. Today, however, this is no longer the case. The cost of designing a commercial aircraft or modern microprocessor is in the hundreds of millions of dollars, more than any building. Many types of design are as complex, resource intensive and time consuming as architecture. In this electronic age, building construction is viewed as "low-tech". However, there are several aspects of building that pose unique issues, not encountered to the same degree in other product areas. These differences make the issues and solutions of information technology in building different than those in other industries.

1.3.1 FIXEDNESS OF THE FABRICATION TECHNOLOGY

One of the differences between engineering fields is the prespecifiability of the fabrication technology. Process plants—for refining oil or making fertilizer—are largely made up of piping connecting a small number of processes. Piping systems have an easily predefined small set of parts from which they are fabricated. Certain kinds of precast concrete structures have a similarly well-defined kit-of-parts, consisting of panel, beam, column and wall types and a small number of joint types. In electronics, the basic units are a small set of components and transistors, which are combined into large hierarchical structures to carry out the different functions of, say, a modern computer. In each of these fields, it is fairly straightforward to identify the set of components (or parts) and the attributes needed to address their physical form, possible composition, function and performance. We can say these areas involve "kit-of-parts" design.

On the other hand, modern buildings in most parts of the world can be designed to have a very great range of possibly applicable construction technologies. An architect is free to use a variety of structural systems, mechanical systems, external cladding, interior partitioning, glazing, and so forth. Each one of these systems has a set of primitive units, with relevant attributes, which are composed according to a set of rules and joining conditions. However, these different systems can be mixed in almost infinite combination. Many are economically competitive.

In an engineering field with such a rich set of fabrication technologies, it is difficult to identify *a priori* the technologies that will be utilized in a particular project. It is also impractical and undesirable to load all the possible fabrication technologies and all their parts and attributes into a single CAD system or building model. Such generality comes at a high cost, involving irrelevant objects, attributes and options. Such a design or construction planning tool would be huge and cumbersome. It is more practical to add information about different technologies as a design proceeds. This implies an extensible computer-aided building design system, allowing the addition of construction technologies and other information as design or construction proceeds.

Most areas of manufacturing also have had very stable fabrication technologies, but this is changing, with new forms of metal forming and rapid prototyping and new types of plastics and composite materials. A richness of technology is being created in manufacturing similar to that found in the building industry. Hence, if the construction industry resolves how to deal with its range of fabrication technologies, it is likely that similar solutions will be adopted in certain areas of manufacturing.

1.3.2 RELATIVE COST OF DESIGN

At one end of the design cost spectrum is computer chip design, where a company might spend hundreds of millions of dollars to design a new processor chip that will cost only several hundred dollars when mass produced. The ratio of design cost to product cost reaches a million-to-one. Of course, the company's investment is returned through quantity; ten or hundreds of millions of chips are expected to be sold. In the middle of the spectrum are consumer products—cameras, copiers, video equipment. These sell in the thousands or few millions, but cost only a few hundred dollars. Their design cost is typically a few million. Here the ratio of design cost to product cost is on the order of a hundred-to-one. At the other extreme, we have buildings. The quantity for this product is something less than one (many designs are never built). The product cost can range from several hundred thousands of dollars to hundreds of millions, similar to the costs of a commercial aircraft. However, the design cost for a building is almost always less than ten percent of the construction cost. One would think that the very low investment in design would result in both more expensive buildings in terms of construction and operating costs (probably true) and very conservative "cookie-cutter" designs with little variation (which is not generally true). From this perspective, it is surprising that buildings can be designed that are interesting and sometimes innovative. When compared to other design fields, it is truly impressive that an architect can undertake the design of a building for a tenth or less of its construction cost.

These cost distinctions, however, make a huge difference regarding the level of planning, analysis and exploration of alternatives that a design team can practically undertake. Building practice becomes more easily understood when we consider the steel, wood product, and concrete research institutes that carry out large pre-engineering studies of products, facilitating their movement into "standard practice". At the same time, large-scale construction is conservative regarding details, relying on those that have been shown to be effective in other projects.

1.3.3 NO TERMINATION OF DESIGN AND REDESIGN

In contrast to most products, building design is of such a scale and complexity, and serves such complex functions, that as needs change, it is usually easier to remodel the building (change its design) rather than move to a different building. Few of us consider redesigning a car to make a coupe into a sedan as the family grows. But we often add a room in a house for that purpose. The point is that most manufactured products have a specific fixed design that is not expected to change. In contrast, one expects that a building will be remodeled, in some cases almost continuously, over its lifetime.

This difference means that certain designs that are more adaptable may be superior to others that provide more specialized amenities. It suggests that most buildings need to be highly adaptable, and may perform better in the long run if they can adapt to different uses and conditions. Except in certain cases, such as temporary construction, the idea of planned obsolescence is not meaningful to buildings. In building, the design goes with the product, allowing it to be re-engineered and refurbished for possibly a thousand years. No other type of large-scale product has the same range of adaptive use.

1.3.4 DISPERSION OF THE INDUSTRY

The construction industry is made up of many small competitors, rather than a few large companies. For any given building project, a unique set of contractors, subcontractors and design professionals are likely to form the project team. With a new set of actors for each project, it is difficult to deviate much from the standard modes of communication and planning. This is one of the reasons that computers and electronic communication have been rather slow to become integrated into building construction. This structure is in marked contrast to other product areas, which have varying degrees of vertical integration and where a small number of material suppliers, part fabricators, product manufacturers and distributors all work together.

While most other fields whose products are three-dimensional have converted to 3D modeling and work conventionally using 3D CAD, that transition is just beginning in architecture. Of all the design fields dealing with 3D products, only the building industry remains largely 2D and drawing based. The reliance on 2D representations is not because paper-based representations are better. It is because there are no mechanisms to initiate changes of such wide impact. In other industries, organizations develop long-term working relations that allow special forms of communication, specifically for exchanging CAD data. Because the building industry is composed of many small organizations, many call the construction industry fragmented. This suggests that the organizations best able to take advantage of CAD may be those somewhat vertically integrated, such as design-build organizations.

Current forms of contracting also greatly limit the consideration that can be given within any building lifecycle phase to issues arising in other lifecycle phases. For example, construction issues are not fully studied during design, while operational issues are not fully considered during design or construction. These limitations restrict the building industry's implementation of improvements that have been regularly applied to other design fields. There have been limited efforts to package projects and services in new ways, treating architecture as if it was a large product, like a truck or airplane, sometimes called a "turn-key" project. In these cases a team will provide a building as a "design-build" package, in some cases even managing the project for a period of time. Such efforts are attempting to capture the efficiencies of concurrent engineering, as more regularly achieved in other design fields. Such practices, however, are currently rare, and face many legal as well as practical problems.

1.3.5 ECONOMIC STRUCTURE OF CONSTRUCTION

An important aspect of construction in the US and to varying degrees in other countries is that it is relatively easy to enter, both in terms of training and in terms of capitalization. A person can move from being a carpenter or concrete layer to a large-scale contractor by hard work and good business sense. Construction in the US is an important industry providing social mobility. In most countries, there is no equivalent in the construction industry to a General Motors, or an AT&T, with major capital investments, portfolios of patents of special technologies and trade secrets. Probably the closest example is the "big five" construction companies in Japan: Obayashi, Kajima, Shimizu, Taisei and Kumagai Gumi. These firms are among the few that have large-scale information technology departments and that research and develop future technologies.

The aspects of the building industry that make it good for easy entry and social mobility also make it slow to address major changes. Low capitalization means that investments in technology are low. There are no central players who have strong leverage for realizing change even if they find economic benefits in doing so. Contractors find it hard to explore or innovate in areas involving automation when the data they receive from architects do not support it. At the same time, architects cannot move from drawings to a database representation of buildings if the bank loan executive, the contractor and the subcontractors have their business experience set around drawings. And none of these parties has the capitalization to justify research or development efforts to work out any problems in making such a transition.

Each party in the building industry—the architects, engineers, building inspectors and other governmental regulators, bankers and loan providers, and subcontractors—all have been trained to rely on paper-based representations of buildings to carry out their current work. All must change in unison, yet there is no one to coordinate such a change. Such a climate makes change in the construction industry different from technological change elsewhere.

1.4 THE COMPUTATIONAL CHALLENGE FOR THE BUILDING INDUSTRY

The current situation has been the result of a long evolution from Vitruvius, quoted at the beginning of the chapter, to the current time. Innovations have been adopted that are both organizational—such as the recent development of fast-track scheduling—and also technical—such as the use of finite element modeling and the electronic storing and transmission of drawings. However, all the changes have been incremental, adopted first in special cases, then slowly absorbed into the wider practices of the industry as a whole.

The potential benefits of computer use in construction are suggested by the applications listed as examples in each phase of the lifecycle. Certainly, there are many more yet to be developed, some of which are surveyed in later chapters. However, the limitations that make using such applications difficult are not likely to be resolved quickly. The limitations are both organizational and technical.

The organizational limitations are those associated with the characteristics of the construction industry—its dispersion and low capitalization. These make innovations more difficult, especially innovations as fundamental as the representations used in broad segments of the industry. However, as the individual parts of the construction industry become more automated, the financial benefits of making changes to the information provided throughout the lifecycle will become more apparent. For example, as the building product manufacturers—the curtainwall and store front fabricators, the interior paneling systems fabricators, the exterior stonecutters, the precast concrete fabricators—become more automated, they will begin generating their own intelligent 3D computer models. As architects develop such models for their own design and communication uses, but without passing them on, it will become less of a big step to provide them to the contractors and subcontractors. Such data exchange is happening now in a few special cases. Another scenario involves the procurement of construction products while design is still proceeding, as needed in fast-track scheduling. This requires ever earlier exchange of data between design and contractor. These are only two possible scenarios that can benefit from computerization. Such opportunities will form in parallel in small incremental steps, and eventually major changes will be realized, affecting all parties in the building industry.

The immediate technological challenge is to develop a digital representation of a building project that can first be used for feasibility, then design, then fabrication, and then operations and maintenance, and will enable all parties to do their work more effectively than now. The representation must facilitate use of computer-based tools and digital communication. It has been generally agreed by all people involved in computers in construction that it is necessary to move beyond drawing representations in order to define more complete geometrical representations with integrated material and performance property data—what has been called a building model. The challenge is to specify the structure, content and use of building models and the technologies needed to support them. Reports from this representation should include drawings, as used today, to support an incremental transition. Major questions arise about the development of such technology. How will such models be defined and communicated among both people and computers? How will applications be supported by the building model, especially in terms of the issues associated with the large range of technologies and building types? Will there be one building model, possibly with variations over the building lifetime, or many? These are among the fundamental questions regarding building models and they will be addressed in later chapters.

1.5 SUMMARY

This chapter has outlined the context that exists today in which the various members of the construction industry operate. The introduction of new technology into the building industry is not difficult, if it applies to only one kind of operator. However, technologies that would affect them all—such as a change in the way a building is represented—are extremely difficult and changes will inevitably be slow.

1.6 NOTES AND FURTHER READING

Throughout this volume, the terms computation and information technology (IT) are used almost interchangeably. IT is a term widely used in Europe, but is used much less in the US, while computation is used here in the US to refer to all aspects of information processing. The vocabulary used in discussing design is also important. If one uses the term "design problem," it suggests that design is a problem-solving activity and has a well-structured context and knowledge with which to work. If we talk about a "design task," then it suggests an actor or operationally oriented process. If we use the term "design decisions," it emphasizes that design is a decision-making process, rather than a sequence of actions that are confirmed later. In general, the terminology used here emphasizes "design task" over "design problem" to emphasize the operational nature of design, and to use the terminology of actions rather than decisions.

An early influential report on building data integration was authored by Wim Gielingh [1988a, 1988b]. He laid out both the lifecycle used here and also some of the dimensions on information management that we will address in later chapters.

A feasibility study for a building project takes different forms, depending upon the building type. Here, the economic aspect is taken as the common kernel. There are few thorough resources dealing with building project feasibility. A reference dealing with the financial aspects is Barrett and Blair [1982]. Also, the *Appraisal Journal* has occasional papers on the subject. An example of a useful building feasibility software system, particularly oriented towards institutions such as schools and hospitals, is developed and sold by SARA Systems, Kansas City, MO. When special populations of users are involved, behavioral issues become significant aspects of programming, undertaken during the feasibility stage. There are a number of good texts on the behavioral aspects of feasibility—building programming—for example, by Duerk [1993], Marti [1981], Preiser [1985], and Alexander, Ishikawa and Silverstein [1977]. In some buildings, functional aspects dominate. A prototype software system emphasizing functional criteria is part of the SEED architectural system [Akin, Sen, Donia and Zhang, 1995]. A commercial product is the Alberti System, developed by Nemetschek Systems, Inc.

While there are many books describing architectural design, few address it from an information processing standpoint. One accessible survey is Rowe's *Design Thinking* [1987]. Eastman's early chapter on the use of computer applications provides another perspective [1975a, Chapter 1]. The standard contractual stages for design services are given in AIA [1994]. A survey of recent application types can be found in Mitchell and McCullough's *Digital Design Media* [1991]. Galle [1995] summarizes the current state of the issues involved in building data integration. The current structure of architectural practice was reviewed recently by Cuff [1991] and earlier by Gutman [1988]. The changing structure of practice, resulting in part from information technology, is analyzed by Tombasi [1997].

Technical information about building types is widely available in any well-stocked architectural library. Some example sources include the books by Sherwood [1978], Panero and Zelnik [1979] and De Chiara and Callender [1990]. An early example of a digital reference of design material and cases is described in Zimring and Ataman [1994]. Research on computer applications addressing the environmental performance of buildings is presented in *Building Simulation '97* [Spitler and Hensen, 1997] and in others in this series. A study of building model support for code checking is de Waard [1992].

In the US, R. S. Means and F. W. Dodge provide relative construction cost information to contractors. Bidding practices and strategies are reviewed in Harris and McCaffer [1995, Chapter 11], which deals with units of cost methods. See also Adrian [1973, Chapter 3]. An overview of IT aspects of cost estimating is given in Myllymaki [1998]. An alternative process model for construction planning is presented by Froese, Grobler and Yu [1998]. Partitioning a large building project into separate, biddable projects was a financing process innovated in the US by William Zeckendorf in the 1960s (who incidently gave the architect I.M. Pei some of his earliest projects). For more, see Zeckendorf [1970]. Williams [1989] provides an interesting account of the construction planning for a contemporary prefabricated building in her book on Norman Foster's Hongkong Bank.

Modern construction management is reviewed by Adrian [1973] and more recently by Levy [1994]. The new generation of resource planning tools are called "enterprise resource planning" (ERP) and are presented in many IT professional journals [Caldwell and Stein, 1998].

Coordination among suppliers in construction is considered by Miyagawa [1997] and Sanvido [1984]. Recent capabilities supporting "virtual supply chains" are discussed in Stein and Sweat [1998].

An example of a construction application that would benefit from a building model is described in Alferes and Seireg [1996]. An information study integrating design and construction is Alshawi and Underwood [1996]. Another example that supports new construction technology is Tanijiri et al. [1997].

A survey of facility management is presented by Hales [1985]. Space planning techniques for efficient layout of factories, hospitals or offices have a long history of research and application. See Cinar [1975] and Liggett [1989].

An overview of the large-scale Japanese construction companies is given in Levy [1993]. The organization of the construction industry in the UK is reviewed in Harvey and Ashworth [1993]. An economic view of the construction industry is presented in Lange and Mills [1979]. Some design/build strategies for building are presented in Booth [1992].

Some of the innovative applications and new software developed for building construction can be seen in the various trade magazines for steel fabrication, concrete fabrication, interiors, facility planning, and so forth. In architecture, the work of architect Frank Gehry and his associate, James Glymph, is notable for their application of computer-aided fabrication. The new Guggenheim Museum in Bilbao, Spain is a spectacular example. Their work has been presented in Menges [1998].

The case for making the next generation of standard architectural representation 3-dimensional has been put forward by many authors. For an early paper, see Eastman [1975b]. More recently, see Galle [1995].

A good introduction to control system concepts that can be applied to the building process is Rasmussen, Pejterson and Goodstein [1994]. See also Alting [1994].

1.7 STUDY QUESTIONS

1. Consider certain building industry innovations, selected by you or taken from the following list. Discuss how each innovation might be introduced and become standard practice, given the structure of the building industry, as presented in Section 1.3. Identify which members of the building industry are affected and how they would have to change. Possible innovations: composite material structural elements, a new form of insulating glass that has a higher first cost, provision of microwave network communication within buildings, internet communication of building product information, internet distribution of building designs for bidding.

2. There are many variations in practices within the building industry, within regions, countries and even within different firms. Meet with and interview a local building industry firm—real estate company, architect, contractor—and learn their procedures. Compare their procedures to those presented in Section 1.2 of this chapter. Note how they are different and explore why these practices take the form they do.

3. Given the information flows and tasks presented in Section 1.2, sketch out a computer application that you think is not now on the market but would enhance current capabilities.

4. Identify how the information flows presented in this chapter might change in a design-build organization. Discuss how these changes can benefit the functioning of the firm and also the client.

5. Write a short essay identifying how the structure of operations and processes in the building lifecycle process might be changed in a beneficial way, beyond or different from design-build. Identify who would benefit, who would be hurt by the change, and why the benefits would outweigh the losses.

6. Take any one phase of the building lifecycle and identify performance measurements for that process that could be used for iterative application control system concepts. Identify how this information may be collected and its implication for a building model.

CHAPTER TWO
The Evolution of
Computer Models in Building

We envisioned even then the designer seated at a console, drawing a sketch of his proposed device on the screen of an oscilloscope tube with a "light pen", modifying his sketch at will, and commanding the computer slave to refine the sketch into a perfect drawing, to perform various numerical analyses... . In some cases the human operator might initiate an optimization procedure... . The different powers of man and machine are complementary powers, cross-fertilizing powers, mutually reinforcing powers.

Steven Coons
"An Outline for the Requirements for a Computer-Aided Design System"
AFIPS Conference
1963

2.1 INTRODUCTION

Credit for developing the world's first interactive computer-aided design (CAD) system belongs to Ivan Sutherland, who in 1963 developed special graphics hardware and a program called "Sketchpad" as for his Ph.D. dissertation. He conceived Sketchpad as a drawing assistant; it cleaned up rough drawings by straightening and connecting lines, constructing geometric patterns, and in other ways anticipating a designer's intent. Sutherland not only wrote the software, he also had to build the display and interaction hardware. The above quote by Sutherland's thesis advisor discusses the initial plans for CAD systems. Sutherland continued to make contributions to CAD and computer graphics and co-founded a high-performance computer graphics company.

Sutherland's thesis introduced a new generation of technology supporting design and fabrication. However, it took six years to develop the technologies that would allow CAD to become a production tool. Several different technologies were involved. The display technology in Sutherland's thesis was new and needed to be redesigned in order to be mass produced. Also, demonstrating the first version of Sketchpad required dedicated use of one of MIT's campus mainframes. This was not feasible for practical use, even in a large company. Small, cheap computers were needed. And, of course, the various pieces had to be integrated and new software developed.

It took six years, to 1969, before Computervision Corporation built and marketed the first commercial CAD system, a system for constructing drawings defined by textual commands and displayed on a plotter. It relied on one of the first mini-computers. In 1970, they introduced an early CRT display tube allowing direct interaction. By the early 1970s, new CAD companies were forming monthly: Applicon, Calma, Autotrol, Intergraph, CADAM and others. In terms of the media attention and entrepreneurial enthusiasm, the emergence of CAD in the 1970s was similar to the emergence of the personal computer in the 1980s and multimedia in the 1990s.

This chapter begins by reviewing the main threads of research supporting interactive computer-aided design and similar applications. It presents a historical review of the display, hardware and software technologies. Although once they were an inhibiting context for the development of complex applications in building, these technologies now enable development. We consider CAD here for its use not only during the design phase of a building project, but as a means to represent and manipulate a product specification throughout a building's lifecycle. Thus we consider CAD in its broadest sense, ultimately being concerned with the full representation of a building and the products that comprise it, whether used during design, construction planning, operation or management.

Computer-aided design systems have evolved according to three general but distinct paradigms. One paradigm is as a *geometric editor*. Another more recent paradigm looks at a CAD system as an *environment for developing discipline-specific applications*. Most current CAD-system vendors address work within both of these paradigms, with varying success. Last, some design systems have been developed to support domain-specific applications but with the addition of a well-defined central model for representing building information. Early efforts based on this paradigm did not accept the CAD-system-as-graphic-editor paradigm, instead focusing on the need to develop what is now called a *building model*. It is to this third paradigm that commercial CAD is finally turning today. In the second half of this chapter, three early groundbreaking efforts in building modeling are reviewed.

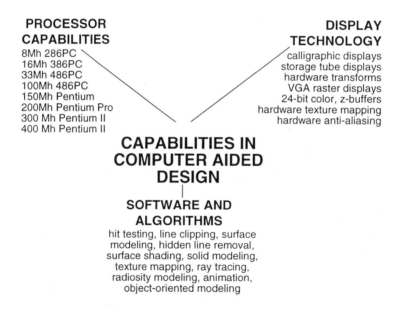

Figure 2.1: The three technologies that computer-aided design is based on and that determine its capabilities.

2.2 CAD TECHNOLOGIES

Early computer-aided design was very technology intensive, beyond the capabilities of most computer processors, early displays and systems capabilities of the day. CAD has consistently been a technology-intensive application, "driving" the technology with its potential market for high-end capabilities. Today, hardware technology has largely caught up with these demands, allowing interactive 3D CAD to run—slowly—on everyday desktop machines.

Computer-aided design is dependent on three different types of technologies, diagrammed in Figure 2.1:

(1) *display technology:* CAD motivated the first graphic display monitors; teletypes that printed on paper were used at the time. Evolution in display technologies has allowed us to move from low resolution monochrome static displays, incapable of displaying much detail or to quickly update a display, to large-scale monitors and high performance graphics boards supporting increasingly fast dynamic color displays. Graphic boards incorporate important display and processing capabilities, such as high resolution, wide color ranges, z-buffering, line drawing algorithms, anti-aliasing, transforms, surface shading and texture mapping. These technologies are the basis for all high quality, dynamic imaging capabilities now commercially available, including those for games.

(2) *processor capability*: The earliest CAD systems relied on command-line interaction, where the user typed in each operation as a command. This was mandated because early display and processor capabilities were not able to support the processing required for graphical user interfaces. Object dragging and 3D capabilities only became possible on high-end CAD systems when processor speeds became fast enough to support these operations. This threshold was reached with the introduction of the Motorola 68000 and the Intel 386. Still today, a complex 3D model can slow any processor and display device to a crawl. However, the transition time to PCs has been constantly decreasing and high-end PC-level machines are now competitive with traditional workstations. Today, real-time interaction is the target capability that drives high performance processors and display technology. The next generation of interactive capabilities is being developed directly for high-end PCs.

(3) *software capabilities*: Early software issues dealt with low-level issues such as how to quickly draw arcs and thick lines. Then came issues of perspective views and hidden line removal. Later came the technical issues of quick shading based on various lighting models, editing operations on surfaces and then on solid models. Recently, the issues in software design have included real-time rendering—especially with mapped textures—and new forms of 3D manipulation and editing. Another developing research area is trying to take advantage of image coherence between animation frames, instead of computing each frame individually. Parametric geometric modeling is another software area that is allowing new products to be introduced, supporting innovative capabilities.

Beginning in the mid-1950s, computer applications were written to automatically calculate engineering formulas that had previously been calculated manually. Computer programs also supported material procurement and billing services. It was the representation of geometry, however, that suggested a major role for computers in the design and manufacture of products. The earliest peripheral devices that could deal with geometry were plotters. They generated a drawing from a description consisting of a list of line segments, defined as pairs of X-Y location coordinates. Initially, arcs and text were generated using software that converted text fonts into the short line segments needed for each letter. In this sense, the earliest CAD systems represented a product simply as a list of line segments and text strings located within some coordinate system. Early display technology followed an approach similar to that used by plotters. An electron beam in the display traced lines, curves and text on the phosphor of the display tube. The excited phosphor lit up for only a short period; hence the drawing had to be constantly refreshed. This sort of display was called a *calligraphic display*. Calligraphic displays required their own processor to continually refresh the drawing and their own memory to hold a list of display commands. Thus, their cost was very high, initially over one hundred thousand dollars. The earliest CAD systems were all based on calligraphic displays. For some details regarding the initial development and dates of CAD-related technologies, see Figure 2.2.

Later, CAD companies developed and used several other display technologies, including storage tubes (principally sold by Tektronix) and plasma displays. In the late 1970s, *pixel-based bitmap displays* became available and within a few years grew to dominate the display market. They were initially limited because they required a large amount of memory (by the standards of the day) to hold the display image. While they required a special display processor, the technology was very simple compared to *calligraphic displays*. Pixel-based bitmap displays also used display monitors very similar to those used for television.

TIMELINE

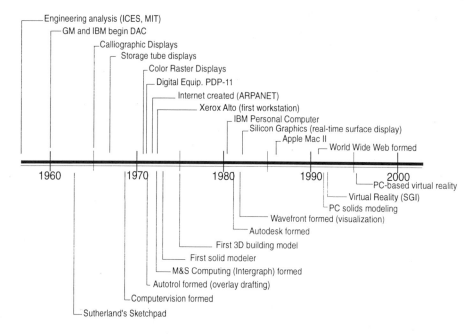

Figure 2.2: A timeline of major technological developments affecting computer-aided design.

Computer-aided design was an early interactive application that required very large amounts of dedicated processor time to respond to a user's commands. In the 1970s, personal computers did not yet exist. The development of mini-computers priced at 20 to 40 thousand dollars in the 1970s made CAD a practical possibility. Such mini-computers, using a simple timesharing system so two to four people could use the same machine at once, barely allowed interactive execution of a CAD program. The most popular early mini-computer for CAD systems was the Digital Equipment PDP-11, with 32 thousand words of 16-bit memory and a processing speed of about 200 thousand operations per second. This is much less than a first generation PC! Nonetheless, these machines offered a huge increase in the number of computing cycles per dollar continuously available to an individual.

From its beginning (1970-1975), CAD software was developed in an evolutionary manner, with only a limited analytic or formal base. Research was intense in the development of algorithms for drawing curves, panning and scaling images or parts of images, and so forth. CAD companies proposed different data structures for representing geometrical data and each company introduced enhancements to gain market advantage over its competitors. If a company's innovation was successful, others soon copied it. Those that did not evolve in this way soon lost market share, in classic Darwinian fashion. Some features that evolved in this way were (i) symbols, introduced first as purely graphic entities, then later as data-loaded ones; (ii) layers, as a means to structure data; (iii) user application languages; (iv) associative cross-hatching and associative dimensioning; (v) 3D wireframe and surface modeling; (vi) solid modeling; and recently (vii) integrated rendering modules and (viii) parametric solid modeling.

2.2.1 PERSONAL COMPUTERS

Up to the middle 1980s, CAD systems were developed for mini-computers or time-shared mainframes. Unix or VMS (the proprietary Digital Equipment Corporation operating system) were the main operating systems available. Systems running on these operating systems are generically called *workstations*.

Even before the introduction of the IBM Personal Computer (PC), a few efforts were made to develop personal drafting systems. The Apple II, though limited to 64K word of memory, was the first popular, personal computer and some early drafting systems were developed on this machine. The IBM-compatible PC, introduced in 1981, and the Apple Computer Macintosh, introduced in 1982, were big steps forward in terms of processing power, with 640K and 256K words, respectively. Simple CAD systems were quickly developed for these machines. Initially, the capabilities of these systems were toy-like, unable to include what were considered minimal features for an effective CAD system, even for 2D drafting. However, the cost reduction was dramatic. Until the advent of the personal computer, a CAD workstation cost on the order of $30,000, which only a few organizations in the building industry could afford. Many buyers of workstations had to use multiple shifts of operators to justify their cost. Operators were specially trained to operate the machine constantly; even the occasional unskilled use of these expensive machines by an architect or engineer was too inefficient.

Several companies began developing CAD software for the IBM PC and the Apple Macintosh, including Autodesk, VersaCad, Summagraphics, Microstation and others. Initially, these systems were also very limited, with poor graphics, little storage and even less speed. With the introduction of the Intel 80386 and the large screen Macintoshes in the late 1980s, desktop CAD truly became available in the $5,000 range. The evolution of the Intel family of processors is shown in Figure 2.3. As the cost of CAD systems dropped, their potential market exploded. A CAD system could be purchased and used possibly only a few hours per day, without fear of wasting an expensive machine. Architects and engineers could use it without the supposed efficiency of a CAD operator. The required investment was of the same order of magnitude as a good typewriter (considering inflation). Today, the capabilities of CAD are available to anyone with access to a PC.

Processing power was not the only demand CAD systems put on PCs. They also had to improve display and graphic capabilities. Neither the PC using EGA graphics, nor the early Macintosh with its small monochrome display, was adequate for serious CAD use. Effective CAD use became practical only after display processors and monitors attained SVGA resolution (800 x 600 pixels), at least 256 colors and fast line and arc drawing.

As personal computer processing speed increased, more of the capabilities available on workstations migrated onto them, such as 3D surface and solid modeling, dynamic "rubber-band" editing of graphic entities, and shaded rendering. Thus, while these capabilities have been gaining widespread use, they are not new. The late 1980s and early 1990s have been a period in which the CAD market has consolidated and expanded.

As a result of these technological advances, many limitations in using computer-aided design have disappeared. Costs of systems are low and their speed and versatility continue to increase. Development of new forms of computational support for building is no longer limited by system costs or the quality of displays. Rather, imagination and the ability to change the status quo—that is, the intellectual capabilities within the field—are the primary limitations to future progress. The generic capabilities produced thus far have not really begun to address the specific needs of architecture or construction. More specialized capabilities are needed that respond to the representations, procedures and practices of the building industry.

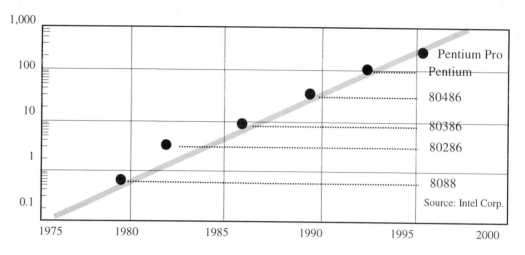

Figure 2.3: The logarithmic increase in computer speed for the family of Intel microprocessors. The vertical scale is MIPS, or millions of instructions per second. Desktop computers are now typically 100 MIP machines.

2.3 CAD AS A GEOMETRIC EDITOR

Many application areas have been developed for building use. However, the one most central to all building lifecycle phases, as identified at the end of Chapter One, involves the representation and manipulation of building geometry, as defined in a CAD system.

A CAD system is a very large computer program, written in a programming language such as C or C++, whose size can range from 30,000 lines of program code to several million lines. In order to deal with such a large and complex system, developers organize their software into functionally based modules. That is, a CAD system has a design. Like other aspects of the CAD industry, CAD system design has evolved incrementally.

A typical design for a CAD system with traditional functionality is shown in Figure 2.4. This is not the organization of any specific CAD system, but rather a structure similar to many different systems. Each of the boxes shown corresponds to a software module, which has multiple inputs and outputs, defined by arrows. Some of the arrows are labeled. Thus we see that there is a *user interface* module (now typically Windows or MOTIF, the Unix graphical user interface) that receives all user input and passes it on to a *command processor*. The command processor packages up individual mouse selections and sorts them to properly define a *graphical operation* that the system can understand. The *interaction facilities* provide high-level feedback to the user, such as rubber-banding, stretching and real-time coordinate readout that are not part of the base user interaction. Operators are of two types. One type consists of the given primitive operators of the CAD system, defined in the figure as *graphical operators*. These are part of the base-level CAD system. The other type consists of sequences of base-level operators called by *application code* and interpreted by the *application language*. The graphical operators operate on *graphical primitives*—lines, curves, text, as well as on any selection or layer of the drawing. The graphical operators and the data structure for representing them may be organized together as objects, or may follow more traditional organizations. The operators typically apply to an in-memory display list, where most operations take place. These are mapped back to a disk file format, typically either after every operation or during "save" operations. IGES and other report generators typically operate on the in-memory version of a project description.

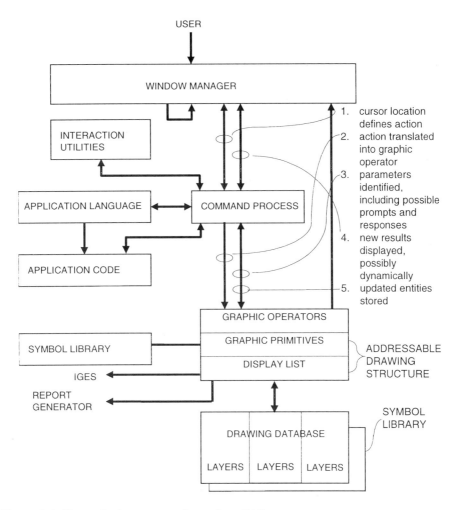

Figure 2.4: The typical structure of a modern CAD system.

This structure suggests how the functionality of CAD systems is organized. It suggests how a CAD program could be "ported" to another machine—namely, by recompiling the software code and, if necessary, restructuring the user interface and database modules. It also shows where errors may occur. Later, we will refer to this figure to explain how some of the evolutionary features and current capabilities arose.

2.3.1 CAD SPECIALIZATION

Initially, CAD systems were not differentiated and were sold in all markets that relied on drawings, such as electronics, manufacturing, building construction, petrochemical plants, road building. However, some companies began developing special operators and features needed in particular markets. Other companies that could not develop their product fast enough to provide the capabilities desired in ALL user markets began to specialize in one or just a few markets, leading to product differentiation.

One of these specialization areas was sheet metal manufacturing. The aircraft and automobile industry realized early that drawings were not the main object to be represented, but rather the development of surfaces. Surfaces could only be poorly represented in drawings, requiring interpolation between a large number of sections drawn through the surface. However, the mathematical definition of a surface provided all points on the surface. Because of their completeness, surface representations could be used to guide the fabrication of sheet metal components with great accuracy, using numerical control machining that was developed in parallel with CAD. In 1966, Steven Coons, Sutherland's Ph.D. advisor, who taught engineering drawing at MIT, developed the first curved surface modeling and editing techniques. It was his vision that is quoted at the beginning of this chapter. In tribute to his pioneering work in the field, the modeling techniques Coons developed came to be called "Coons' patches". Another early pioneer was Pierre Bezier, who, while at Renault, developed surface modeling techniques called Bezier surfaces, which are still used today. Surface modeling was important because it was the first example showing that the objective of CAD was to represent the object being designed, not a drawing of the object.

Special-purpose CAD systems were developed in other areas as well, including (i) electronics, where each step in the design and fabrication of a circuit board or silicon chip has been highly automated; (ii) shipbuilding, which also relies on specialized operations for hull design and analysis, steel plate definition and welding, and bulkhead design and fabrication; and (iii) geographical information systems (GIS), which involve large mapping databases. Initially, these specializations were treated as application areas and special features were simply added to a general-purpose CAD system as *application code*, as shown in Figure 2.4. Eventually, they became separate from standard CAD systems because they required unique representations and base operations that were not supported by general-purpose CAD. However, as all CAD systems matured, some capabilities that were once only available in special-purpose CAD systems have actually migrated back to general-purpose CAD. Parametric surfaces, solid modeling and kinematics, for example, are currently moving from special-purpose CAD to general-purpose systems.

2.3.2 SOLID MODELING

In the late 1970s, 3D wireframe drawings were introduced into general-purpose CAD systems as an early 3D capability. This was a incremental step, adding 3D vertices, new editing operations, 3D coordinates and a multiple view display allowed both users and developers to begin working in 3D. Soon after, some simple surface modeling capabilities were introduced to general-purpose CAD systems, allowing full 3D representation. It became possible to represent surfaces because pixel display systems could show colored shaded areas in addition to lines. The display of shaded surfaces greatly expanded the need for a large range of colors. Older pixel displays supported only 4 or 16 colors. New algorithms were soon developed to enhance the capabilities for rendering smooth shaded surfaces, casting shadows, properly displaying surfaces that are partially blocked by other surfaces, and other needs.

Modeling an assembly of complex parts required defining the multiple surfaces of each shape by specifying each surface, trimming and adjusting them so their edges matched. This was a very tedious task. Research groups at several universities worked on the problem of making surface modeling simpler, and on defining higher level editing operations. In 1973, two independent efforts developed related approaches. One came from Stanford University, where Bruce Baumgart designed and implemented a shape modeling program that defined an object by the set of surfaces that bounded it. The program had operators that, given two shapes, generated the union, intersection or difference of the shapes. These operations are called the *Boolean operators* and are demonstrated in Figure 2.5. At about the same time, Ian Braid at Cambridge developed a similar system. Both programs included new graphic operations that combined two solid shapes, each defined as a volume enclosed by a set of surfaces, and returned another solid shape. Because the result was always a closed shape, iterated operations were allowed. New data structures for representing solids were developed. These were added at the graphic objects level of a CAD system, as shown in Figure 2.4. Because a solid was represented by the surfaces that bounded it, this form of solid modeling was called a boundary representation or BRep. Later, other forms of solid modeling representations were developed.

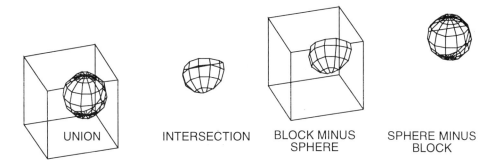

UNION   INTERSECTION   BLOCK MINUS   SPHERE MINUS
                                  SPHERE   BLOCK

Figure 2.5: Examples of solid modeling operations.

This work was soon followed by a formal analysis that determined the properties needed to guarantee that a set of surfaces was volume enclosing and to make solid modeling an abstract algebra. Not any set of surfaces is sufficient to represent a volume-enclosing region. A set of well-formedness rules is needed to distinguish a well-formed BRep from any set of surfaces. Among these are topological properties requiring:

(1) all edges to be two-connected; that is each edge on a face must coincide with exactly one other edge on a face

Geometrically,

(2) all faces must have consistent orientation

(3) no faces can self-intersect

Many volumes have holes through them, such as a wall may have for a door or window. A volume may also be hollow, with another shape inside that bounds the volume. For example, if all the constructed parts of a building were combined by unioning them together (with doors closed), the rooms would remain outside the construction volume but be interior to the building. In such cases, the outside set of surfaces enclosing the building would define a closed volume, called a *shell*, and the interior set of faces enclosing the room volumes would define other shells. All the shells together bound the solid part of the building. Both of these conditions impose additional constraints regarding orientation:

(4) edges bounding a surface with holes must be consistent with regard to the side of the edge adjacent to the surface

(5) the surfaces defining a set of shells making a volume with hollows must have consistent orientation

Other well-formedness conditions have also been defined.

These capabilities had several benefits. For one thing, they allowed users to take high-level design actions, similar to sculpting, that automated all the detailed actions involved in cutting multiple surfaces that in surface modeling had to be dealt with manually. These high-level actions also met the properties of a formal algebra, and thus could be concatenated, allowing the automatic computation of complex shapes. They were intuitively much simpler than surface modeling, where each type of surface had its own editing style. In the late 1970s, solids became commercially available, though not in stable form until the mid-1980s. They are now available in most CAD systems developed for the Architecture, Engineering and Construction (AEC) markets.

For most industries, there were many benefits to be gained from moving from 2D to 3D, whether it involved surface or solid modeling:

- Any number of drawings could be automatically plotted as reports—by cutting sections or making various isometric or perspective projections from a 3D model. All drawings extracted from the same 3D model were guaranteed to be consistent with each other. No other means have been developed so far that can guarantee the consistency of a set of 2D projections supposedly describing a single 3D shape or assembly.

- A 3D model could be used in many ways that a drawing could not. Among the many uses were 3D visualization, automated machining and manufacturing, and as input to various forms of analyses. Except in special cases, these uses were not possible from a digital drawing representation.

The use of 3D modeling has allowed many advances in certain industries, especially in manufacturing, aerospace and process plant design and construction. For example, it has facilitated the integration of design and analysis applications and facilitated automatic fabrication and assembly. The building industry must still make the transition from paper or electronic drawings to 3D modeling to realize these benefits.

2.3.3 PARAMETRIC MODELING

More recently, a new generation of geometric modeler has been introduced, based on *parametric modeling*. These systems move away from the conceptually straightforward notion of a geometric editor. In earlier solid and surface modeling systems, the user entered a sequence of parametric definitions to define a shape or editing operations on a shape. Examples of parameters might be the three dimensions and location of a solid box, the location and diameter of a sphere, an edge that needs to be filleted and the radius of the fillet or the cross-section of a shape and the centerline that the cross-section sweeps through. In general-purpose geometric modelers, these parameters are evaluated and a set of surfaces corresponding to the resulting shape is generated. Further operations, including union, intersection and difference, may be applied to the current shape. No history is retained of the operations used to generate the current shape.

In a parametric modeling system, a shape is defined through a similar set of construction operations. Each operation is defined with its parameters. But in addition to evaluating the resulting shape as a set of volume-enclosing surfaces, the modeler retains the set of input construction operations and parameters. These operations form a set of operators and operands in the form of an algebraic expression. This expression can be later edited—graphically or textually—and the shape automatically regenerated. Such editing capabilities are very useful in design, where the geometry of a product is refined over time. Shapes may be refined by changing their existing parameters or by adding additional operations to the model.

An example of a simple parametric solid modeler is shown in Figure 2.6. A menu of primitives and operators is shown at the top. In practice, they would be presented to a user graphically. The algebraic expression of primitives and operations defined up to now is shown in the middle of the figure, along with the parameters. The layout of the primitives, without the application of the operators, is shown at the bottom left. The evaluated shape is shown at bottom right. The algebraic expression in the center of Figure 2.6 is called the *unevaluated representation* of the parametric model. The bottom-right figure is called the *evaluated representation*. In Figure 2.7, a sample editing operation is shown. Suppose that the user selected one of the points defining the front roof plane and changed its Z-value. Upon re-evaluation, the new evaluated model would appear on the right. Very complex parametric models can be defined, with each of the original defining operations available for editing.

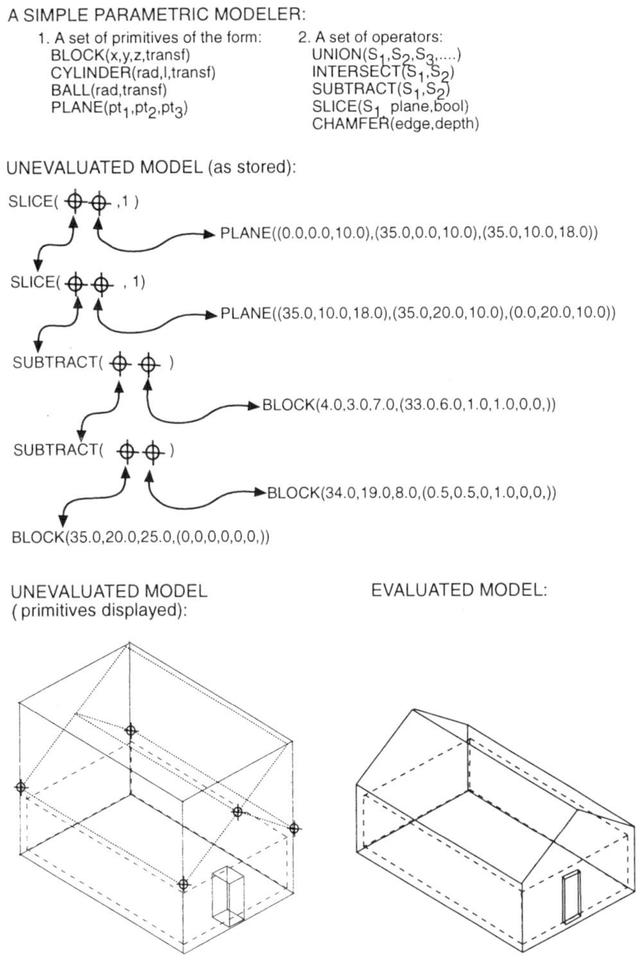

A SIMPLE PARAMETRIC MODELER:

1. A set of primitives of the form:
   BLOCK(x,y,z,transf)
   CYLINDER(rad,l,transf)
   BALL(rad,transf)
   PLANE(pt$_1$,pt$_2$,pt$_3$)

2. A set of operators:
   UNION(S$_1$,S$_2$,S$_3$.....)
   INTERSECT(S$_1$,S$_2$)
   SUBTRACT(S$_1$,S$_2$)
   SLICE(S$_1$ plane,bool)
   CHAMFER(edge,depth)

UNEVALUATED MODEL (as stored):

SLICE(⊕ ⊕ ,1)
→ PLANE((0.0,0.0,10.0),(35.0,0.0,10.0),(35.0,10.0,18.0))

SLICE(⊕ ⊕ , 1)
→ PLANE((35.0,10.0,18.0),(35.0,20.0,10.0),(0.0,20.0,10.0))

SUBTRACT(⊕ ⊕)
→ BLOCK(4.0,3.0,7.0,(33.0,6.0,1.0,1.0,0,0,))

SUBTRACT(⊕ ⊕)
→ BLOCK(34.0,19.0,8.0,(0.5,0.5,0,1.0,0,0,))

BLOCK(35.0,20.0,25.0,(0,0,0,0,0,0,))

UNEVALUATED MODEL
(primitives displayed):   EVALUATED MODEL:

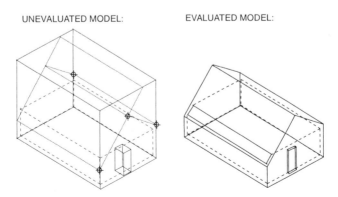

Figure 2.6: A simple unevaluated parametric model and its evaluated result. The parametric model is defined and stored as a set of parametric primitives and the sequence of operations modifying or combining them. The primitive shapes are shown bottom left, the evaluated model is shown at bottom right.

UNEVALUATED MODEL: EVALUATED MODEL:

Figure 2.7: An example of an editing operation performed on the parametric model defined in Figure 2.6.

Development of a wide variety of tools allowing direct manipulation of 3D forms has been the important benefit of solid modeling. Solid modeling technology allows people to create and edit shape directly on the computer, providing an effective alternative to working with 3D physical models. In all likelihood, solid modeling will become the dominant representation for building geometry in the next few years.

2.4 CAD AS A PLATFORM FOR APPLICATION DEVELOPMENT

In the previous section, we discussed the use and evolution of CAD as a geometric editor. Another view of CAD is that geometry is only one of many attributes involved in the representation of a product. This view takes the position that representing a product as geometry alone is very incomplete. To the basic representation of geometry, it adds other properties, such as material and performance properties, and to the structure of that representation it adds relations between entities. Material and performance properties of objects are defined as *attributes*, represented as text, numbers or compositions of simple values (such as RGB color values).

Mathematically, connectivity and other such relations are known as a system's *topology* and are most commonly represented as a graph. Representing structural systems requires the prescription of elements that are "connected" to others for the transmission of stresses. For a piping system, the connectivity of the piping system must be prescribed for flow analysis. Similarly, in architecture, the connectivity of rooms is needed for dealing with privacy, accessibility and to check fire-exit codes. Wall boundaries are needed for energy analysis. Many types of relations are required in different contexts. There are prescribed means with which most CAD systems carry attributes and define topological relations between elements.

This view considers a general-purpose CAD system as a platform upon which to add domain-specific capabilities. The CAD system provides the low-level capabilities needed to support special-purpose functionality. In addition to representing geometry, it must also be able to represent attributes and support creation, selection, editing and deletion of attributes. It must also be able to represent topological networks and support them with editing capabilities. Also, a CAD system that supports application development has to provide an application development language, allowing the definition of specialized entities with their associated geometry, attributes and topological relations and providing special operators to edit and manipulate the special objects. Most CAD systems today have made these extensions, supporting customization.
Several additional subtle capabilities are required to support open extensibility:
- providing ways for new operations or applications to be added to the base system, in terms of menus and graphical user interface
- providing a means to associate attributes, topology and geometry together; the most common means was to associate attributes and topology to the geometry using the CAD systems symbol capability
- ability to block certain basic operations, so that they cannot be applied to application-defined objects, destroying the internal consistency built up in the special application operations
- easy interfaces to load external data, such as material or performance properties of entities
- means to interface existing "legacy" applications to these specialized CAD models, allowing data to be extracted from the CAD system to run on the application, or for loading the results of another CAD program into this one

CAD systems that provide an application development platform allow implementation of some powerful and useful applications. While a geometry editor carries no special knowledge beyond geometry about a product, the extended applications built on top of them are able to incorporate many types of domain-specific knowledge. They can incorporate design or construction rules, checks about legal compositions and other product information of central importance to the product field. For a given type of product design, these capabilities can be immensely useful and begin to realize the image of a CAD system as a design or engineering assistant.

The current generation CAD systems were first developed in the early 1980s, with only partial cognizance of the need to support application development and before the development of modern object-oriented programming techniques. Most current CAD systems evolved from electronic drafting systems, using the paradigm of geometric editor. Today, the main vendors are attempting to restructure their systems to support object-oriented development. They are integrating Internet and Web-based communication, so as to provide a future-oriented application development platform. At the same time as they develop new technologies for application development, CAD vendors need to support current customers who desire only incremental improvements. The challenge is to maintain one's current customer base while at the same time rebuilding the foundation code supporting all system operations.

2.5 EARLY EFFORTS AT BUILDING MODELS

Even though a single CAD system may support the effective development of multiple add-on applications, it is not likely that these applications will operate together. The different applications may each rely on different definitions of the objects representing a building, different relationships among objects and different rules of how objects are composed. In fact, most applications built on top of the current main CAD systems are not compatible with the other applications on the same platform. This is the natural result of each application having its own representational needs and its own concepts and semantics about how a building is defined.

In the early days of architectural CAD, however, a few systems evolved from assumptions quite different from those associated with geometric editors. That all the separate computational tools needed to design and build some facility should be integrated around a central representation of a building is not a new idea. In the mid-1970s, a number of independent efforts were made to develop integrated systems, based on a single building model supporting a suite of applications. These systems were developed before the existence of platform-based systems and developed from scratch an integrated system supporting building design. They were based on the assumption that the basic task of design was to develop a specification of a building, supported by applications to define the various components and to analyze the behavior of the compositions. These efforts each defined a single, coherent representation of a building around which all applications should be built. More than one of these eventually evolved into commercial CAD systems, but as a result of this different heritage, offered different capabilities and provided different bases for future application development than standard CAD systems.

Most of these efforts were British. Three early building modeling systems were funded by Her Majesty's Health Service, the national health agency: the OXSYS CAD system from Applied Research of Cambridge, and the CEDAR and HARNESS hospital design systems. A fourth major effort was funded by the Scottish Housing Authority and developed by a group at the University of Edinburgh. Three other efforts were made in the US. An important but largely forgotten early effort was a building model supporting design using the Techcrete precast concrete construction system. Techcrete was developed by architect Carl Koch and Associates and the building model was developed by Theodore Myer of Bolt, Beranek and Newman. Another research group was at the University of Michigan; their best known model was called ARCH-MODEL. ARCH-MODEL went through several versions, each expanding on the original earlier capabilities, and eventually evolving into a Macintosh-implemented CAD system. It was based on a consistent model of geometry—using solid modeling and a relational database for storing non-geometric data. The other effort was the author's work at Carnegie-Mellon University, where three different building modeling systems were developed and tested: BDS, GLIDE and GLIDE-II.

Here, three of these early efforts are reviewed: (1) the council housing effort by Aart Bijl at the University of Edinburgh; (2) the OXSYS CAD system developed by Applied Research of Cambridge, a commercial firm, but as a research project for the national health service; and (3) the CAEADS system developed by the author's group at Carnegie-Mellon University along with the University of Michigan, sponsored by the National Science Foundation and later by the US Army Corps of Engineers. These have been selected because accessible documentation on them still exists and their impact on later efforts.

In the following discussion, an attempt is made to reconstruct the ideas and issues involved in the three efforts, based on recollections and discussions with the developers near the time they were finished and the reports on this work that still exist. It is useful to consider them with respect to the timeline shown in Figure 2.2, which shows what hardware and software capabilities were available at the time. These early efforts responded to building modeling issues in an intuitive fashion, addressing the problem at hand by defining the needed concepts as they were recognized, often anticipating concepts that were more formally developed only much later. The systems surveyed here, plus the other systems mentioned above (accessible through references at the end of this chapter) have provided some of the foundations for current integrated building model efforts. The supporting drawings and images have been taken from original sources. Many are of low quality, but are presented to indicate the functionality provided by these systems.

2.5.1 SCOTTISH SPECIAL HOUSING AUTHORITY HOUSING SYSTEM

An early housing and architectural design system was developed by the Architectural Research Unit at the University of Edinburgh, led by Aart Bijl. It was funded first by the Scottish Special Housing Authority (SSHA) and then later augmented at the national level by the Department of the Environment. Funded research began in July, 1969, and continued into 1973. It was undertaken in two major projects: the first focusing on housing unit design, the second addressing site planning for housing estates. The period in which this project was undertaken can be discerned from the computer equipment shown in an early photograph from the first project's research report, which is reproduced in Figure 2.8.

Figure 2.8: Reproduction of the hardware configuration used in the first phase of the SSHA Housing Design System: a Digital Equipment PDP-7 with 8K memory and 340 display.

The goal of these two related projects was to develop practical tools. In these early days, the distinction between prototype and production software was not as clear as it is now. Thus it was planned that the hardware, and the software system developed on it, would be passed over to SSHA and they would continue using the hardware, maintaining and further developing the software. The SSHA was evidently a very forward-looking government unit. It had previously developed very organized quantity surveying methods. Part of these methods was a well-structured set of "units of construction" and "units of work", providing an important, rational foundation for developing the CAD applications.

Both systems involved definition of a design development process that could produce the desired product specifications and that could be realized on a computer. Both reports included early process analyses of the various operations needed to develop a design.

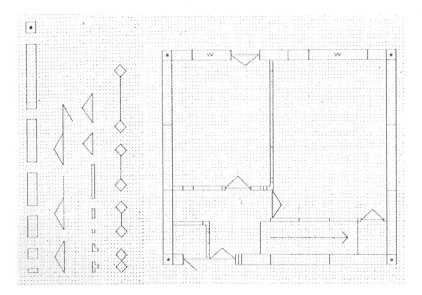

Figure 2.9: An example of a floorplan and symbols used for walls, doors and windows.

2.5.1.1 Floorplan Design System

The first phase was conceived as an intelligent floorplan layout system. The elements used to define the floorplan were units such as wall, window, door, cabinet and stair (see Figure 2.9). The plan elements were known as "blobs", which were symbolic entities that were selected, specified and located to compose a housing unit. The symbolic entities were kept very simple to reduce display time. Each blob referred to a detailed data structure that more fully described it. Each blob had associated unit geometry, materials and labor codes, joining codes and other information.

As the blobs were assembled, the spaces bounded by the arranged walls were identified so that they could be later analyzed. Also, their junctions had to be indicated and assigned the appropriate detailing. The SSHA floorplan system was one of the earliest building models that explicitly modeled both the solid, constructed part of building and also building spaces.

The 2D data structure of the SSHA building model is shown in Figure 2.10. A wall component carried material and other information. It is shown as a component bounded by surfaces that were shared with rooms and wall-ends called "leaves" that abutted junctions. Both components and rooms could be traversed to capture their geometry. This structure supported a number of advanced operations. Because materials were associated with each building element, they could be retrieved and displayed, along with the part of the room perimeter they described (see Figure 2.11). Each of the joint conditions defined in the plan could also be displayed. From the room perimeter, the program automatically computed room areas and the enclosing surface areas for each room. This procedure was useful in computing material areas and volumes. Based on the glazing area on outside walls and its location, and on inside wall boundaries, the iso-contours designating sunlight levels could be derived (see Figure 2.11).

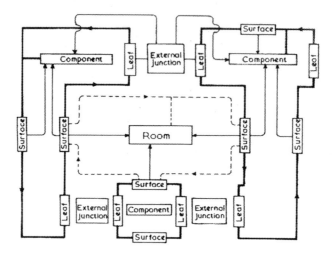

Figure 2.10: An abstract representation of the SSHA building model.

The system also had the capabilities to evaluate beam and joist sizes and positions (see Figure 2.13) and to calculate heat transfer through exterior walls. It is assumed that these calculations were all made by small analysis routines written especially for this system. No reference is made to external or commercial applications that were interfaced.

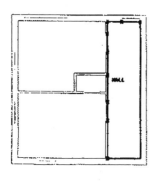

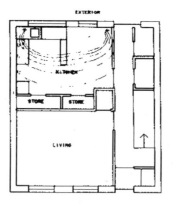

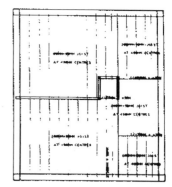

Figure 2.11: Automatic computation of room perimeter.

Figure 2.12: Automated floor joist sizing and layout.

Figure 2.13: Generation of natural light contour levels.

The house unit design system was similar but predated by a decade or more the architectural design add-on packages offered by CAD vendors and third parties. It relied on interaction with the user through an early display monitor—initially a calligraphic display, later a storage tube display. The system relied on simple menus consisting of keywords on the display screen. It took advantage of the well-defined building unit codes already created and used by SSHA but was developed to support standard on-site construction. It was quite successful in showing at an early period the potential capabilities of CAD systems using two-dimensional geometry.

2.5.1.2 Site Planning System

The second stage of the SSHA Building Model was a site planning system for developing the site layout and cost estimation for a housing estate. It defined the contour and ground plane conditions, then assigned housing and garage units, roads, drainage lines, landscape areas, footpaths and retaining walls to the site. These elements were each defined in any order, and their borders were rectified into tight packing polygons afterward. The primary purpose of the system was not for designing the site plan, but rather for entering an already designed plan and getting a quick, detailed cost estimate for it. Editing operations were provided so that the specification could be changed and cost updates computed. The site plan system included an early terrain model. It supported topographic definition using either contours (given a Z-value, enter a string of X-Ys), or elevations along a survey line (sequence of X-Y-Z). Both initial elevations and finished elevations were entered, from which cut-and-fill volumes were derived. Elevation interpolation was performed by software acquired from the University of Michigan.

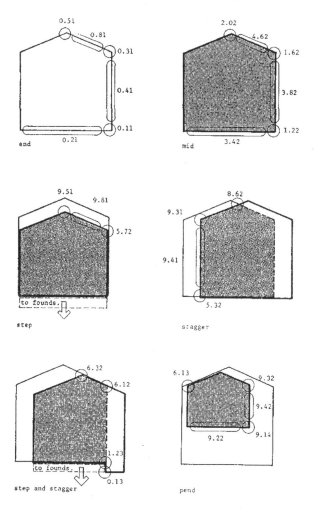

Figure 2.14: Building unit end conditions involve various conjoint, subjoint and overlapping conditions.

This CAD system prestored the house-type library and inserted it into the site. Each house unit was located on a site by an origin, angle and elevation for each optional level. The house definitions involved several complex issues. The system defined a house type by its perimeter, as a sequence of vertices. However, house plans allowed for split levels so that each of the possible split-level regions had to be defined separately. In addition, party walls were described, because the detailing of party walls, particularly due to changes in elevation, were a significant cost item. The party-wall condition was defined by accessing the two adjacent walls. The different cases are shown in Figure 2.14. Party walls were also defined by an array of vertices, defining the vertical perimeter of the wall. Also, sleeper foundation walls were also defined at split levels or as part of the basic type definition, again as a line of vertices. Garages were defined similarly, but with only one level and no sleepers.

Foundations were a significant cost item and the height of foundation walls and the merging of shared foundations were significant considerations in later cost calculations. Simple estimates were made of the bottom level of each foundation wall, where steps were located, and depended upon whether the steps were vertical or sloped. Given the merged foundation walls, with top and bottom elevations, routines estimated their cost.

Party walls were analyzed to determine if they were mutual, stepped or staggered. These conditions identified whether the boundaries of the wall are inside, align or intersect each other (see Figure 2.14). Thus their conditions correspond to special cases of the 2D polygon intersection operations. Each condition required its own set of details. All sections of the perimeter were analyzed and the appropriate detail with its associated unit cost was looked up and applied. The areas of shared wall were also identified, and by subtraction so were the portions of party wall on the exterior. Each type of wall had its own cost unit.

Roads and paths were entered by their centerline and width. Vertical alignments were computed from the finished elevation and users could adjust these if desired. Road elevations were interpolated and smoothed automatically. The result produced a new cut-and-fill volume. This volume of added or subtracted soil was computed, using elevations at grid points. An example of a layout is shown in Figure 2.15.

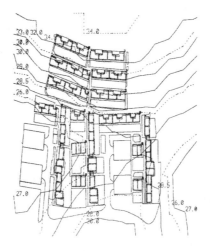

Figure 2.15: An example of a system-generated site plan, with buildings, roads and contours. The original contour lines are shown as dotted.

Landscape areas, paved areas, ramps and outdoor steps were defined by a polygonal border and a type classification. Fences, drainage lines and retaining walls were all defined by polylines and associated attributes. All the polygonal areas were rectified so as to be consistent and provide complete coverage. House and road areas were fixed and landscaping borders were adjusted to them, and the remaining adjustments were averaged. The output of the program, after all the entities were defined, was a bill of quantities and their material or detailing code. These were then entered into a quantity survey program to derive a cost estimate.

The site plan system had to deal with many geometrical operations, such as set operations on polygons, calculating the volume of cut-and-fill operations, and the entry of various geometrical types. These operations were all defined and implemented by the project team. Today, libraries of routines are available for most of these capabilities. The SSHA program is not very different from current site planning applications now available, though the current systems have enhanced 3D graphics. This program could display plans and sections, but not 3D images. Both the site planning and housing programs incorporated significant amounts of knowledge about house design and construction practices, about construction detailing, adjustment of roads and paths to terrain, and foundation planning. They were pioneering projects, directed toward production use. The system was used in production for a period of about ten years.

Later, the project leader, Aart Bijl, felt that the kind of task-oriented knowledge implemented in routines in the two projects was severely limiting, both in its generality and in its conception of architecture. The routines were fragile and could not deal with the variety of conditions often encountered in different design contexts. He later undertook a different direction, developing systems based on logic programming with semantic networks. Later, in a book reviewing all of his efforts, Bijl presented a set of points regarding what would have to be addressed if intelligent computer applications were going to support creative architectural design as we now know it.

- Many aspects of a design developed on paper are ambiguous and are open to multiple interpretations. One reason for this is that most design problems are ill-defined and the designer must decide which aspects of a problem should be keyed upon and which ignored. Allowing these multiple interpretations is an essential aspect of architectural representations, whether paper- or computer-based, because it allows human designers to develop new interpretations of existing information, much like a scientist forms new hypotheses.
- The rationale for many design decisions is often tacit. That is, a creative designer does not always make decisions based on well-defined and explicit knowledge; knowing why a design decision was made may be akin to knowing why a soccer player made a specific move—it probably just seemed the right thing to do. If the soccer player held back to rationalize a decision, the context in which it was initially made may disappear.
- Design is an evolutionary process. The values embedded in a design evolve as the design is studied, often through making trial designs. As the values change, the programmatic aspects also evolve.

These points are directed toward the use of computers in the design phase of the building lifecycle. They reflect upon various efforts to automate aspects of design. At the same time, they are general truths that apply to many actions undertaken by people in response to real-world conditions and requiring creativity.

2.5.2 OXSYS BY APPLIED RESEARCH OF CAMBRIDGE

Applied Research of Cambridge (ARC) was a commercial unit that grew out of the Centre for Land Use and Built Form Studies (now called the Martin Centre), a research unit associated with the School of Architecture at Cambridge University. ARC was formed in 1970 to commercialize the research developments initiated within the Centre. An early research project taken on by ARC was the development of a CAD system to support hospital design, using system building techniques. The tool was to support hospital design based on OXSYS, a prefabricated building system developed by the Oxford Regional Health Authority to facilitate the construction of hospitals after the Labor government in the UK had nationalized the health care system.

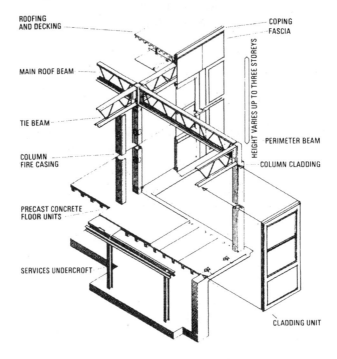

Figure 2.16: Isometric of the Oxford method of hospital construction.

The Oxford Method was a post-and-beam and slab system of construction, based on a Tartan Grid. A Tartan Grid is an orthogonal structural grid, with smaller grid lines on either side of the structural grid. This grid is used for laying out walls and fittings on walls, or in section, for laying out ceilings and finished floor elements. The construction system consisted of metal post and beams, with concrete slabs and various types of external cladding. Mechanical elements, fittings and most of the components of hospitals were predefined. An exploded isometric characterizing the OXSYS system is shown in Figure 2.16.

The CAD system was first developed on the ICL Atlas mainframe computer at the Cambridge CAD Centre, an Industrial Research Establishment initiated in the 1960s by the UK's Department of Trade and Industry to support advanced technology applications for industry. The overall system was written in FORTRAN and heavily overlaid in memory. (This means that the program swapped in and out various program segments—a tactic used before virtual memory.) The overlays were managed by OXSYS-O, the OXSYS operating system (later renamed BDS Overlay System (BOS)). The overlays consisted of different application modules. An overall diagram of the resulting environment is shown in Figure 2.17.

The OXSYS building system was based on a predefined set of building components. Thus all building elements could be defined in a library, which ARC called a Codex. ARC updated and extended the Codex as the OXSYS building system was refined. The Codex was extensive, including structural elements, cladding components, partitions, slabs and ceiling elements, as well as room types, interior fittings and fixtures, and mechanical equipment, all organized into families and subfamilies. The Codex was not based on any standard, but was pragmatically organized according to types within this building system. Early publications suggested that it could be cross-linked to standard categories, such as the European Sfb standard (somewhat like the CSI standard in the US). Publications indicated that there were Codex elements for assemblies, as well as for parts, which could be used for schematic design. However, it is not clear if there were hierarchical links from an assembly to its parts.

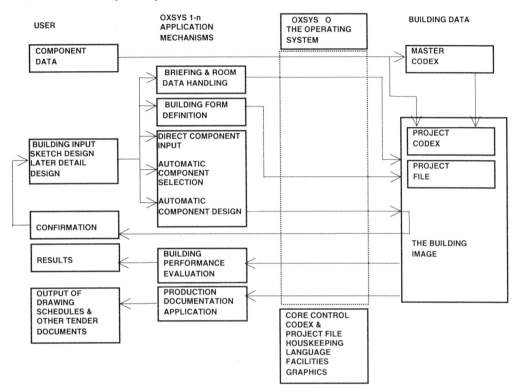

Figure 2.17: The system architecture of the OXSYS system, developed by Applied Research of Cambridge, Ltd.

Each element was described in terms of its dimensions, descriptive text giving its name and general use, weight, function-related properties, environmental characteristics and also the graphic codes required to draw the element in different views. Geometrically, each element was defined by its projections onto a 3D rectangular box. Up to twelve projections or views could be stored for each component. Poor quality images of these graphic codes for three elements are shown in Figure 2.18, taken from an early product description. Because the total Codex for OXSYS was so large, a smaller version was required for definition of a particular building project, called the Project Codex. It was similar in some ways to the symbol libraries and project symbols (WBLOCKs and BLOCKs in AutoCad®) found in most contemporary CAD systems.

Within the Project Codex, elements were classified into families, subfamilies and components. They were spatially defined by an orthogonal bounding box specifying their X-Y-Z spatial extent, and a list of allowed penetrations. Other information included adjacencies, material properties and various performance characteristics such as maximum span or load. Whenever possible these specifications were precomputed. Attributes, including the bounding box dimensions, could be marked as either fixed or variable. Given a required level of performance, the components could be automatically selected using the component selection routines in the system. Alternatively, they could be selected and laid out manually.

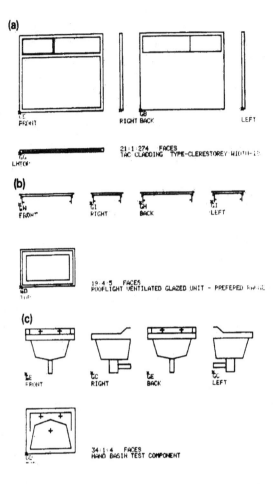

FIGURE 2.18: The multiple geometric descriptions for three different objects used in OXSYS.

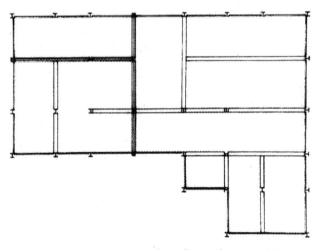

Figure 2.19: Steelwork automatic selection based on column positions.

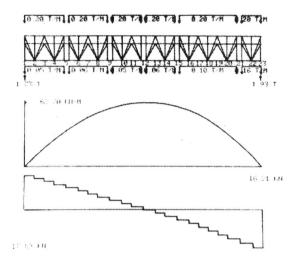

Figure 2.20: Automatic analysis of steel beams.

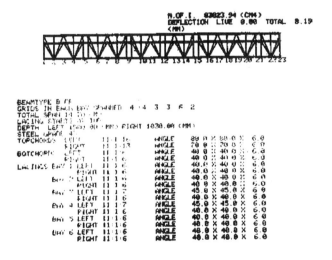

Figure 2.21: Automatic sizing of steelwork.

A project was the composition of individual components. It carried the spatial transformation and parameters of each component in the layout. New components could be added at any time and existing ones could have their location or parameters changed to suit different conditions. Some components were variable in length. Upon placement, spatial conflicts were checked using the box-level geometry. An "allowed penetrations" attribute was used to identify permissible conflicts. Connections were checked with the adjacent elements in the Tartan Grid. A variety of applications could then analyze a composition directly or extract data and reformat it to generate a drawing. Figures 2.19, 2.20 and 2.21 show three structural applications: the first application was for selection of elements, the second for analyzing shear and moments in a truss element, and a third for automatic sizing. The applications interfaced to this model were custom developed and not general-purpose ones.

Document generation was treated as a separate application. It retrieved the graphical data stored with each part, locating and formatting it to generate plans and sections. Examples, including a plan and two sections of a building layout, are shown in Figures 2.22, 2.23 and 2.24.

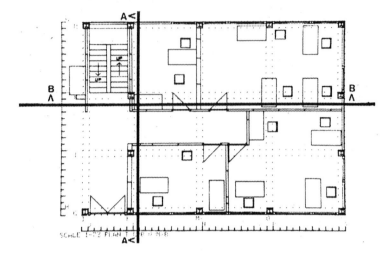

Figure 2.22: Plan showing layout of cladding partitions, fixtures and furniture (Tartan grid is dotted).

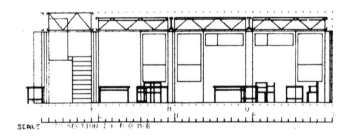

Figure 2.23: Section through B-B in plan above showing roof and first floor structure.

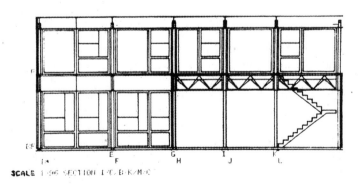

Figure 2.24: Section through A-A showing partial elevation and section.

In addition to an elaborate Codex for building parts, OXSYS included a well-defined structure for representing building spaces, based on spatial sets. It was organized in hierarchies defined using the relations shown in the left column of Figure 2.25. The examples show that for two spaces, any subset may be defined using the spatial set operators. The structure was developed based on the recognition that functional zones are not always consistent with other types of zones, like HVAC zones. Thus multiple, spatial hierarchies were supported. Different zone hierarchies could be defined, then reviewed and overlaid. The zones were entered by a user and the spaces were then tabulated and compared against the building brief. An example is shown in the right column of Figure 2.25. The building is partitioned with two different sets of zones. The resulting subzones are shown in the fourth figure down. The ownership of the derived zones by the two higher level ones are shown at the bottom.

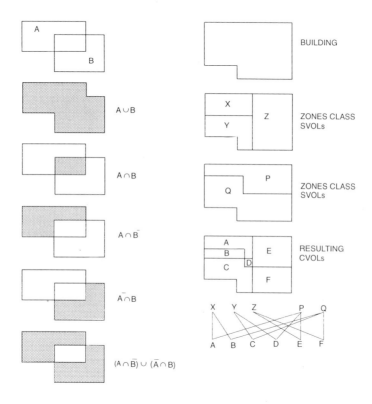

Figure 2.25: The union-intersection of space zones allowed classification of all spaces.

An important thrust of the OXSYS CAD system was to address the combinations of details arising in the assembly of Codex parts. It involved a fairly complex, rule-based system of checks for whether particular assemblies were allowed (all coded in FORTRAN). The developers went through a "knowledge elicitation" process, acquiring the rules that were implicitly part of the building system from the OXSYS architects. The rule-coding process was difficult because the Oxford System of building kept evolving, even to the extent that the size of the grid was changed. As a result, the rules had to be constantly updated. In use, many rules appeared to be "fragile", in that they reflected recommendations of the system's architects but did not apply in all cases. Project architects who used the system often needed to work around different rules or make additions or changes to the system. In retrospect, its developers felt that the OXSYS CAD system was a means for enforcing the conventions of the system architects on the project architects. These issues are still a concern today where expert systems are designed to enforce certain types of design decisions.

The file management in the BOS system facilities addressed several important issues about data security that still are not resolved in many current CAD systems. One feature was that as data was changed during a session, the changes were written back to disk on duplicate copies, without updating the original. Thus the initial state could be retained until an update was complete and the user accepted the update. At this point, the original file segments were overwritten. The second feature was that BOS maintained an interaction log of all user commands. In case of a crash or hardware glitch, the previous operations could be re-executed, thereby ensuring that the work done was secure.

The OXSYS CAD system was ported to Prime mini-computers and tested on several hospital projects. The initial Prime implementation was restricted to 64K words of memory. Later, a larger virtual memory version of the Prime operating system became available.

Associated with the project database was also a database defining the brief for the project, which allowed allocations within the final design to be compared against targets included in the brief. OXSYS also supported concept design, by allocating spaces corresponding to each of the spaces specified in the brief in a certain configuration.

The system was marketed commercially beginning in 1978 under the name BDS—for Building Design System. It was an early example of an object-based system that incorporated multiple views of an entity, some being graphical and others not. The BDS system defined entities abstractly, then generated views of them. This approach to design was later refined in the commercial GDS system and in other British systems. GDS was originally conceived as a post-editor for OXSYS-generated drawings, to deal with annotation and representational conventions too tedious to do parametrically. It rapidly became apparent that it could do the whole drawing production in a way more familiar to architects than through Codex creation. Many of the BDS concepts and implementation strategies, especially at the system level, were incorporated into GDS.

BDS was eventually withdrawn from the market when full 3D and solid modeling geometry began to be integrated into production systems. However, it was a strong example of object-based modeling and the development and integration of multiple applications. It included design rule checking, implemented in a somewhat more general way than was realized in the SSHS work at Edinburgh. It dealt explicitly with the issues of producing a full product specification for construction. The OXSYS CAD system served as a model for at least one other commercially marketed CAD system, initially called RUCAPS and later SONATA. SONATA included full 3D modeling and other capabilities beyond those that were initially developed in BDS. There are probably some people still running SONATA today. A similar object-based system is STAR which is now marketed primarily in Europe. Apparently support for the last eight installed BDS systems was withdrawn in 1986.

The OXSYS and BDS systems were based on a single, integrated model of a building that supported all design tasks as well as an integrated suite of interdisciplinary applications. While OXSYS and BDS did not support user-defined parts, some of their descendents did. In an unpublished review of the work undertaken by ARC, Ed Hoskins, Director of ARC, made several comments that are still relevant today:

• BDS was based upon the assumption that information technology could have a catalytic effect on the design process, which in retrospect was somewhat naïve; an established practice will not reorganize itself around a new, foreign and sometimes inflexible set of procedures.

- BDS's economies in production documentation were inadequate to justify significantly increased resources to the design phase of building production.
- While the facilities provided in BDS were truly interdisciplinary, BDS did not adequately support multiple concurrent operation.
- The predefined parts did not give the designer the freedom to work out new details that were unanticipated in the original system.
- A building component-based model is inadequate if it cannot support the open-ended needs arising from new, special details and from distinct or problematic site conditions.

2.5.3 GLIDE-II AND CAEADS

During the 1970s and early 1980, the CAD-Graphics Laboratory at Carnegie-Mellon University, directed by the author, was active in the development of several early building models. The Carnegie-Mellon group was not concerned with developing finished solutions, but rather with developing a computing environment that would facilitate the easy implementation of integrated systems for architectural design. The work of the group had two foci: (1) the development of solid modeling for use in buildings, and (2) the integration of solid modeling with database and other capabilities needed to generate an environment for advanced CAD system development.

In 1974, the group developed an early, solid modeling-based, building model with permanent file store, called Building Description System (another BDS). In 1977, an expanded team developed GLIDE (Graphical Language for Interactive Design), an interpretative language with built-in solid modeling capabilities with permanent storage of global variables. GLIDE allowed users to define new parametric primitives and attach attributes to them. Its operations included the spatial transforms, spatial set operations and Euler operators required for defining new parametric shape primitives. The language and permanent storage provided a simple database structure for defining and storing building schemas with complex geometry. New schema structures could be added after the database was loaded with values. Some fairly complex building models were defined. GLIDE was implemented on a PDP-10 time-sharing computer and was written in BLISS—a proprietary Digital Equipment system development language. This work was funded by the National Science Foundation.

The insights and capabilities of this earlier work led to funding by the US Army Corps of Engineers to develop a portable engineering database system that could run on various hardware systems. This was difficult, in that certain system facilities such as graphics were built up on a particular machine configuration, and there were no standards for graphical interfaces and databases at that time. However, a new system, called GLIDE-II, was designed and implemented, defined as an interpretive programming language with permanence, with a backend file system for storage. Within GLIDE-II, a building schema was defined that supported interfaces with a set of applications, some specially defined and some generic, the latter having to be interfaced through translators. The integrated system comprising the environment, applications and schema was called CAEADS, for Computer Aided Engineering and Architectural Design System. CAEADS included a predefined, general-purpose building model and interfaces to a set of applications, developed in the 1980 timeframe. Below, GLIDE-II and CAEADS are both reviewed.

2.5.3.1 GLIDE-II

GLIDE-II was developed as a portable database language, written as a bootstrapped extension of Pascal (portable meaning that it could be configured on multiple machines, in a semi-commercial manner). Like its predecessor GLIDE, it was an interpretive programming language with permanent storage of global variables. Syntactically, it was a superset of Pascal and built by extending a P-code interpreter, the original way to easily implement Pascal. Language extensions supported features needed for product modeling. Also, the language was implemented as an interpreter, so that each global statement was executed as soon as it was completed. The implementation was designed to run on Digital Equipment VAXes.

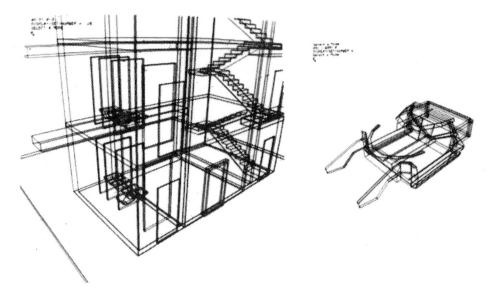

Figure 2.26: Two graphic images from GLIDE: A 3D building layout and a portion of an automobile chassis.

In GLIDE, solid modeling capabilities were built into the language as primitives. In GLIDE-II they were written in the GLIDE-II language and stored in a library that could be used or replaced with another. This allowed the geometric modeling capabilities to be extended or modified. The Euler operators for defining shape topology, various shape and surface definition and transform operators were all written as procedure calls, then used to define high-level solid modeling operations. This potentially allowed new surfaces to be added as abstract data types—as a data structure and a set of operations for manipulating the new surface type. Some screen images of geometric shapes generated in GLIDE are shown in Figure 2.26. In GLIDE-II, the Pascal Record could represent a geometrical shape, an object, or complex relations. An object could be defined without a predefined fixed shape or a set of attributes; these could be defined later by the user, resulting in a custom object and shape. Alternatively, all of these could be predefined and only an instance ID and transform need be entered by the user. Other variations also could be supported, including instances defined by a few parametric values.

A project was partitioned into multiple separate name spaces, called *Frames*. Frames were similar to directories in a file system, each being named and organized hierarchically. They held both Records and operators and allowed structuring of code making up a building model in a hierarchical database schema.

Legacy applications could be interfaced to the building model, but other operations could be defined and added by application developers who supported end users. Developers could, for example, define design tools to solve layout and specification problems. If a tool design did not fit the current need, it could be revised. The use of a common environment supported the smooth transition between normal use and system development. This smooth transition, where designers and engineers start as users of high-level tools, but then have the opportunity to incrementally redefine existing tools or develop new ones, was an important concept behind the development of

GLIDE and GLIDE-II. User interface capabilities were limited. All commands were entered textually, with entity references given using an entity name, possibly with subscripts.

GLIDE-II's Frame and Record structure encouraged work in schema organization. It was recognized that in construction, a single fixed schema was not practical. Many different construction technologies could be used to solve the same problem (steel, concrete or wood). A semi-hierarchical structure, into which detailed type definitions and operations are loaded from stored libraries as one designs, was explored as a schema definition strategy. Thus a high-level Frame for "structural system" could be defined early with general attributes regarding cost, for example. Later, the detailed schema below the structural system could be loaded with different entities and procedures corresponding to a specific type of structural system. The attributes of a higher level Frame need not know what entities or procedures computed its aggregate properties. The multiple hierarchies formed to support this type of performance and information aggregation were called "abstraction hierarchies".

Frames offered no clear way to add new types of performance after the schema was initialized, since doing so required adding new attributes to existing Frames and Records. Other language extensions were needed to resolve this limitation. Frames did not support inheritance; they were more like modules in Modula-2 (the unsuccessful successor language to Pascal). Instances of Record types were created within a Frame.

A fundamental issue in programming languages is the notion of *type*. Type specifies what operators may be applied to what objects or variables. Until to recently, an object or variable could only be of a single type. Polymorphism is the computer language concept of an object or variable having more than one type. If new attributes were added to an object, how could the computer know which objects possessed the added attributes and which did not? A type of polymorphism called *type union*s was developed in GLIDE-II so that existing entities could have new semantics added to them. Thus a wall or other high-level data object could have new variables appended to receive additional values. A set of appended variables could correspond to a class of performance—for example, for a structural or acoustic class. Similar appendages could be associated with other Records in the Frame tree, which would allow new, detailed properties to be assigned, then aggregated up to the top-level Frame. The motivation for type unions was to allow new performance properties to be added to a Record after it was created. This type of extensibility involved three different levels of complexity:
1. unions of types, allowing a type to be defined as the union of other types
2. unions of types that apply to newly created instances (without values), so that all new instances would have the newly added data fields
3. unions that extended an existing instance already loaded with values
Only the first two levels of capability were implemented. The needed capability in design databases to add new data fields after an object is instanced is still missing from most commercial systems today. The type union feature has close similarities to some forms of implementation of object-oriented systems existing today (see Chapter Four for a more detailed discussion of inheritance).

The GLIDE-II file system also supported design *Alternatives*. At any point in time, an Alternative could be initiated that would redirect the results of all operations to a new file. Work then would proceed, using the data in the new file as primary data, with earlier opened file(s) as the source for entities not in the new file. Secondary Alternatives could be created from an earlier one, resulting in a tree of Alternative files. An example of such a structure is shown in Figure 2.27. Object searches within the GLIDE-II runtime environment were directed up the Alternative tree from the current leaf node. The system would return the first object match encountered; thus an Alternative could overwrite or append new objects or values to an existing model. A user could return to an earlier branch and create other Alternatives, creating multiple branches in a parent-child relation.

Once a base design or Alternative had child Alternatives, e.g., branches, it could not be modified. At any time, an Alternative could be merged into its parent, which then resulted in its siblings being thrown away.

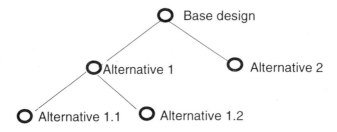

Figure 2.27: An example of an alternative tree in GLIDE-II. If Alternative 2 in merged into the base design, Alternatives 1 and 1.1 are thrown away.

GLIDE-II allowed very flexible development of a building specification. But this flexibility came at significant cost. While the original GLIDE could create an object in a fraction of a second, GLIDE-II took several seconds. Some layout procedures would create hundreds of objects at a time, resulting in performance that was not really interactive, but rather batch operation oriented. Like GLIDE, GLIDE-II wrote directly to disk, without transaction management. Thus if an error occurred, the project might become corrupted. Multiple copies of a project were used as a (crude) recovery mechanism.

2.5.3.2 The CAEADS Building Schema

GLIDE-II was used as a platform for developing a significant building model for the Army Corps of Engineers. It consisted of the Frame structure shown in Figure 2.28, consisting of five Frames. It consisted of three Frames below the project level, for program, building and geometry. As can be seen in the schema, geometry was not emphasized.

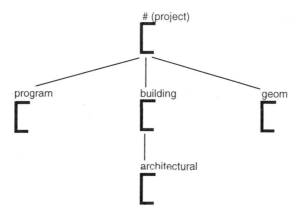

Figure 2.28: The simple Frame structure implemented for the CAEADS project.

The schema within the building Frame is shown in Figure 2.29. It consisted of a hierarchical structure that decomposed a building into systems and components. Systems and parts were defined at multiple levels of abstraction. Performance properties were aggregated upward from parts to assemblies. Boxes correspond to Records with the non-geometric attributes or relations they carry listed inside. Lines identify relations between Records.

Within the building lattice, the top-level Frame was thought to correspond to the project. It included properties important to the overall project as project goals. These were aggregated cost, area, number of units or rooms, and so forth. In addition, it held Boolean flags, each signifying completion of a particular subsystem. These flags served as the top value for each subsystem hierarchy. The top Frame also carried references to its component Frames and Records. At each lower level, additional details were added. The structural hierarchy might be decomposed into aisles, bays and foundation, then these into their subcomponents or, for mechanical systems, into subsystems and zones. There are, in fact, many ways to do such a decomposition. The resulting structure was an early example of an *abstraction hierarchy*.

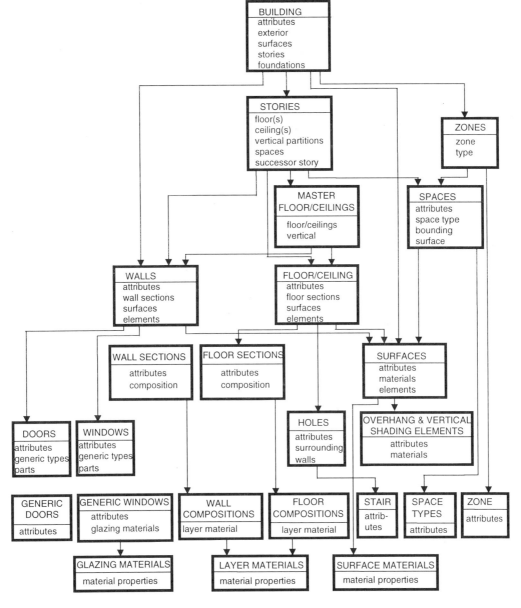

Figure 2.29: Diagram of a building model schema developed within GLIDE-II, incorporating different levels of abstraction, used in the CAEADS project. Different technologies could be plugged into the high-level functional requirements.

The main structure implemented was the spatial one. A building was decomposed into walls (exterior and interior), floor/ceiling structures and spaces. The relations among these were those needed to support updating and editing. Other hierarchies were used to organize information about different functions, where a function corresponded usually to a building subsystem. Each subsystem had a global definition, which, in order to be satisfied, was decomposed into a hierarchy of subcomponents and analyzed. Managing these systems led to a series of applications whose results were aggregated over subsystems, then systems, up to the overall project Frame. We describe the various objects in some more detail below.

The building object was the root of the model. It defined a single space-enclosing structure. It also provided the entry point and access path references to the rest of the design.

The story Record was used to represent stories as well as all floor-ceiling levels between floor/ceiling constructions. A story was stored in a linked list, starting with an inclusive story (the aggregation of all floor-ceilings on a story) and then linked together each floor-ceiling level contained in that story. Each member in the linked lists had a back-pointer to the inclusive story of which it is part. A story references all the walls and spaces that were part of it, i.e., were bounded by that particular floor/ceiling combination.

Zone was a part of the building that was defined as one major spatial unit with a certain purpose. It was usually the combination of a set of adjacent Spaces (see below). For example, a thermal zone could be defined as a part of the building that is controlled by one thermostat. The characteristics needed for describing a zone were specified by the zone type.

Space was the volume bounded by a closed loop of wall surfaces, the upper surface(s) of the floor(s) and the lower surface(s) of the ceiling(s). A Space belonged to a story, is of a given (programmatic) type and has a user-defined Activity.

Wall was a vertical partition within the building. Each wall lay on one straight line, defined on the X-Z plane (horizontal plane). The body of the wall was bounded by two parallel surfaces, two intersecting walls at each end and one or more floors and ceilings. The two endpoints, as determined by the intersection of the wall's line and intersecting walls, were labeled 1 and 2 and were used for determining the wall's direction. The direction was used to distinguish the right-hand and the left-hand side of the wall. A wall could be ended with a line perpendicular to the line on which it lies if an intersecting wall does not exist at that end. The right-hand or the left-hand surface could be divided into segments not lying on a common straight line if the wall had more than one wall section (such as resulting from application of a wainscot or tile on a part of the wall).

The wall composition was defined as a set of wall sections (WallSection) and a list of layers that reference materials in the architectural catalog. Each of the layers held the elements enclosed within it and WallSection held the elements that were in all layers (used in spatial conflict checking).

Floor/ceiling was any building partition that was not vertical. A floor/ceiling was a body bounded by two surfaces, the lower one and the upper one, and a set of perimetrical surfaces, which when combined formed a closed loop. The upper and lower surfaces could lie on planes not perpendicular to the X-Z plane, and not necessarily parallel to one another, as long as the planes did not intersect one another within the perimeter boundary. The perimetrical surface belonged either to a wall (the common case) or is derived through an intersection between two floor/ceilings.

`Hole` was a body composed of a space and its bounding walls. It extended vertically over several levels and cut through at least one floor/ceiling, e.g., a stairwell. `Element` was an object that was embedded within another object. In walls, for example, these could be windows and doors, or piping and ductwork.

Material and material properties were all defined in libraries accessed separately.

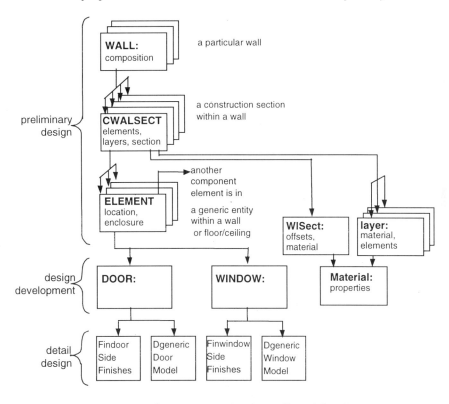

Figure 2.30: The GLIDE-II data structure for the wall model in CAEADS.

Each design specialty was assumed to have its own hierarchy (this one being the architectural hierarchy), with a subset of the properties needed to manage the performance of concern. It was the different functional areas that led to components being part of multiple hierarchies. While each specialty could be treated as a hierarchy, all of them together required a partial ordering or semi-lattice.

Walls could be defined at different levels of detail, depending upon the stage of development. A wall's structure is shown in more detail in Figure 2.30. Initially, they were defined as centerlines. Their topological connectivity with other walls was defined, so that if one wall was moved, all those attached to it would grow or shrink. Doors and windows and other elements were located within the wall. The detailed wall definition was a solid model, which was parametrically defined so that its shape changed in relation to the walls to which it was connected and to the location of the floor and ceiling surfaces. Its construction was defined in layers, consisting of a structural core material with finish layers defined on both sides.

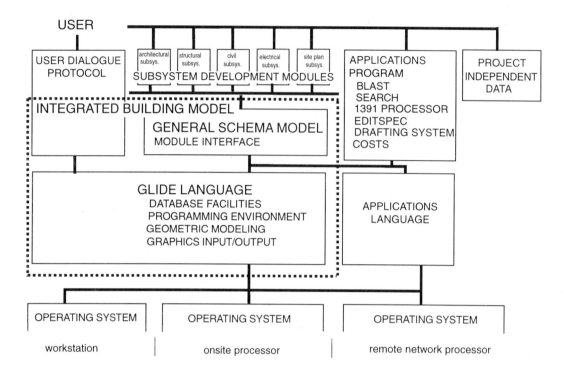

Figure 2.31: The system architecture of CAEADS.

The CAEADS schema was defined as a collection of abstract data types. Thus operators were defined in each Frame for the objects defined within it. A limited form of object inheritance was utilized that did not include procedural inheritance. There was more focus on defining parametric shapes and embedding complex relationships than determining a single fixed model.

CAEADS was a large system organized with several modules, as shown in Figure 2.31. The General Schema Module and Integrated Building Model were implemented in the GLIDE-II Language, the Applications Language. The schema supported various applications and this was assumed to be an extensible set. The application modules were assumed to involve layout and simple configuration testing, as well as interfaces to external applications. The University of Michigan developed the habitability application and the interface to BLAST, a dynamic load, energy analysis program. Of those shown, only the 1380 specification, the habitability application and portions of BLAST were fully integrated. A CAD drafting system was purchased and although the plan was to integrate it into the environment to support quick revisions, the integration was never completed.

The overall user configuration was that shown in Figure 2.32. It was conceived so that several users could access the building model in time-sharing mode (which today would be considered client-server mode). They would first define the building model instance and then later extract data from it as input to the various applications. Some applications would be executed centrally on the model itself, while others would eventually run at remote sites. In practice, CAEADS had only file locking as a means to control access, limiting access to a single user at a time.

Graphical interaction with GLIDE-II was through an Evans and Sutherland Picture System, an early line of high performance, vector-based, real-time graphical hardware. While geometric models could be developed on that system or on the VAX machine on which GLIDE-II ran, proper communication between them was not fully operational.

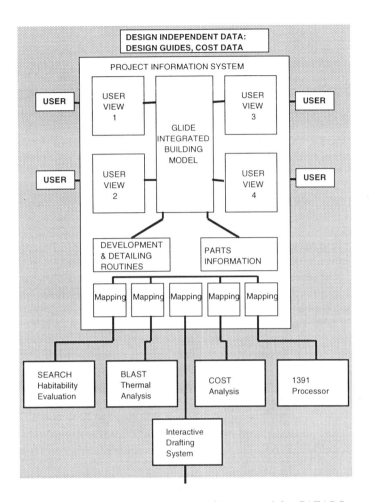

Figure 2.32: A general diagram of the user architecture of the CAEADS system.

The funding for CAEADS ended in the Fall of 1981. Later GLIDE-II was made available in the public domain and was picked up by several commercial firms. It was developed into at least one commercial product (unrelated to buildings).

The GLIDE-II and CAEADS effort involved a significant amount of explicit systems-level development. It was one of a few efforts to develop computer languages supporting design. Some of the language features developed in GLIDE-II were later incorporated into object-oriented databases. Because it specifically dealt with a programmable database model, CAEADS involved an explicit building model schema that had some of the characteristics of current building models, including a high-level core building representation, with extensible, construction technology-dependent subschemas. The work on GLIDE-II development focused on the kinds of language/database features needed to support building model extensibility, a continuing research issue. GLIDE-II also was based on modern solid modeling geometry. Thus each shape was a parametric solid, which could be resized during design by changing its attributes.

The sequence of efforts in developing software facilities for building modeling by the Carnegie-Mellon team identified a number of open-ended research issues:

- One issue was how to develop appropriate transaction management methods allowing concurrent access to and updating of a building model by multiple users.
- Another issue was how to specify and manage relations during design. While the meaning and use of attributes to define the properties of building objects were reasonably well understood, the definition of relations was not.
- Finally, a third issue was how to set up intermediate object abstractions for the designer's use. Design works through a lattice of abstractions, moving incrementally to finer levels of detail. At the bottom level are the physical components making up the design, which are well understood. However, those at the intermediate stages are highly varied and reflect different design strategies. For example, spaces may be defined before walls, then walls as centerlines, then with a thickness. How should these different abstractions be organized?

2.6 SUMMARY OF THE MID-1980S

The projects reviewed in this chapter span a period from the late 1960s to the early 1980s. While the hardware used in CAEADS was far superior to that used in the SSHA system, other aspects moved at a slower, even negligible pace. All approaches represented a building as a set of components with associated attributes—what is now called an object-based representation. The initial building models integrated special-purpose applications, while later ones began to integrate applications that were stand-alone. All were designed to address both design and construction-level details. The schemas of the SSHA and OXSYS systems were mostly fixed for a particular type of construction technology. The CAEADS system supported schema extensions during design and thus was potentially more general, but it supported extensibility only in a particular top-down order. All systems relied on single hierarchies of predefined objects, organized into multiple levels of abstraction.

By the early 1980s, it was apparent that a building's description required multiple, overlapping hierarchies, one for each method of analysis during the design phase. Also needed was a way for users to define the various intermediate levels in the hierarchy—corresponding to intermediate abstractions—as well as the bottom-level components. In architecture, different designers, or even the same designer doing a different type of project, might use different abstraction hierarchies, though in structures, acoustics or energy, the levels in the hierarchy were somewhat more regular. All the systems here predefined the levels of abstraction, but this met with resistance from many architects, who developed their own abstractions unique to their design process.

Another general issue was how to deal with side effects. Locating a window also created a potential light source for a lighting program. Placing mechanical equipment in some location also created structural loads. The functions that an object affects depend upon the behaviors being considered. All these systems worked well if a fixed set of functions was built into the model. However, extensibility called for more powerful procedures and was recognized by all three research groups as an unsolved problem.

Each of these efforts developed a building model as part of a custom designed design environment, with applications operating on the building model. At the time of their development, there were few external design applications, and thus the need to integrate multiple external applications was not heavily considered. The CAEADS project was the only one to interface with standard applications. The interface, however, was one way, unidirectional from the model to the applications. It was recognized that some applications create data that would be useful to other applications, but managing the identity of application data, identifying which model instances were modified, was an unsolved problem.

These three efforts were developed using very limited computational capabilities, in regard to processing and display capabilities. CAD systems in the early 1970s were typically limited to 32K words of memory with much less than one MIP speed. Displays could only deal with a few thousand vectors per refresh (without filled polygons). And solid modelers and geometric modeling capabilities had yet to be commercially developed. When viewed in this light, these early capabilities were ambitious, providing examples that even today are not easily replicated.

2.7 THE CURRENT TRANSITIONAL STATE

The easiest way to utilize a new technology is to use it to replace existing manual tasks. The framework and goals of the task are well understood and only the task methods need to be revised. Both the marketing staff of CAD companies and the early users recognized CAD's potential in production drafting and the systems evolved to better respond to this particular task.

From this rationale, based both in terms of how firms could effectively utilize this new tool and how CAD companies could market them, initial, crude systems were refined to support electronic drafting. The features needed to effectively support drafting emerged through market competition. In other areas, electronic calculation of manual, analytic methods were replaced with computerized ones in engineering. Rendering also was automated, first by computer-generated perspective line drawing, then later by visualization programs supporting shaded color definition and lighting simulation. In large offices, such tools could be selected for use by a department head or project leader and were often purchased and paid for by a single project. Such adoption on a task-by-task substitution basis allowed computers to be adapted without challenging basic practices or the way people had been educated. It also allowed the effectiveness of computers and the decision to use them to be assessed, at least roughly, against doing the same task in the traditional, manual way. As a result, productivity has been a continuing justification for the purchase of a CAD system.

Early automobiles were called "horseless carriages". This mental concept implies that automobiles were thought of as an augmentation of the existing carriages of the day. The revolution in highways and transport came later. In the same way, computer-aided design and drafting (CADD) is the horseless carriage equivalent of computers in building. Most of the traditional tasks involved in manual design and engineering have been automated—drafting, 3D rendering, engineering analyses—in horseless carriage style. In construction, the process of adaptation of computation is not much different. Construction firms have automated their accounting systems, like all businesses, and use electronic databases for cost estimating. Critical path scheduling and PERT charts were used in project management before computers became available, but are now done electronically. Other tools are available for earthwork planning, for managing storage and preparation areas within the construction site, and for automatically defining and laying out concrete formwork and defining the material list for the formwork. That is, packages now exist for most of the tasks of construction management that were previously done manually.

Not all CAD efforts were directed at "horseless carriage" models of building design in the 1970s and 1980s. A number of efforts attempted to redefine the process of design and, in some cases, of building. The three systems reviewed here are examples of the more radical efforts. Many judgments of the time considered these systems intellectually interesting, but too radical to be accepted by current practice. Hoskins' comment about being naïve in believing that a new tool could lead to radical new processes is instructive. Convincing a customer of a new tool's worth also required teaching the potential customer a new way of designing and operating, something that was often resisted and very expensive to the seller. Twenty-five years later this reality has not changed.

Productivity assessment is based on a comparison of doing similar tasks, using different media and technologies. Productivity is not an easily applied argument when significant changes are made in the methods of production, because its effectiveness cannot be easily compared with previous procedures. It can only be assessed at some large aggregate, such as a total project, after a long learning curve.

Of the two paradigms of CAD presented at the beginning of this chapter, it is fair to say that the dominant usage in the building industry is still based on using CAD as a graphics editor. The primary representation for architecture, civil engineering and construction work associated with buildings is a set of drawings. In the effort to support sophisticated users, most CADD systems also support application development. However, appending these to a general-purpose CAD system has proven a complex undertaking, with changes being required to the CAD system. Only applications in some specific areas have been successful, such as piping, electrical and mechanical systems. Almost all of the applications are integrated at the level of platform only. That is, while graphics can be exchanged, attributes and relations among objects cannot be passed among applications in different domains.

Now, in the late 1990s, with current markets becoming saturated and the price of a CAD system dropping, major CAD companies in the building area are looking to advance the market to another level. Efforts are being made to develop new platforms from the bottom up for application development. These include the newest application development tools and libraries. A new paradigm for building modeling is emerging, based on 3D and solid modeling, object-oriented languages and databases, effective graphical user interfaces and Web-based communication. It is also recognized that they must provide means to coordinate and exchange data between separately developed vendor-produced applications.

While the application software part of these environments may vary, there are important reasons to develop a public and sharable data description of the building.

- Users are likely to want to use applications developed by different software vendors. It is unlikely, possibly undesirable, that any one CAD vendor should supply all building applications. Thus, there is a need to have a common framework that can exchange data across vendors.
- Building product suppliers, such as those providing information in Sweet's Catalog and Architect's First Source, are ready to provide their product information in a format that can be readily used to produce designs and support various forms of analyses. They need to have a standard format to put this information in so it is accessible by all.
- With the downsizing of most design firms and the move to collaborative team design, there is an increasing need to exchange design information electronically in an easy but reliable manner. No such process exists today unless all members of the design team agree on a single CAD system.

The challenge before us is

> to develop an electronic representation of a building, in a form capable of supporting all major activities throughout the building lifecycle.

This challenge is the focus of *building modeling*, and is justified by its support of more significant changes in current practice. Its advocates claim that building modeling facilitates changes to the process and conceptualization of how buildings are designed, how they are constructed and operated. Visionaries in the field point out many new possibilities: electronic walk-throughs of candidate designs; rapid development and simulation of a design's appearance and performance, allowing refinement and possibly optimization during design development; planning and simulation of construction while still in the design phase; development of computer-aided manufacturing production methods of building components; and new methods of coordination of work crews on the construction site.

Building models will only be realized when a significant professional community needs to use information represented in a building model to design, construct or manage buildings—buildings that are more complex, easier to construct, and more easily maintained than is possible with current technology. That is, building modeling must prove its advantage in the marketplace. The user community will have to move from current practices, based on a paper-generated information trail, to electronic ones. This applies to architects, civil engineers, building product suppliers, contractors, building regulation organizations, and building operators, among others.

It is quite possible that the development of a broad-based, sophisticated user community that can perceive the advantages of building modeling could happen only after electronic drafting and simple add-on applications have first introduced the techniques and practices of computers in building. That is, the learning curve in building may entail incremental steps from current practices to future ones. Even now, the challenge is not as much the goal itself as it is making the incremental transition.

2.8 NOTES AND FURTHER READING

The major publication from Sutherland's thesis was [Sutherland, 1963]. There have been very few papers describing CAD system architectures. Two are Sedgewick [1974] and Eastman [1990].

The work of Coons was the first to focus on modeling an object, rather than a drawing of an object. The original report was Coons [1967]. Coons' patches are a restricted form of Hermite surface that is no longer widely used in commercial CAD systems. For a review of Bezier's work and contribution to CAD, see Bezier [1993]. The volume incorporating the Bezier paper presents a history of the basic concepts in computer-aided design.

The seminal papers on solid modeling are Baumgart [1974], [1975], and Braid [1973] and [1975]. Early formal work on solids was presented in Requicha [1980]. A survey of early work is presented in Baer, Eastman and Henrion [1979]. The standard text for solid modeling is Mantyla [1988]; a more general survey for surfaces and solids is Foley et al. [1991, Chap. 11].

The three different efforts in building modeling surveyed include much material never presented in published form, though some were presented at conferences. The author has drawn upon these documents and on technical reports and personal correspondence with the original researchers.

The Scottish Housing Council system developed by Aart Bijl's group is presented in internal reports that may not be widely available [Bijl, Renshaw et al., 1971], [Bijl, Shawcross et al., 1974]. A published report is by Bijl and Shawcross [1975]. An early review of this work is given in Bijl [1979]. Bijl's later work, reflecting on these early efforts, is presented in Bijl [1989].

The BDS system developed by Applied Research of Cambridge is reported on in Hoskins [1973], [1977], [1979a], [1979b]. This section benefited greatly from comments and internal papers provided by Ed Hoskins and Paul Richens.

Three different building model environments were developed by the author's group at Carnegie-Mellon University. The first, Building Description System (also BDS), is reported in Eastman and Baer [1975] and Eastman, Lividini and Stoker [1975]. The second system, Graphical Language for Interactive Design (GLIDE), was presented in Eastman and Henrion [1977] and Eastman [1978]. The third system, GLIDE-II, developed for the US Army Corps of Engineers, is presented in

Eastman [1980b]. Some of the system-level issues are reviewed in Eastman [1980a]. Treatment of some complex relationships embedded in a model are given in Eastman [1981]. All three systems are reviewed in Eastman [1993b]. This paper includes sample code written in GLIDE-II. Some aspects of the CAEADS wall model are presented in Yasky [1980]. Detailed documentation of the CAEADS model schema is in Yasky's Ph.D. thesis [1981]. The transaction management issues growing out of the CAEADS effort, and a conceptual solution, were reported in Kutay and Eastman [1983].

Other historical surveys are Mitchell [1977] and Galle's annotated bibliography [1994]. Other significant early efforts included the CEDAR [Thompson, Lera et al., 1979], [Thompson and Webster, 1978] and HARNESS [Meager, 1973] hospital design efforts. The Karl Koch and Bolt, Berenek and Newman system, supporting the Techcrete construction system, was reported in Myer [1970]. Harold Borkin's group at the University of Michigan undertook important early efforts, but few journal articles were ever produced. Two are Borkin et al. [1981] and Patricia McIntosh's Ph.D. thesis [1982].

The line of development initiated in the 1970s and presented here has continued to the present time. Some of these efforts include the KAAD modeler by the team of G. Carrera and G. Novembri at the University of Rome, and Y. Kalay at the University of California, Berkeley [1994]. It incorporates its own design applications and focuses on the representation of design knowledge and goals. SEED (Software Environment for Early phases in Design) is a project at Carnegie-Mellon University by Flemming et al. [1995] that also incorporates special applications and focuses on new methodologies for the early design stages. Another knowledge-based modeler that incorporates default knowledge and some knowledge of presentation formats is the CAD/CG system by Watanbe [1994]. The ICADS system by Pohl and Myers [1994] incorporates multiple actors, implemented as expert systems, that automatically apply their expertise in the development of a design. Another knowledge-based system that incorporates heterogeneous applications but provides a consistent user interface is KNODES, by Rutherford and Maver at University of Strathclyde [1994]. The IBDS developed by Fenves and his colleagues at Carnegie-Mellon University was another important example of integration, focusing on a suite of special applications around a common representation [1994].

2.9 STUDY QUESTIONS

1. To better appreciate the historical evolution of CAD systems, one can look up old copies of trade journals, such as *Computer Graphics World* or *Cadence,* and examine old articles and advertisements, for example, in the 1975-85 timeframe. Select some now common capabilities, such as integrated visualization, or development of an application language, or attributes associated with a graphic entity, and try to plot a history of these features, identifying when they first became available, how they spread to many CAD systems, and how they evolved over time.

2. For a specific application area, such as energy analysis or construction of formwork, identify the entities and attributes needed to describe the objects used in that application. Also, identify some of the topological relations.

3. As a study of the history of computer-aided building design, take an early research or commercial system and trace its evolution, in terms of the principal parties involved, the special concepts it used, and the innovations it made. Systems that might be studied are the CEDAR and Harness hospital systems in the UK, the commercial Techcrete system, and the ARCH-Model program from the University of Michigan in the US, and the early commercial architectural CAD systems, such as TriCad, Calma, and GDS.

CHAPTER THREE
Early Product Exchange Standards

There are a great many existing specialized languages and programming systems for many of the individual areas which must be covered by a Computer-Aided Design System, but each of these languages and systems has its own restrictions and interwoven computational complexities so that it would be completely impractical to attempt to integrate such systems in a straight-forward manner. Furthermore, such a brute force approach would not satisfy the Computer-Aided Design requirement in the first place, since there would be little or no cross fertilization between the various systems... . In fact to postulate the existence of a closed system for Computer-Aided Design as we mean it is completely and absolutely contradictory to the very sense of the concept ... the very nature of the system must be such that its area of application is continuously extended by its users to provide new capabilities as the need arises.

Douglas Ross and Jorge Rodriguez
"Theoretical Foundations for the Computer-Aided Design System"
AFIPS Conference
1963

3.1 INTRODUCTION

The need to transfer data between engineering applications has existed since before the development of the first CAD systems. The statement quoted above was in a paper on information modeling in design, at the same conference session where Ivan Sutherland introduced SKETCHPAD, the first CAD system. Ross went on to develop System Design and Analysis Technology (SADT), the early popular process modeling language used by many US industries. About the same time, ICES (Integrated Civil Engineering System), one of the larger efforts developing a suite of engineering analyses in the late 1950s, identified the problem of data exchange and proposed some initial methods for dealing with it.

The need for data exchange is an inherent one, arising from the collaborative nature of design. Some examples of conditions leading to the need for data exchange include

- translating data between two different CAD systems used by firms jointly working on the same project, but using different CAD systems
- extracting data from a CAD system for input into an analysis application, such as structural or thermal, without intervening manual translation, thereby saving significant time and avoiding the probability of translation errors
- loading the output from some external application that generates design results into a CAD system or other general design application
- loading data describing an already constructed building in an archival representation into a CAD system format for planned renovation work
- reading in details or building parts from a parts catalog represented in some general format for inclusion in a CAD project, and possibly for use in analyses
- extracting base information from construction drawings from which to generate shop drawings

   • deriving bills of material automatically from construction drawings
Such needs arise continuously throughout the lifecycle of a building. They are encountered most densely, however, at the beginning of the lifecycle, during design and construction.

3.1.1 METHODS OF CODING DATA EXCHANGE

In the middle 1970s, CAD system companies learned of the need for data exchange from their customers, and some developed means for reading data into and writing the data out of a CAD project file. A variety of low-level methods are technically possible:

a) writing parts of a project out to a file, translating into a textual format, and conversely, reading such a text file into a project, and parsing it and executing commands to create the described objects, while the system is running

b) writing a standalone application that reads the file format in which project data are stored on disk, interprets it and copies the project data, then writes it to a file in a text format, for reading into another application using method a) above

c) providing subroutine calls that can be executed from other programs to extract data from the file format in which projects are stored on disk and create objects within a different application

d) providing subroutine calls that can be executed from the CAD program, allowing writing of data in the desired format or alternatively

e) adding the application to the CAD system environment, and calling it from the CAD system so that both are running at the same time

These alternatives vary according to where the data are residing when accessed and what program is doing the accessing. The data may be loaded into the runtime data structures used within the CAD system or sitting on a project file. Control of data access may be in the CAD system, in a standalone application, or in the application receiving the data.

Today, all of these methods have been widely used. In the first case, Intergraph offered its users the SIF format; Autodesk offered its customers the DXF format and other systems have their external formats. Methods b) and c) require means to read data on a project file, independent of the CAD system. That means that all the entities on the file are read, making accessing a subset of the data difficult. Method d) is supported by providing a library of subroutine calls that are embedded in the application; the CAD system, however, provides the querying (selection) capabilities and allows calling the application on the selected entities. The quality of the external file formats and embeddable subroutine libraries provided by different CAD vendors varies greatly.

Proprietary formats provided by CAD system companies are controlled by the companies and can change at any time. They are developed for the interests of the vendor and not necessarily the user community at large. As a result, continuous efforts have been undertaken to develop non-proprietary standards, developed by national or international standards bodies.

3.1.2 NEUTRAL FILE APPROACHES

An important concept that emerged at the time the first CAD data exchange standards were being developed was that of a neutral file. The neutral file is a response to the combinatorial problem that arises when several engineering applications need to exchange data.

Suppose we have two applications, A and B, and we want to exchange data in both directions between them. If we develop custom translators, using any of the methods listed in Section 3.1.1, we will need to write two of them: A-to-B and B-to-A. If we have three applications, however, we need to write six translators: A-to-B, A-to-C, B-to-A, B-to-C, C-to-A, C-to-B. If we have four applications among which we need to exchange data, there is the need to write twelve translators. That is, in order to add a single application, one has to write 2*N translators, where N is the existing number of applications in the suite, diagrammed in Figure 3.1. This makes the development of a large suite of interconnected applications very expensive.

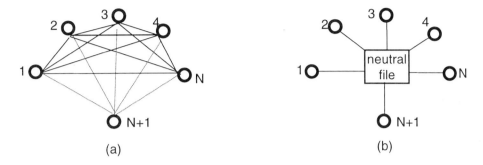

(a)   (b)

Figure 3.1: The normal method of data exchange (a), and an approach using a neutral file (b). Each edge between two applications corresponds to two translators.

The important recognition is that if a common, shared data format can be defined, from which all applications can exchange data, the number of translators needed for any one application is two—one to write to the shared data format and one to read from it. The shared format is what is called a neutral file format. See Figure 3.1(b). This recognition has been one of the major concepts applied to all efforts to develop data exchange standards.

In this chapter, we review the most common, early data exchange methods and standards. In Chapter Five, the current efforts in data exchange using ISO STEP are reviewed. In order to allow comparison and evaluation, we use a set of increasingly complex examples, described in Section 3.2. The first of the exchange methods to be considered is DXF, as only one but certainly the most widely used proprietary exchange format. The principal standard exchange format in the US, IGES, is reviewed next, so that the differences between DXF, IGES and the newer standards are clearly shown. The rest of the chapter is devoted to reviewing these two standards, and developing criteria for the evaluation of product data exchange methods and tools.

3.2 EXAMPLES OF EXCHANGE DATA

In order to articulate the various translation capabilities presented in this and later chapters, we will rely on an increasingly complex set of examples:
- an arc
- a planar 3D surface, with a surface pattern
- an arbitrary 3D solid shape, as might be used for a building part
- a model of a wall, supporting multiple views and applications

Each of these examples is described below in terms of its data and semantic requirements.

We must also become increasingly precise regarding concepts and terms. Throughout the rest of this chapter, and the rest of the book, we will be using the following terminology:

entity: refers to a graphical or physical element within a CAD system, data model or the physical world. It may be a class or an instance, as defined below.

entity class: refers to an abstract grouping of elements having similar structure and properties. An entity class may be a generic concept, such as "wall" or a specific type of product, such as Pella Window Model W48X36.

entity instance: refers to an individual entity of an entity class. Entity instances may be counted and their count will correspond to the number needed to realize a project.

3.2.1 ARC

An arc is a continuous portion of a circle within a Cartesian coordinate system. Thus, an essential definition of an arc is the circle of which it is a part, and the part, minimally defined by two angles that, from the circle's center, define the boundaries of the arc. See Figure 3.2. The two angles partition the circle into two parts and the part of interest must be designated. To accomplish this, the two angles are typically ordered and a direction implicitly imposed, or a flag may designate the order (clockwise or counterclockwise). Thus, the minimal information for representing an arc is a coordinate point, a radius, two angles and a direction flag. There are many variations used in different CAD and drawing systems, such as an implied direction—always clockwise—or use of three points—for center, beginning point, endpoint and implied direction. In the last case, there is a constraint that the distances from the center point to the two bounding points are equal. Some representations also use an arc representation for circles, designating a circle if the start and endpoints (or angles) are the same. Arcs can be edited by revising any of their parameters, for example, their centerpoint, radius or two angles.

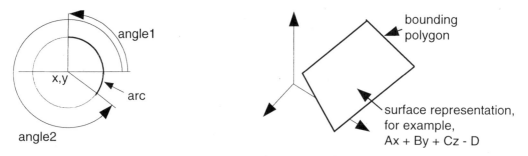

Figure 3.2: The representation of an arc.   *Figure 3.3: The representation of a plane.*

A minor issue is that we have defined an arc as pure geometry. We haven't said anything about its line weight, color, the style of line (solid or dashed) or whether it has an arrow at its ends. These are all presentation attributes of a line, e.g., how it should be presented in a particular drawing. We will not consider presentation attributes here, nor of surfaces later, as these are large topics of representation in themselves. Consider, for instance, a line of varying width, which is outlined with an embedded shading pattern within the outline. A geometric arc is the simplest example we will consider.

3.2.2 PLANAR 3D FACE

In Chapter One, we argued that three-dimensional geometry is a necessary step in realizing effective building models that are useful throughout the building's lifecycle. The simplest form of a fully 3D entity is a planar face, as diagrammed in Figure 3.3. There are two parts to a planar face: the surface that it lies on—here a plane—that has infinite extent, and the bounded region of interest. Nonplanar faces rely on the same general strategy: definition of the surface, then of a polygon on the surface that trims the surface to an area of interest. Faces are the basic entity used in most 3D modeling, for example, for modeling construction parts and equipment and for modeling the buildings in rendering packages.

The first issue is the representation of the face's surface. There are many representations for a planar surface; most curved surfaces have control points that can be specified, so they are planar as a special case. The representation given in Figure 3.3 uses the *implicit* form,

$$Ax + By + Cz - D = 0$$

which specifies the conditions for any point lying on the plane. The implicit form can also be defined using vectors,

$$(A,B,C) \cdot (x,y,z) = D$$

An *explicit* vector form also can be defined by specifying a point **c** and two vector directions **a** and **b**. Two intersecting vectors uniquely define a plane:

$$p(u,v) = c + au + bv$$

In the implicit representation, the surface ranges to infinity. For special cases, the surface parameters in the explicit representation can also define a bounded region. In it, u and v are range parameters that can be caled to specify a parallelogram region. Alternatively, the point and two vectors can be used to define a triangle. For triangles, the points define both a unique surface and the shape of the triangular region. Many surface modeling applications rely on triangulated faces for the simplicity of representation and use.

Suppose, however, that the shape of the face in which we are interested is not a parallelogram or triangle. Suppose it is a polygon, such as is defined by the wall shown in Figure 3.4. The wall has five sides, plus a door and window; that is, it is a polygon with holes (the door can be treated as a notch instead of a hole, if we assume there is no threshold). Either a much more complex polygonal boundary is needed, or else the wall face needs to be decomposed into a set of, say, triangular faces. Some possible decompositions of the wall polygon are also shown in Figure 3.4.

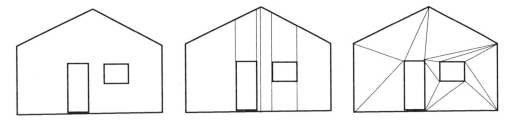

Figure 3.4: A complex polygon and two alternative decompositions of it, first into a set of quadrilaterals and then a set of triangles.

The general representation of a planar face consists of a surface and one or more polygons located on the surface. In Figure 3.4., at least three polygons must be used: one outside and two inside polygons, which must not intersect one another. In most surface modeling systems, only one outside polygon is allowed. Some surface modelers don't allow nested polygons, in which case the inner ones must be defined by connecting them to the outside, making a seam in the surface, as shown in Figure 3.5. (The separation between lines connecting the outside loop to the inside one are exaggerated to make the structure of the single polygon more apparent.)

Figure 3.5: Approximation of a polygon with holes.

Figure 3.6: The set of surfaces making a solid model of a wall with door and window.

Figure 3.7: A simple solid shape with a cylindrical plan centerline.

Some additional issues must be noted. The bounding polygon may be linear or curved. It may have more than a sequence of points and may include various curves or approximated curves. Also, most graphic systems rely heavily on the normal of the surface (the vector at 90 degrees to it), which defines an orientation of the face. A planar face has a single normal, while curved surface faces have many, possibly one for each pixel in which it is displayed. Normals are derived from the surface definition.

Additional issues involve the ability to edit the face; these are the most subtle semantics of a model. Any representation of a face should allow the face to be spatially transformed: translated, rotated, scaled, mirrored, viewed in perspective, as a whole, without worrying about its internal structure. More elaborate editing, for example of the surface or the bounding polygon and the coherence of the model with regard to editing, will depend upon the representation and operators used.

3.2.3 SOLID SHAPE

A solid model is the representation of a 3D shape as an enclosed volume. The solid model may define a shape that corresponds to a solid object or to a space that has a bounded shape and volume, such as a room. Solid modeling was surveyed in Chapter Two. An important capability of solid models is that plans, sections and orthographic or perspective projections can be automatically derived from a solid. Thus, they can be used to generate the geometry for a set of drawings. Currently, this is the only reliable way to guarantee that all the geometric views of a shape—for example, a set of plans and elevations of a building—are consistent, i.e., that they represent a single building.

We define two alternative solid models: (1) a solid shape for a wall, with a door and window, as shown in Figure 3.6, and (2) a curved wall, as shown in Figure 3.7.

3.2.3.1 Wall with Multiple Views

The most detailed example for our review builds upon research in building modeling in the CAEADS project of the 1980s, described in Chapter Two, to provide the detail definition of a wall entity and its components. Functionally, a wall is a boundary between two spaces, one of which may be the external environment. The dimensions upon which the boundary acts include people access, sound, air flow, light, heat, smell and vision. While walls are occasionally designed with explicit performances in all these dimensions, usually they are not articulated except where special conditions exist. Special cases include the netting and screen walls used in hot climates and tents, and the special barriers created in modern clean rooms, where computer chips are built. In addition, a wall is often a structural element, providing part of the load-carrying capability of the building's structural system. Last, a wall is sometimes used as a chase area, for routing other systems.

Given a schematic wall entity, several types of wall construction may be specified. These can be considered abstractly in terms of generalized classes. A *core wall type* is composed of a core that provides its support and layers on both sides that modify its performances. The core may be wood or metal studs, brick, concrete or other types of construction. A core wall is diagrammed in Figure 3.8.

Such a wall may have several different cross-sections. The different cross-sections result from openings and segments. *Segment* entities provide the main part of the boundary and provide for its structural rigidity. Each wall segment is a homogeneous piece of layered construction, with the same core and built-up layers on both sides. A wall is decomposed into segments satisfying these conditions. A segment is further decomposed into three other entities: a structural *core*, made, for example, of a configuration of wooden or metal studs, brick or rubble, optional *insulation* that fills interstitial space within the core, and various *layers* of material that are applied to either side in some order. Layers may include plaster, gypsum board, paneling and even paint or wallpaper. The insulation and layers modify the boundary properties of the core. *Openings* modify the properties of the wall according to light, sound, access and so forth. Thus, an opening may be specialized into a door or window. Openings are defined by a *hole* and a *filler*. *Pass-thrus* use the wall as a routing chase.

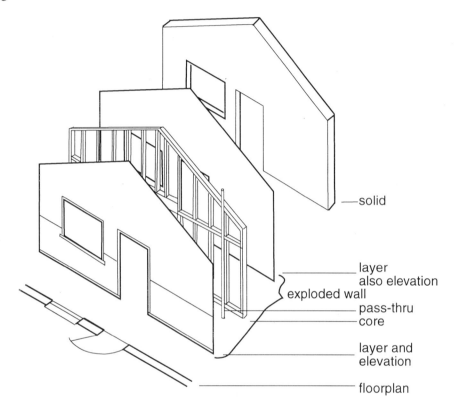

Figure 3.8: An isometric drawing of a core wall, with two openings, layers defining different segments and a pass-thru. A solid model of the aggregate wall is at the rear.

Each of the wall's entities has one or more descriptions, as needed for dealing with the wall over its lifecycle. Four different geometric views of the wall are used. For schematic design and later for facility management, the wall is represented in a floorplan view. For finish construction, it is represented in two elevations. The wall as a whole is represented as a solid model, with properties. Each of its components also has a geometric definition. The wall segment is geometrically defined with a vertical polygon showing the region of the wall it is to occupy, a left and right offset from a centerline to the two finish surfaces on both sides, and a derived area. Openings are also defined with a vertical polygon, an area and a symbol defining the type of filler. The vertical polygon areas are used to derive the energy flows through the wall sections and openings. Pass-thrus are geometrically defined as a centerline and also a solid model, the latter used in order to check spatial conflicts. The insulation and layers are defined by thickness, the core offsets, and optionally, a set of solid models of its components, for spatial conflict testing and construction.

In addition to the wall geometry, we consider one form of analysis, in this case, analysis of a wall's energy resistance. A segment's energy flow is based on its area and the heat resistance of its core, insulation, and layers. For openings, the resistance is the r-value of the window times its area. In a real-world application, it is likely that contextual issues would lead to other forms of analysis, such as lighting, structural, or acoustical, that would have to be added as design proceeds.

The wall has been defined in an integrated fashion. However, most users will not need the entire wall description, relying instead on different subsets. Each of these subsets of the complete wall description is called a *view*. A person interested in building the wall will be interested in its geometry and materials and this person's view will consist of these descriptions. An energy specialist will be interested in the thermal properties of the wall's materials and their surface area. Such partial descriptions are also used by computer applications, for example, for generating or evaluating part of a building design, or for surveying its materials. Such views are of great importance to all aspects of building modeling because it is through different views that all users and all applications interact with the model.

A variety of integrity relations must be satisfied between the descriptions of the components and the wall assembly if it is to define a realizable wall. Here, integrity focuses on the consistency of the various representations. First, we consider geometrical consistency. The segment and opening polygon definitions must be spatially disjoint but fully cover the wall elevations. Their areas must be consistent with their polygon descriptions. The core and layer thicknesses must add up to the thickness of the segment. The solid model of the wall must be consistent with the union of all the segments. Pass-thrus must be spatially disjoint from openings but be subjoint to the wall geometry.

In terms of topological relations, we assume here that every wall is adjacent to two spaces (one possibly being the outside environment). The wall is bounded on all sides of its perimeter: to walls it abuts, to one or more floor surfaces and one or more ceiling surfaces. A special case is a wall that abuts what might be called a NULL surface—resulting in an open edge. A wall has at least one opening or segment, though there may be many. Pass-thrus are optional. Each segment has exactly one core, but can have any number of layers on each side. Here, the geometry and thermal properties are all assumed to always exist for each part, on a one-to-one basis.

Editing operations need to maintain the consistency of the wall data. Thus, there needs to be an overall structure for the wall, allowing all its parts to be moved together, as in the solid model example above. If one of the geometric properties of any part of the wall is changed, the other parts need to be updated to be made consistent with the change—for example, the core structure needs to be changed if the door is moved. If materials assigned to the layers change, then the thermal properties and thicknesses also need to be updated. There are very many such relations, of which only a few are noted here.

These four examples—the arc, the planar surface, the solid and the wall—will serve as test cases for the different modeling and exchange formats reviewed in this and later chapters.

3.3 DXF

From the middle 1980s onward, PCs grew to become the dominant platform for architectural- and building-oriented CAD systems. Initially, many systems vied for leadership. After heated competition, AutoCad®, developed by Autodesk Inc., became the largest selling CAD system in the architecture, engineering and construction (AEC) market worldwide, followed in the US by Bentley Systems' Microstation®. In Europe, other systems competed for a significant piece of the market, such as Nemetchek's Allplan®. In the late 1980s, the use of AutoCad in conjunction with the older, more powerful workstation-based CAD systems became common. Sometimes joint use arose when a project was encountered that involved different firms using different CAD systems.

In other cases, firms may have migrated to AutoCad, but still maintained files developed on another system, or alternatively, they may have used multiple systems simply because of the different functionality each provided.

Autodesk's internal file format is the DWG format, kept confidential until recently. In the late 1980s, they responded to the demands for data exchange by defining a public, external file format called Data Interchange Format (DXF). DXF was defined in both a textual and a binary format. It was limited to the kinds of graphic entities and conventions that were incorporated in AutoCad. Because DXF was initially based on a limited set of assumptions, particularly regarding associativity, translators could be easily developed for it. Several software consulting companies started writing custom translators between DXF and workstation CAD systems. Later, other CAD system companies responded to user demands and developed their own translators into and out of the DXF file format. These translators supported exchange between different CAD systems, with varying degrees of reliability. Today, DXF has become a *de facto* data exchange format for 2D information, even used to exchange data between non-AutoCad systems. It is limited in 3D, and IGES (discussed below) is used more commonly for this use, especially for curved surface modeling. Other CAD companies offer external file formats for moving data in and out of their project files and most also supply DXF translators, both in and out.

3.3.1 DXF FILE FORMAT

The DXF file format is described extensively in Autodesk documentation. Only a quick overview is presented here.

Group code range	Following value	Group code range	Following value
0-9	String	100	Sub-class marker
0	Entity type	140-147	Floating-point
5	Handle	170-175	Integer
6	Line type	210-239	Floating-point
8	Layer Name	210	X-extrusion dir.
10-59	Floating-point	220	Y-extrusion dir.
10	X-coordinate	230	Z-extrusion dir.
20	Y-coordinate	999	Comment
30	Z-coordinate	1000-1009	String
50	Start angle	1010-1059	Floating-point
51	End angle	1010	X-pt view target
60-79	16-bit integer	1020	Y-pt. view target
62	Color	1030	Z-pt. view target
90-99	32-bit integer	1060-1079	Integer

Figure 3.9: The value types defined within the various group code.

A file corresponds to an AutoCad® *drawing*, which may be used to define part of a single drawing or several. The file format consists of five sections: a Header section, a Tables section, a Blocks section, an Entity section, an Extended Entity section and an Object section, followed by an end-of-file marker. All sections of a DXF file are composed of groups, each of which occupies two lines in the file. The first line of a group is a group code, an integer code, and the second is the group value, in a format depending upon the group code. Group codes are typed by their ranges, as shown in Figure 3.9. A few commonly used group codes are shown under the range they are within. Each code has a particular interpretation. For example, codes 10, 20 and 30 are x, y, and z coordinates; 62 is a color; codes 210, 220, and 230 are the x, y and z cosines of an extrusion direction. All codes 1000 or above are for extended entity data.

The Header section provides general information about the drawing, such as the units of measurement used, rotation directions of angles, formats for dimensions, text and grids. As of Version 12, there were 137 status variable settings in the header, each having a value code and receiving a value. An example is shown in Figure 3.10. The file is partitioned with grayscale hatching to indicate groups. The uses of the first four group codes are shown in the figure in a larger *sans serif* font.

```
0
SECTION
2
HEADER
9
$ACADVER
1
AC1009      ACAD version
9
$INSBASE    insertion base
10
0.0
20
0.0
30
0.0
9
$EXTMIN
10          lower bounds
4.998122
20          of drawing
2.113167
30
-0.81644
9
$EXTMAX
10          upper bounds
11.133136
20          of drawing
6.162342
30
1.0
```

Figure 3.10: Header section of a DXF file.

```
0
SECTION
0
TABLE       table section
2
STYLE       style type
5
1C
0
ENDTAB
0

ENDSEC
0
ARC
8
0
5
1
10
6.020374    center - x
20
5.140089    center - y
30
0.0
40
1.022252    radius
50
220.163257  start angle
51
86.339878   end angle
```

Figure 3.11: Entity definition of an arc.

```
0
3DFACE
8
0
5
A
10
10.216935   first point
20
4.27148
30
-0.2608
11
8.087602    second point
21
4.571953
31
-0.20591
12
8.039898    third point
22
2.293276
32
-0.81644
13
10.040557   fourth point
23
2.265458
33
-0.799889
```

Figure 3.12: Entity definition of a 3D Face.

The Tables section defines the parameters for format entities that are used within the drawing. These are VPORT (viewports), LTYPE (linetypes), LAYER (drawing layer table), STYLE (text styles), VIEW (defining view or camera locations and angles), BLOCK_RECORD (block reference table), UCS (coordinate systems used in the drawing), APPID (applications that generated data), and DIMSTYLE (dimension styles). These are used in defining formats for the Entities section. The Blocks section (separate from the BLOCK_RECORD table) defines the symbols used in the drawing. It has the same format as the Entities section defined below it. These are grouped, however, as Blocks. The Entities section carries the graphical entities. The number of graphical entities has grown over time with AutoCad®'s features. The entities include LINE, ARC, CIRCLE, POINT, TRACE, SOLID, TEXT, SHAPE, INSERT (of a Block), ATTDEF (format of text attributes), ATTRIB, POLYLINE, VERTEX (defines components of polyline), 3DFACE, VIEWPORT, 3DSOLID, BODY, SPLINE (surface), IMAGE (bitmap), HATCH and others. Each entity is followed by an 8 Group that identifies its layer, followed by the specific parameters associated with that type. Notice that attributes are a kind of entity. They are defined in terms of a text string that is associated with a Block, and its various presentation formats. The Extended Entity data is a general means by which applications may carry additional data with Blocks. The coding of Extended Entity data is generally proprietary to the application.

The Objects section is new with Version 14. It defines nongraphical objects. Some are predefined and others can be defined for use by applications. Objects are similar to entities, except that they have no graphical or geometric meaning. All objects that are not entities or symbol table records or symbol tables are stored in this section. The root object in this section is an Object Dictionary, which defines the objects of interest to an application. The Object Dictionary refers to the other objects below it and, in like fashion, the other objects can refer to subobjects. Each object is given a unique Handle ID, which is used by other objects to refer to it. The predefined objects include DICTIONARY, IMAGEDEF (references to image files), MLINESTYLE (custom line styles) and SPATIALFILTER (a region of space of interest).

The new Object extension extends DXF in a significant manner, allowing it to represent application objects and provide means for accessing them. It will be interesting to see the use of the Object section within the new generation of applications being developed by Softdesk, Autodesk's AEC application subsidiary.

3.3.2 EXAMPLES OF REPRESENTATIONS

The codes for representing an arc in DXF are shown in Figure 3.11, preceded in the figure with a TABLE entry. The arc has x and y center coordinates, a radius, a start angle and an end angle. The rotation direction is defined globally for all arcs. A separate representation is used for circles. The planar bounded surface is characterized here as an AutoCad 3D face and shown in Figure 3.12. Its coding is with four points, each with an x, y and z coordinate. Any surface information, including the surface normal, must be derived from these points. Clearly, this works only for planar faces. No boundary may be added to describe the portion of the plane of interest; it is always a rectangle, or if two vertices coincide, a triangle.

AutoCad offers several ways to define 3D shapes. For example, one can extrude a 2D polygon and cap both ends, making a closed bounded volume. Starting in Version 13, it has included 3D solid modeling in the Designer optional application, using a third party software library called ACIS, developed by Shape Data Ltd. Since ACIS is written as a separate program and integrated into AutoCad, it translates its representation of shape into and out of the AutoCad DWG binary format. The geometric description is held together with a Block to make the total solid model. In addition, an ACIS model carries a large amount of extended entity data.

Figure 3.13: The two alternative views of a solid in DXF. The left is wire frame, while the right is surface meshed. Both are carried in the DXF format.

There are several, alternative ways to partially represent the multiple-view wall example. An application could be developed that carried the various views together as nested Blocks; that is, a plan view, two elevation views, and a solid model view, carried within an outer Block. DXF would have to carry the multiple levels of decomposition also as nested Blocks, in order to carry along the various segments, windows, doors, layers and pass-thrus. Attribute entities could be carried within the Blocks or as Extended Data. Special operations would have to be developed to maintain the consistency and correctness of each entity instance as the wall was edited. Subsets of this representation have been developed for walls by AutoCad value-added resellers (VARs), using similar strategies. In these cases, a Block becomes a kind of data structure in which the needed structures are added. An operation can open a Block, manipulate its parameters, then package it back up, protecting its internal structure from inadvertent manipulation. Topological relations can only be represented in attributes. In this case, relations between entities can be defined if each one carries an application ID marking the application that created it, and has an attribute value which carries the application ID of the entity to which it is related.

It is apparent that the structures required for an application manipulating a multi-viewed wall represented in DXF would result in a highly customized data structure that would have to be carefully communicated and agreed upon by other applications wishing to use the resulting data. In practice, that is the way applications now are developed in AutoCad, with the result that most such applications cannot communicate with each other.

3.3.3 ASSESSMENT

For drawing files, DXF is quite compact, taking only a small amount of memory space. The type codes for all entities are carried in place just before the data values. This allows the writing of simple parsers for interpreting DXF files; all information can be processed in a single pass over the data. It also allows arbitrary mixing of entities.

Like AutoCad itself, DXF did not support references between objects (relations or topological links) until recent versions, when the need for this became obvious. All references between entities require that they are made forward in the file, so that all entities referred to are already defined before they are referenced. Again, this facilitates easy interpretation of the data. In more complex formats, intermediate tables must be created to hold cross-references between entities. There has been consistent effort to keep the structure of DXF interpretation so that it can be done in a single pass of the file.

No organizational mechanism exists for assessing conversion processors to or from DXF. Vendors may make whatever assumption they wish—for example, converting all 2D entities to polylines or converting crosshatching to line segments. It is thus incumbent on users to test and evaluate a DXF exchange processor to see if it meets their needs. Firms wishing to use DXF in production applications usually try out a range of test cases between the source and target applications, to see how the two conversion processors behave.

DXF was originally developed for, and serves well as, a two-dimensional exchange format. It is limited in three dimensions to representing three-dimensional surfaces and solids in the same way that AutoCad does, implicitly as polylines, in terms of the points that bound them. These surfaces can only be evaluated precisely if they are planar. One surface, Edgesurf, provides a true spline representation, but its use is limited. In practice, AutoCad facets all curved surfaces; any curved surface mapped into DXF is approximated in this manner. More accurate geometry may be carried in curved line segments, and these can be used as surface or shape generators, but must be updated with each operation. Another aspect of the DXF format is that surface normals, which are used extensively for computing surface shading and for hidden surface elimination, are not carried within the DXF format, frequently resulting in problems when translated 3D models are displayed.

Because of the heavy reliance on DXF among users and the support by a wide variety of vendors, there have been efforts in the UK to make it an official standard. It was endorsed by the NEDO Committee on CAD Data Exchange in Building for the exchange of unstructured 2D drawings. Similarly, there has been a wide range of efforts to develop standards for layer semantics and naming. In the US, these have been led by a joint committee of the American Institute of Architects, the American Society of Civil Engineers and the Construction Specifications Institute. Similar efforts have been undertaken in the UK and in Scandinavia.

Up to Version 14, the primary means to structure data in DXF was by using Blocks. Attributes associated with a shape, or attributes for which a shape has not yet been developed, must begin with a Block. Blocks can be opened and manipulated through special operations, then closed at the end of the operation. The treatment of attributes is left to the application developer, usually in the Extended Entity data associated with a Block or as an attribute entity carried by the Block, resulting in each application developer relying on their own conventions, which are seldom publicly documented. Topological relations also require *ad hoc* treatment. In practice, DXF is used for exchanging 2D and faceted 3D geometrical data.

The introduction of Objects provides a rich new alternative for developers to package geometric, attribute and relation data into a particular object representation needed for an application. Naming conventions and the Dictionary Object are used to clearly define the objects used in some application. These resolve a long-term problem of application developers: that the structures defined by the application could be operated on by the standard AutoCad operations in ways that violated the semantics of the application structure. For example, a wall layout application might maintain the connectivity relations of how each wall abuts others at its ends. However, if the standard MOVE operation available in AutoCad or any other CAD system is applied to a wall, then these relations are likely to become invalid, because the operation does not maintain the relations. By having the application operations apply to the objects in a specific Dictionary, and disallowing general operations on Objects in that dictionary, the consistency of the application can be maintained.

Recently, a new organization has proposed making the AutoCad DWG file format a public standard, like DXF. It is now fairly easy to obtain the encoded binary format. This is likely to allow faster translation among CAD systems, but does not resolve the need to develop *ad hoc* structures to package information needed for complex applications.

3.4 IGES

In the late 1970s, the rapid growth of new CAD/CAM applications led to problems of data exchange in large team projects managed by government agencies such as DOD and NASA. Incompatible CAD systems used by team members required custom translators to be built to support the exchange of engineering data for such projects. A NASA administrator, Robert Fulton, pushed an initiative to develop a standard text file interface format. He made a strong public request to each CAD company to agree to develop translators to and from a standard interface format, if one was developed. He then got staff members from General Electric and Boeing Company to expand upon their own earlier internal efforts at developing exchange standards, and draft a standard, neutral exchange file format.

Because it was an initial effort in the area and most of the people involved recognized it as a beginning, it was called "Initial Graphics Exchange Specification" (IGES). The specification was drafted by technical staff at Boeing and General Electric, with input and review by all interested parties. The standard was later transferred and managed by the National Institute for Standards and Technology (NIST). Since 1978, when the initial version of IGES was released, four other versions have been generated, with extended capabilities to deal with 2D and 3D graphic entities, including surfaces, solids and non-graphic attributes. IGES interfaces are a required capability of all CAD systems procured by the US federal government. We examine the IGES format here in some detail.

IGES is an application-independent neutral file format. It was thought that if each CAD system company developed one translator writing to IGES from its internal representation, and another translator reading IGES into its internal format, sufficient means should exist for moving data from any CAD program to any other program. This is a far more effective means to develop translators than to rely on a large number of translators between each pair of CAD or other applications. The unit of translation is always one file.

The fundamental unit of data in the file is an entity, which may be geometrical or non-geometrical. Geometric entities define the physical shape and include points, curves, surfaces and solids as well as relations that are collections of such entities. IGES specifies a model via a *definition space*, a *project space*, and a *drawing space*, allowing the placement of a single geometric entity at various instance locations within a model, then placing multiple views of the model on a drawing using a Transformation Entity. Non-geometric entities allow development of complex structures. They include a View Entity that allows derivation of various two-dimensional projections from either a larger 2D drawing or 3D model. The Drawing Entity allows associating one or more views to be extracted and placed on a drawing in paper coordinates. The Associativity Entity allows several entities to be related and/or grouped. Also, entities may be referred to that reside on a different file.

3.4.1 THE IGES FILE FORMAT

The essential contribution of IGES was to define a file format that all CAD companies would write translators to and from. An IGES file consists of five or six sections:

Flag: an optional section

Start: a human readable prologue to the file describing its contents

Global: contains information needed by the preprocessor and postprocessor to handle the file (Figure 3.14). It includes the digital representation of integer and all floating-point values within the file, model space scale and unit type, maximum values for integers, number of line widths, IGES version that wrote the file, person who generated data and other general information for the translators. This section is in free format; that is, its format with regard to columns is not defined.

Directory: provides an index to an entity and includes descriptive data about the entity in fixed format; Figure 3.15 gives an abbreviated listing of the fields making up the directory entry. One directory entry exists for each entity in the file; most entries carry pointers to an entry carrying needed values; understanding the coding of this table of values is critical for correct generation and interpretation of IGES data. This section is in fixed format.

Parameter Data: provides the specific data for an entity instance, in variable length format.   Most entities are reserved for standard types; a few are open for implementer-defined macros. Some entities have a subclass, called a Form number.

Terminate: defines the number of lines in each section.

Index	Type	Description	Index	Type	Description	Index	Type	Description
1	string	param. delimiter	9	integer	signif. digits S	17	real	max. line width
2	string	record delimiter	10	integer	max. exponent D	18	string	date & time
3	string	product ID send	11	integer	signif. digits D	19	real	min. resolution
4	string	file name	12	string	product ID rec.	20	real	max.coord. value
5	string	system ID	13	real	model scale	21	string	author name
6	string	preprocessor ver.	14	integer	unit flag	22	string	author organ.
7	integer	size of integer	15	string	units	23	integer	IGES version
8	integer	max. exponent S	16	integer	no. line weights	24	integer	compliance std.

Figure 3.14: Parameters used in the Global section.

The Global section includes a parameter delimiter (defaulted to a comma) and a record delimiter (defaulted to a semicolon). These are used to delimit all succeeding strings in the Global section and all succeeding sections of the file. The Global section also gives the maximum and minimum coordinates for the file and the precision of coordinate values.

The entries in the Directory section provide necessary high-level information about an entity, including its transformation, the level or layer it is on, its presentation in terms of line format, pointer(s) to the view(s) it is in, any dependency relations, and its use. It also refers to a Parameter entry for the geometric definition of the entity.

1 8|9 16|17 24|25 32|33 40|41 48|49 56|57 64|65 72|73 80								
(1) Entity Type Number	(2) Param- eter Data	(3) Structure	(4) Line Font Pattern	(5) Level	(6) View	(7) Tranfor- mation Matrix	(8) Label Display Assoc.	(9) Status Number	(10) Sequence Number
#	⇒	#,⇒	#,⇒	#,⇒	0,⇒	0,⇒	0,⇒	#	D, #
(11) Entity Type Number	(12) Line Weight Number	(13) Color Number	(14) Param- eter Line Count	(15) Form Number	(16) Reserved	(17) Reserved	(18) Entity Label	(19) Entity Subscript Number	(20) Sequence Number
#	#	#	#	#				#	D #+1

Figure 3.15: Format of the Directory Entry section. It consists of two, eighty character strings. The key at the bottom of each cell denotes the type of data, where # means integer, ⇒ means pointer, 0 means zero, and D is the literal plus a number.

The heart of the IGES standard is the definition of the various graphical entity types, defined in entry #1 of the Directory. The major entity types included in Version 4.0 of IGES are shown in Figure 3.16. They cover 2D, 3D surfaces and solids. Non-geometric entities are also specified in the Directory and parameter sections. These are shown in Figure 3.17. Special Entities include provision for dimensions, associations, properties and attributes. Dimensions are non-associative. The Associativity Entity supports relations between entities. The Property Entity allows specific characteristics to be associated with a Geometric Entity, as well as constructions for associative dimensions, symbols and network topological connections.

Versions 5.0 and 5.1 added many, new entity types, some to support drawing entities, new dimension entities and boundary representation solid modeling. The additions are tabulated in Figure 3.18. For each entity listed, IGES defines the mathematical parameterization upon which the entity is based, the values of which are carried in the parameter section of the file. Their location in the Parameter section is in entry #2 of the Directory. Care has been taken to provide precise definitions of each parameter.

The Parameter section is the fourth section. The Parameter section begins with the entity type, followed by a sequence of parameter values for that type. The values are in free format, marked by separators. Real values may be defined in decimal or exponent format. A subset of the entities provided is shown below. The next to last eight spaces in each line of the Parameter section is a back-pointer to the Directory section it describes.

3.4.2 ENTITY PARAMETERS
The parameters for a variety of different entity types are presented below. The arc and circle use the same parameterization. Instead of radius and angles, IGES uses three points, for the center,

start and termination. The predetermined direction of rotation between the start and termination points is counterclockwise. (No interpretation is given if the start and termination points are at different distances from the center.)

Entity Type Number	Entity Type
100	Circular Arc
102	Composite Curve
104	Conic Arc

106	Copious Data
	    Centerline, Simple Closed Area
	    Linear Path, Witness Line, Section Line

108	Plane
110	Line
112	Parametric Spline Curve
114	Parametric Spline Surface
116	Point
118	Ruled Surface
120	Surface of Revolution
122	Tabulated Cylinder
124	Transformation Matrix
125	Flash
126	Rational B-Spline Curve
128	Rational B-Spline Surface
130	Offset Curve
132	Connect Point
134	Node
136	Finite Element
138	Nodal Displacement and Rotation
140	Offset Surface
142	Curve on a Parametric Surface
144	Trimmed Parametric Surface
146	Nodal Results (Appendix J)
148	Element Results (Appendix J)
150	Block
152	Right Angular Wedge
154	Right Circular Cylinder
156	Right Circular Cone Frustum
158	Sphere
160	Torus
162	Solid of Revolution
164	Solid of Linear Extrusion
168	Ellipsoid
180	Boolean Tree
184	Solid Assembly
430	Solid Instance

Figure 3.16. The geometric entity types provided in IGES Version 4.0.

Entity Type Number	Entity Type
106	Copious Data
	Centerline
	Section
	Witness Line
202	Angular Dimension
206	Diameter Dimension
208	Flag Note
210	General Label
212	General Note
214	Leader (Arrow)
216	Ordinate Dimension
218	Linear Dimension
220	Point Dimension
222	Radius Dimension
228	General Symbol
230	Sectioned Area
302	Associativity Definition
308	Subfigure Definition
320	Network Subfigure Definition
322	Attribute Table
402	Associativity Instance
408	Subfigure Instance
420	Network Subfigure Instance
422	Attribute Instances

Figure 3.17: The non-geometric entity types in IGES 4.0.

Circular Arc Entity Type (Number 100):

INDEX	NAME	TYPE	DESCRIPTION
1	ZT	Real	Parallel ZT displacement of arc from XT,YT plane
2	X1	Real	Arc center abscissa
3	Y1	Real	Arc center ordinate
4	X2	Real	Start point abscissa
5	Y2	Real	Start point ordinate
6	X3	Real	Terminate point abscissa
7	Y3	Real	Terminate point ordinate

The second entity type described here is called Copious Data. It carries a list of tuples, where a tuple may be a pair, a triplet, or a sextuple (for 2D paths, 3D paths, or a located vector). Its parameterization is given below. Various form flag values in field #15 in the Directory entry identify how the Copious Data Entity is being used. Of these, 11 indicates 2D points defining a piecewise linear curve with a common X-Y plane, 12 indicates 3D points defining a piecewise linear curve, 13 indicates an s-value tuple (a 3D coordinate and a vector) representing a piecewise linear curve.

Copious Data Entry Type (Number 106):

INDEX	NAME	TYPE	DESCRIPTION
1	IP	Integer	Tuple size
			IP=1 x,y pairs, common z
			IP=2 x,y,z coordinate

			IP=3 x,y,z coordinates and i,j,k vectors
2	N1	Integer	Number of n-tuples
3	varies	Real	First coordinate of first tuple
4	varies	Real	Second coordinate of first tuple
.	.	Real	.
.	.	Real	.
.	.	Real	Last coordinate of last tuple

The Copious Data type is used extensively for various forms of polylines or curves, often embedded in other entities. When one entity is embedded in another, the dependency is carried in the Directory Status number, entry #9.

Entity Type Number	Entity Type Description
123	Direction
141	Boundary
143	Bounded Surface
182	Selected Component
186	Manifold Solid Boundary Rep Object
192	Right Circular Cylindrical Surface
194	Right Circular Conical Surface
196	Spherical Surface
198	Toroidal Surface
204	Curve Dimension
213	General Note
214	Leader (arrow)
216	Linear Dimension
218	Ordinate Dimension
228	General Symbol
230	Sectioned Area
316	Units Data
402	Associativity Instance
404	Drawing
406	Property
410	View
416	External Reference
502	Solid Vertex
504	Solid Edge
508	Solid Loop
510	Solid Face
514	Solid Shell

Figure 3.18: New entities defined in IGES 5.0 and 5.1.

Another example entity type is the Plane. It is described as a planar surface and a closed curve that bounds it.

The Plane defines a set of implicit plane coefficients, a pointer to one or more bounding closed curves in the plane and optionally a location for a display symbol located in the plane, to be used as its reference. The Form Entry in the Directory determines whether the plane is bounded or is infinite but with a hole. If the plane is bounded by one outer curve and multiple inner curves, then the different curves are related by a Single Parent Associativity Entity (Type 402, Form 9). All the associated closed curves are physically dependent upon the plane.

Plane Entity Type (Number 108)

INDEX	NAME	TYPE	DESCRIPTION
1	A	Real	Coefficients of plane
2	B	Real	Coefficients of plane
3	C	Real	Coefficients of plane
4	D	Real	Coefficients of plane
5	PTR	Pointer	Pointer to directory entry of closed curve or zero
6	X	Real	XT coordinate of display symbol
7	Y	Real	YT coordinate of display symbol
8	Z	Real	ZT coordinate of display symbol
9	SIZE	Real	Size parameter for display symbol

Arc and line are two basic geometrical entities. Entity type number 110 is a line. It is simply defined by its two endpoints.

Line Entity Type (Number 110)

INDEX	NAME	TYPE	DESCRIPTION
1	X1	Real	Start point P1
2	Y1	Real	
3	Z1	Real	
4	X2	Real	Terminate point P2
5	Y2	Real	
6	Z2	Real	

For most IGES models, transformations are needed to specify the location of various entities within the project space. Transformations make use of a 3x3 rotation matrix and a translation vector, as shown below. The transformation matrix is used to map from definition space to model space, and possibly from model space to *drawing* space.

$$\begin{bmatrix} R_{11} & R_{12} & R_{13} \\ R_{21} & R_{22} & R_{23} \\ R_{31} & R_{32} & R_{33} \end{bmatrix} \begin{bmatrix} XINPUT \\ YINPUT \\ ZINPUT \end{bmatrix} + \begin{bmatrix} T_1 \\ T_2 \\ T_3 \end{bmatrix} = \begin{bmatrix} XOUTPUT \\ YOUTPUT \\ ZOUTPUT \end{bmatrix}$$

This form of transform varies from those used in various CAD systems, but is easily derived. The entity definition for a transform is

Transformation Matrix Entity (Number 124)

INDEX	NAME	TYPE	DESCRIPTION
1	R11	Real	Top row of matrix
2	R12	Real	
3	R13	Real	
4	T1	Real	
5	R21	Real	Second row of matrix
6	R22	Real	
7	R23	Real	
8	T2	Real	
9	R31	Real	Third row of matrix
10	R32	Real	
11	R33	Real	
12	T3	Real	

The wall example will use a solid Block to cut the openings in the wall. The Block Entity is defined as a rectangular parallelepiped in terms of its three X, Y and Z dimensions. It is then located by defining a location (in model space), and two vectors, for the local X- and Z-axes. The X- and Z-axes must be orthogonal.

Block Entity Type (Number 150)

INDEX	NAME	TYPE	DESCRIPTION
1	LX	Real	Length in local X-direction
2	LY	Real	Length in local Y-direction
3	LZ	Real	Length in local Z-direction
4	X1	Real	Corner point coordinates of Block
5	Y1	Real	
6	Z1	Real	
7	I1	Real	Unit vector defining local X-axis
8	J1	Real	
9	K1	Real	
10	I2	Real	Unit vector defining local Z-axis
11	J2	Real	
12	K2	Real	

In order to represent a curved solid, we use a Solid of Revolution Entity, described below. It consists of a closed curve, a coordinate location and vector defining the axis of rotation, and the amount of rotation (between 0 and 360 degrees) about the axis. The rotation is counterclockwise when viewed from the positive direction of the vector. If the curve is closed to itself, the Form value will be 1; if it closes with the axis, the Form value will be 0. One or the other must hold for a legally defined Solid of Revolution.

Solid of Revolution Entity Type (Number 162)

INDEX	NAME	TYPE	DESCRIPTION
1	PTR	Pointer	Directory Entry Number of curve entity to be revolved
2	F	Real	Fraction of full rotation through which curve is to be revolved
3	X1	Real	Coordinates of point on axis Default (0,0,0)
4	Y1	Real	
5	Z1	Real	
6	I1	Real	Unit vector in axis direction Default (0,0,1)
7	J1	Real	
8	K1	Real	

The wall example will use a Solid of Linear Extrusion. It is similar to the Solid of Rotation specification, in that it defines a closed curve and a vector to sweep it through. The first argument points to the closed curve, the second defines an extrusion length, and entries 3-5 define the unit vector and direction of the extrusion.

Solid of Linear Extrusion Entity Type (Number 164)

INDEX	NAME	TYPE	DESCRIPTION
1	PTR	Pointer	Pointer to closed curve entity
2	L	Real	Length of extrusion in positive direction
3	I1	Real	Unit vector specifying direction of extrusion
4	I2	Real	
5	I3	Real	

The closed curve must be planar. The axis of extrusion need not be orthogonal to the closed curve, but it may not be coplanar to it.

In order to do the wall example, several associations need to be defined using the Associativity Definition Entity. This entity points to other entities to which it is related. Optionally, they also may have back references to the Associativity Entity, allowing access from any entity to which it is associated. The primary use of this entity is for developing applications for uses such as those required for the wall example. The Associativity Definition Entity identifies the class of entities in the association and the number of instances within the class.

Associativity Definition Entity Type (Number 302)

INDEX	NAME	TYPE	DISCRIPTION
1	K	Integer	Number of class definitions
2	BP1	Integer	1 = backpointers required
			2 = backpointers not required
3	OR1	Integer	1 = classes are ordered
			2 = classes are a set
4	N1	Integer	Number of items per entry
5	IT(1)	Integer	1 = pointer to a directory entry
			2 = value
			3 = parameter is a value
			-3 = parameter is a pointer
.	.	.	.
4+N1			
.	.	.	.
K*(3+N1)			items 2 to 4+N1 are repeated K times

K identifies the number of classes. Entries 2 and 3 define whether or not backpointers and ordering are fixed for the first class. Entry 4 defines the number of instances in this class, and following it, the value is specified for each instance reference. Items 2 to (4+N1) are repeated for each of the K classes. When a particular set of instances is associated, they are related with an Associativity Instance Entity - Entity Type 402. This entity type is not shown. The relation between Associativity Definition and Instance are made by putting a unique Integer ID in the Form field in the Directory entry of each.

The normal CAD symbol definition and instancing capability are defined using the Subfigure Entity. Like Associativity Entities, there is a Definition Entity and an Instance Entity. The Subfigure Definition Entity Type defines the subfigure name, the number of entities in the subfigure, a pointer to each entity and the level of nesting of subfigures.

Subfigure Definition Entity Type (Number 308)

INDEX	NAME	TYPE	DESCRIPTION
1	DEPTH	Integer	Depth of subfigure
2	NAME	String	Subfigure name
3	N	Integer	Number of entities in subfigure
4	DE-1	Pointer	Pointer to 1st Directory entry for associated entity
.	.	.	
N+3	DE-N	Pointer	Pointer to last Directory entry for associated entity

Each instance carries a reference to the Subfigure Definition Entity and a translation and scale, used to adjust the Subfigure's origin, if needed. The Directory entry carries a full transformation to its model location.

These ten parameter definitions represent just part of the full set. Each release of IGES has added additional entity types to those already supported. The range includes those for exchanging finite element models for structural, fluid flow or similar applications, for geometric and solid modeling systems, and for very limited forms of parametric modeling.

The Termination section consists of one line in fixed format. It consists of four eight-character blocks. The first letter identifies the section (S, G, D, or P) and the next seven characters identify the number of lines in that section, right justified. The last eight spaces in each line in each section identify the section and line number in a similar eight-character format.

As with DXF, a binary format is also provided within IGES, but is not described here.

3.4.3 EXAMPLES OF REPRESENTATIONS

The IGES representation of an arc and bounded planar surface are shown in Figure 3.19. The four sections are readily apparent: Start, Global, Directory and Parameter, formatted in a fixed width font. The arc is Type 100. The first line of Directory Entry tells us that the arc's parameters begin on line 1 of the parameter entries, the structure field is not used, line font is given at `1´ (interpreted as solid), and Level is 3. There is no View, Transformation matrix or Label. The Status Number is zero. The second line shows a default line weight. (If one were actually given, it would be based on a computation using Global Parameters 16 and 17.) There is no color and there are two lines of parameters. No other Directory entries are used. The parameter section puts the arc at zero in the Z-axis, and for the arc parameters, assigns: X1 = 4.9698262032692D0, Y1 = 4.7499999958091D0 as the centerpoint, X2 = 3.6872903948943D0, Y2 = 5.2243589702755D0 for the start point and X3 = 5.9290492707671D0,Y3 = 3.7754202693626D0 for the endpoint.

```
IGES test file generated for example translation data          S0000001
translator version IGESOUT-4.01.                              S0000002
                                                              S0000003
,,7HUNNAMED,18HD:\CAD\ARCFACE.IGS,13HIGES-testcase,13HIGESOUT-4.01,    G0000001
32,38,6,99,15,7HUNNAMED,1.0,1,4HINCH,32767,3.2767D1,          G0000002
13H960711.075755,9.9973665264706D-9,9.9973665264706D0,10HC. Eastman,    G0000003
8Hpersonal,6,0;                                               G0000004
      100       1       1       1       3          00000000D0000001
      100                       2                            D0000002
      108       3       1       1       3          00000000D0000003
      108                       2       1                    D0000004
      106       5       1       1       3          00010000D0000005
      106                       5      12                    D0000006
100,0.000,4.9698262032692D0,4.7499999958091D0,3.6872903948943D0,      1P0000001
5.2243589702755D0,5.9290492707671D0,3.7754202693626D0;        1P0000002
108,1.0000000000000D0,1.0000000000000D0,1.0000000000000D0,    3P0000003
3.0000000000000D0,00000005,0.,0.,0.,0.;                       3P0000004
106,2,5,7.4866291864961D-2,2.2734243109031D0,6.517093972319D-1,    5P0000005
-1.4687028842991D0,3.9879336363363D0,4.8076924796289D-1,      5P0000006
-2.3711110963092D0,3.1595726143533D0,2.2115384819559D0,       5P0000007
-4.3234341751093D-1,1.2849074958964D0,2.1474359216145D0,      5P0000008
7.4866291864961D-2,2.2734243109031D0,6.517093972319D-1;       5P0000009
S0000003G0000004D0000006P0000009                              T0000001
```

Figure 3.19: IGES file that holds an arc and a bounded 3D face, with a closed curve bounding it.

The Bounded Planar Entity starts its parameters on line 3 of the Parameter section (Figure 3.19.). Its Structure, Line type and Level are the same as for the arc. It also has two lines of parameter values, and uses Form 1. In this case, Form 1 means that the closed curve defines the positive region of the face. In line 4 of the Parameter section, the A, B, C and D coefficients are identified as 0.0, 0.0, 0.0.0 and 3.0. (a plane intersecting 1.0 on the X, Y, and Z axes). The parameters point to line 5 as the start of the closed curves and put the display symbol at the origin. Entity 106 (Copious Data) carries the boundary. It starts on line 5 of the parameter entries and has the same line, layers and structure as the arc. The Copious Data has five lines of parameters and its Form value is 12. Here, Form 12 indicates a string of X, Y, Z coordinate triples. The parameters for Entity 106 again indicate tuple type 2 (X, Y, Z triples) and five entries, which are listed in the following lines. The last line is the termination line.

```
IGES test file generated for example translation data              S0000001
translator version IGESOUT-4.01.                                   S0000002
                                                                   S0000003
,,7HUNNAMED,20HD:\CAD\EXTRUSION.IGS,13HIGES-testcase,13HIGESOUT-4.01,  G0000001
32,38,6,99,15,7HUNNAMED,1.0,1,4HINCH,32767,3.2767D1,               G0000002
13H960711.075755,9.9973665264706D-9,9.9973665264706D0,10HC. Eastman, G0000003
8Hpersonal,6,0;                                                    G0000004
     162          1         1         1           00000000D0000001
     162                              1                     D0000002
     106          2         1         1           00010000D0000003
     106                              2        12           D0000004
162,3,0.32,0.0,0.0,0.0,0.0,0.0,1.0;                                1P0000001
106,2,5,1.0,0.0,0.0,1.0,0.0,5.0,1.30,0.0,5.0,1.30,0.0,0.0,         3P0000002
1.0,0.0,0.0;                                                       3P0000003
S0000003G0000004D0000004P0000003                                   T0000001
```

Figure 3.20: IGES file that holds the curved wall. The wall is defined as a vertical rectangle, swept through an angle of rotation about the z-axis.

The solid shape, defined as a surface of revolution, is shown in Figure 3.20. The Solid of Revolution is a Constructive Solid Geometry Entity, which is a class instance type of entity. In use, a Solid Instance Entity (Entity Type 430) would also be used, carrying an associated transformation matrix. The Instance Entity is omitted here. (There is also a Surface of Revolution (Entity Type 120), but it does not define a complete solid shape.)

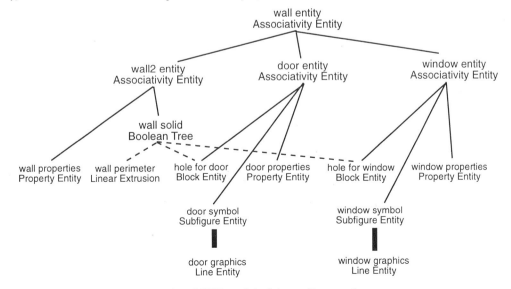

Figure 3.21: The structure of an IGES model of the wall example.

IGES has been augmented to support data exchange among intelligent applications. The most notable is the IGES neutral format for 3D piping. This standard supports the exchange of three-dimensional piping and related equipment between one application and another. It was developed under the US Navy's SEAWOLF program and extended with support by the Navy-Industrial Digital Data Exchange Standards Committee (NIDDESC) and representatives of the process plant industry. It relies on IGES Version 5.1. The 3D Piping exchange format supports applications in a particular domain and is called an *application protocol (AP)*. An AP has a specific scope, further defines how particular entities will be used, and provides conformance requirements and tests.

Our wall test case would also be defined as part of an application protocol (AP). Its scope would include schematic design and design development. A partial representation of the wall, structured as needed by an application during design, is diagrammed in Figure 3.21. It consists of a set of Constructive Solid Geometry solid shapes: the perimeter of the wall elevation, and solids for sub-tracting out the openings. The wall also has an associated Property Entity to carry its attributes. The CSG tree is a simple form of parameterization, allowing the different openings to be easily moved. The opening entities are structured to consist of the Block that defines their hole, an asso-ciated subfigure that is located in the opening, and a Property Entity for attributes. This allows an application to modify them together, maintaining their consistency. The subfigure relation is identified by a heavy line.

The wall model here is incomplete; it has no way to incorporate pass-thrus or to deal with complex walls with multiple cross-sections, as supported by the segment concept defined earlier. (It may be possible to define such a wall in IGES.) At the same time, the door and window information —geometric and property—are grouped together, allowing the data for these entities to be easily manipulated together. After design, the wall CSG tree would probably be evaluated and the Blocks used to subtract openings would be permanently located. The Associativity Entity of IGES allows multiple reference to a single entity and could be used to define the abutting relations between wall edges.

Notice that the definition of such a wall would be carried completely in the IGES specification. IGES provides no means to document user-defined entities. A few higher level translators have been written, most notably for 3D piping models.

3.4.4 IGES TRANSLATOR STANDARDS

The first implementation of IGES processors began in 1981, with public demonstrations at national shows like AutoFact and the National Computer Graphics Association Conference. Files were exchanged for test drawings between different CAD systems. In general, the translator from a CAD system to the IGES neutral file format is called a preprocessor and the translator from IGES format to a CAD system is called a postprocessor. An initial test involves assessing whether a preprocessor generates a "legal" IGES file. Some tools have been developed to parse and test an IGES file for syntactic correctness. Once correct, there are a variety of tests that one can perform on a set of IGES processors. For testing a single pair of processors associated with a single CAD or engineering application, the best test involves a loop where the preprocessor is used to convert application data into an IGES file and then the postprocessor to convert it back. An end-to-end test uses the preprocessor of one application followed by the postprocessor of a different receiving application. The round-trip test does two end-to-end tests, one from application A to application B, then another back from application B to application A.

When a translator is written, implementers encounter differences between the DXF or IGES entity definitions and what their CAD system or application provides. They must make adjustments to the definition of their own entities to fit the neutral, standard entities. In practice, IGES implementations consist of a relevant subset of the entities with which a system can deal.

An essential aspect of documenting the functionality of IGES pre- and postprocessors is to define the corresponding entity maps. An entity map identifies for a processor (either pre- or post-) the correspondence between application entities and IGES entities. The map table should identify: (1) the from-entity-to-entity correspondence, (2) whether the created entity is one-to-one or one-to-many, (3) whether the map is exact or approximate and (4) what entities are not mapped. Such a map table is recommended documentation for every IGES processor. The map tables describing the processors used in a desired translation will quickly allow identification of problems in relation to the translation needs.

In general, the testing steps are to
1. Establish the requirements: document operating system and application versions, estimate expected file sizes and frequency of translation, identify critical issues (accuracy, format requirements, etc.).
2. Establish test cases: one or more test cases should be developed to assess the effectiveness of the pair of processors used. The test cases should be typical of the data expected to be transferred, run on the same hardware, and cover all the entities likely to be encountered in practice. The test case should be documented so that it can be recreated if necessary on the preprocessor application.
3. Apply test procedures: the test sequence should record all error or warning messages created by the processors. The neutral file should be inspected and copied, so that later errors can be traced back to one of the processors applied. Plotting both the starting application dataset and the final received dataset expedites the search for inconsistencies.
4. Analyze test results: includes comparison of before and after plots, identification of entity conversions and looking at the "inspect" description of the beginning and end entities to identify numerical differences. Two plot files may look the same but the translation may be unacceptable, depending on the scale of the plots and the tolerances that are required.
5. Document results.

DXF and IGES are work tools currently used everyday by the CAD community. Anyone who has needed to exchange data between one CAD system and another has probably used one or both of them. While both have capabilities to represent attributes and relation, especially IGES, they are used today essentially as geometric translators. That is, they work well if the translation task is the transfer of geometry only. They quickly break down when dealing with even slightly more complex structures: associative dimensioning, or smart crosshatching (that automatically rehatches if the perimeter is edited). There are no agreed upon conventions for what complex entities like walls are or should be; hence, few translators bother to deal with them at all.

3.4.5 DISCUSSION OF IGES

IGES relies on a part definition followed by parameters, allowing parameters to vary for each entity, if needed. DXF treats some of the part parameters as global parameters. Like DXF, IGES defines the format of each entity externally (in a manual) and only represents in a translation file the parameters for each entity. The usefulness of both interface mechanisms depends upon one's requirements. The results of the combined use of some preprocessor and postprocessor cannot be guaranteed beforehand, but only determined by testing.

Complex entities can be defined and translated using, for example, Subfigure and Associativity Entities. From the beginning, IGES supported definition of complex entities composed of simpler ones, with piping, electrical circuits and complex mechanical parts as test cases. The IGES developers recognized that complex models must be interpretable by both humans—for example, to test them—and also by machines. This resulted in a fixed format layout that expanded the memory consumed. Continuing assessments have applauded IGES's versatility and the rigor of its entity specification. Critical comments, however, have been directed at its verbosity and the difficulty in tracing back through a set of translation files to identify when entities change or disappear.

3.5 LIMITATIONS OF NEUTRAL FILE FORMATS

Much has been learned from the IGES and DXF data exchange experiences. Early versions of IGES, in particular, had many weaknesses that were corrected in later versions. A major review of product data exchange efforts in the late 1980s by Bloor and Owen identified the following broad range of issues plaguing effective data exchange, and suggested means to address them:

drawing office conventions: Different offices use different conventions to organize drawings in a single project, for example, with regard to the use of multiple views, layers and different coordinate systems and symbols. These issues can be resolved through cooperation among the parties involved, agreeing on conventions and standards, or by structured conversion between the conventions.

medium incompatibility: The use of conflicting types of storage media, such as tape, floppy disks or networks, was an early issue. Medium incompatibility has been largely resolved by technological evolution and with the growing availability of CD-ROM and other relatively high density formats.

functionally different CAD systems: Mismatches may arise when there are not adequate representations in the sending or receiving CAD system. This is a fundamental issue reflecting the different functionality of engineering applications. Often the functional differences are based on different intended uses, for example, geometric representations for analyses vs. those for different types of design. The varying functionalities rely on different representations that may be ambiguous or incompatible with each other. This problem is endemic and is not likely to go away.

numerical approximations: Conversions between different surface representations are not always defined and may result in approximations and numerical problems. The same problems apply to solid representations. Reduction in the number of surface types may be possible with the growing use of non-uniform B-splines (NURB) for representing all types of curved surfaces. However, like the functionality of CAD systems, this issue is also likely to remain endemic to the field.

vendor cooperation: Vendors have not been enthusiastic about providing means to transfer data out of their systems, because it allows users to mix-and-match different CAD systems for their functionality, rather than being limited to a single vendor. It is usually not until users are organized that their demands are recognized and vendors begin to produce data exchange preprocessors and improve them.

explosion of data: This issue drove early data exchange efforts and reduced the initial acceptance of IGES. As hardware costs have reduced and performance increased, file sizes have become much less of an issue.

informal specification: IGES specifications were defined in words. In contrast, most modern computer languages are defined more formally, for example, in Backus-Naur Form (BNF) or its derivatives. Some European efforts at formal specification of exchange languages were undertaken that showed the benefits of a more rigorous approach.

complexity of format: The data structures used by different vendors are complex. The physical file layout of IGES added to these formatting issues, overlaying several levels of complexity. This is not the case in databases, where there is a separation between the structures used by end-users to organize a schema and the physical file organization used to implement that schema. The complexity can be ameliorated by making a similar separation in exchange tools and with the development of better tools for defining structures used in data exchange.

limitations of coverage: CAD vendors rightly counter the claim of their lack of support of exchange standards by noting that the entities provided do not cover the functionality their systems offer. Two common examples are parametric modeling and the resistance of the UK CAD Centre to support IGES for its database-implemented process plant design system. This claim imposes on the standards groups the obligation to swiftly support newly developed entity types.

mismatched subsets: A major problem encountered in IGES was that vendors did not implement all of the entities in the specification, primarily because they did not incorporate them in their system. But if a sending CAD system and its preprocessor supply them, then they will be dropped by the receiving system. The proposed solution for this has been to define subsets of the overall specification that vendors may adopt.

mappings: Without strong guidance, implementers may make arbitrary decisions regarding how to map an entity into a neutral file entity. A vendor entity may be optionally mapped into a variety of different neutral file entities. This may be due to inexact specification of the entity. It also arises from mismatched coverage. The proposed solution is to better define neutral file formats in a more formal manner that includes their intended use.

timeliness: The problems of data exchange have grown and solutions to them have been slow in coming. There has been significant lag in the forming of the standards groups and the time taken to generate standards, then getting the vendors to implement them. It is not expected that these issues will go away.

As can be seen, some of these problems address the learning curve required to adopt new technologies. This issue will exist with any change of practice. Some of the issues are organizational, dealing with the ability of user groups to get CAD suppliers to provide needed functionality, despite vendor reluctance to support standards that do not benefit them. However, even if the CAD suppliers responded, some issues would still not be resolved. These are conceptual and intellectual issues that only research in building modeling can address. The items in the above listing that include such issues are varying functionality of different building applications, numerical approximation, informal specification, limitation of coverage, mismatched subsets, and mappings. These deserve a more detailed examination.

3.6 TECHNICAL PROBLEMS OF DATA EXCHANGE

At various points in this volume, we step back and take stock of current open problems. Here, the reader is encouraged to focus on the technical issues of data exchange between a pair of applications, a critical issue in moving to an electronic representation of buildings.

3.6.1 FUNCTIONAL DIFFERENCES IN BUILDING APPLICATIONS

One kind of exchange involves the transfer of application data describing a building project to another application with similar functionality. This most commonly occurs when multiple firms are involved as architects, perhaps when one architect is responsible for schematic design and design development, while another firm is responsible for construction documents. Such a pass-off is never clean and never done just once. Usually multiple pass-offs are required, allowing the firms to work in parallel. Exchanges might also be required if the input data used in one application—an energy analysis or structural analysis application, for example—are used in other similar applications to verify or check the results for consistency.

While it is fairly straightforward to translate a drawing in one CAD system so that it can be read into another and plotted to appear the same as the original, if the second CAD system needs to be able to edit the graphic information, then effective exchange becomes much harder. All CAD systems are different internally—responding to different priorities regarding speed of operations, ease and sophistication of editing operators and range of graphic entities supported. Thus each CAD system makes different assumptions about how to convert its data to an external format, or how to enter an external format into it.

These issues are notoriously problematic for smart entities, such as associative dimensions, crosshatching that will automatically update when its border is edited, and any presentation issues that rely on front-to-back ordering. In all of these cases, editing of the translated objects often identifies unanticipated behavior. Another example is shown in Figure 3.22, indicating a polyline incorporating two fillets. The polyline is translated to another CAD system (via DXF or IGES) that does not have arcs as part of polylines. To compensate, the translation approximates any arcs encountered with a linear approximation made up of short line segments. Such adjustments to many of the graphic entities in both IGES and DXF translators are extremely common. If the non-translated polyline is later edited, it will behave differently than the translated one. If such a problem arises only occasionally, it is manageable. If it occurs several hundred times in converting one drawing, it is not. Both IGES and DXF are widely used for exchange of graphic data, but because of these limitations, the majority of them involve static, non-editing uses.

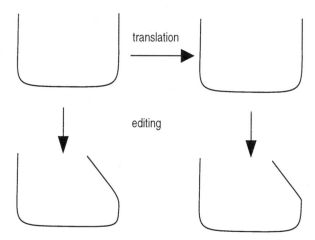

Figure 3.22: After translation, the effects of edits may be very different. Tangency was maintained in the original fillet, but not in the translated one.

The other general type of data exchange is between two applications with different functionality. One scenario may involve merging the results of a design application into a CAD system; another involves extracting information from one application to use in another. These cases must deal with changes of representation, for example, from a 3D solid model of a beam to its centerline for structural analysis, or from a full polyhedral wall representation to a centerline and a height attribute for energy analysis. These application interfaces require data conversion of entities and also the filling in of missing data. More complex capabilities are needed here than can be provided by either DXF or IGES.

A detail problem is found in the translation of attributes associated with some entity. In general, CAD systems associate attributes with geometric entities. The first problem involves the widely different assumptions made by different CAD systems regarding how attributes are associated with or attached to geometry. They may be

- associated with a symbol bounding the geometric entity
- associated with the primitive components making up a geometric entity, such as the lines embedded in a polyline or 3D shape
- carried as extended data within the entity

Because of these different ways of associating attributes, the translation between one application and another must essentially be developed as a custom interface. Such differences in the use of attributes made people realize that effective translation of them was not easily handled in IGES. A much more comprehensive approach to translation was needed to deal with these issues. Difference in functionality is a deep problem in the exchange of data between applications.

3.6.2 NUMERICAL APPROXIMATION

Related to the issue of functional adequacy is numerical approximation. At the bottom level, it arises when one application uses a different numerical representation than another. While Real numbers used to have many different representations, currently there are few. REAL and DOUBLE are both representations of real values, with different precision. Intergraph and Microstation have a unique representation today, relying on a fixed-point, real value representation (on disk only). If one is translating data between systems that use a different real value representation, then the values will change some, the acceptability of the change depending upon its uses. The tolerances in building seldom make numerical approximation of real values an issue; it can, however, be a major issue in manufacturing.

Another issue is tolerances. Especially in solid modeling, the allowed deviation of a polygon bounding a surface from the surface itself can lead to computation errors. Also, a solid model will have tiny cracks where two surfaces join. The allowed tolerances for these cracks vary in different CAD systems. As a result, a solid in one system will be considered an unrelated set of surfaces in another.

The other type of numerical approximation arises when one entity type used in one application is converted to another type for a second application. For example, an increasing number of CAD systems are using non-uniform B-splines (NURBS) as their general curved surface representation, because of its generality and ease of editing. However, other systems, such as AutoCad, do not support NURBS. Conversion of a NURBS surface to a set of faceted surfaces results in loss of precision and change of type, which are not easily recovered. Sometimes these conversions can be made with greater or lower tolerances, for example, based on the exponent of fixed-point numbers, or based on the number of planar surfaces used to approximate a NURB surface, and put under user control. They point out the need to distinguish exact translations which require format changes, possibly with information lost because it is not carried in the receiving application (such as variable line widths and color) from conversions between different types that require approximations. Analysis of different conversion methods, with control parameters of the conversion, is a fruitful direction taken in some conversion programs to make these issues manageable.

3.6.3 INFORMAL SPECIFICATIONS AND MAPPING

Early on, many problems associated with pre- and postprocessor implementation in IGES were due to incomplete specifications. All of the entity types in IGES include guidelines for doing the conversions correctly. For example, the IGES documentation for the Plane Entity (type 108), probably the most common 3D entity, says:

> "The case of a bounded portion of a fixed plane is indicated by the existence of a pointer to a closed curve lying in the plane. This is parameter five. The only allowed coincident points for this curve are the start point and terminate point. Setting this value to zero indicates the case of an unbounded plane.

The case of a bounded portion of a fixed plane minus some portion(s) of that plane, such as those shown in (Figure 30) [shows a polygon with several interior "holes"], are expressed through the use of a Single Parent Associativity (Type 402, Form 9), where the outer closed curve defines the parent bounded plane and each internal closed curve defines some child bounded plane to be subtracted from the parent." (page 86)

The intended rules—that the closed curves lie on the plane, and that the only coincident points are the beginning and termination points—are expressed in words to guide the person writing a translator. Second, several conditions that may or may not be allowed are not addressed:

- the implied or explicit normal of the surface
- if any of the bounding curves are self-intersecting
- if any pair of the interior closed curves can intersect each other
- if an interior closed curve can intersect the exterior bounding curve
- how to define an unbounded plane with holes within it, for example to use as a mask

Thus programmers writing translators are given guidelines that cover some of the less common conditions but are on their own for others. Receiving applications are also on their own in addressing them. As a result, IGES translators between two applications are successful when they have been hand-tuned to support each other. Generally, translation to and from a neutral file is typically ninety percent correct, requiring recreation or correction of a portion of the data.

A related issue is that a preprocessor may generate the graphics needed for a model, but not its underlying structure. For example, it may generate the geometry for the wall example, but not correctly translate the Associativity: the Wall, Door and Window entities might be defined in the correct location, but the location of the Door and Window may not be relative to the Wall. The properties of each may not be associated with the higher-level entities.

In such cases, it becomes apparent that the semantics needed for effective translation are more than can be provided in a single file format. The file format defines the content of each field. The relations between fields are also required to allow correct interpretation. More formal and rigorous specification methods are needed for defining the representation and intended use of an entity.

3.6.4 COMPLEXITY OF FORMAT

Both DXF and IGES—when used to develop applications requiring a complex structure of geometry, attributes and relations—suggest the cognitive limits of what people can understand and manage. They require solid professional programming skills and should be undertaken by software organizations, certainly not the end-users of these tools. At the same time, database technology suggests that simpler approaches are possible.

Similarly, the ambiguity of mappings from known entities in an application to some unknown entities in DXF or IGES—based on some decision of the pre- or postprocessor writer—is inadequate. Mapping tables should be an integral part of all processor generation.

3.6.5 LIMITATION OF COVERAGE

Currently, neutral format files or models assume that they will provide explicit coverage for a set of entity types that is consistent with or approximates the entity types encountered in CAD and other engineering applications. For a vendor, it only makes sense to define interfaces with the entities in their own system. Entities not supported will tend to be dropped. One solution is to specify subsets of entities that must be supported. Another alternative, however, is to support some forms of type conversion within the neutral file or model itself. This would allow the user to define how to convert "foreign" entities into the current application. Some CAD systems, for example, allow selection of certain mappings so that the DXF file they export is appropriate for the use intended.

A more formal analysis of types may also alleviate this issue. The concept of inheritance has been well developed in the theory of programming languages and databases. We will review inheritance in the next chapter. Inheritance structures potentially allow vendors to identify the level of information they choose to make available and also allow standards bodies to incrementally respond to the need to support vendor-specific entity types. This issue is addressed in detail in Chapter 10.

3.6.6 MISMATCHED SUBSETS

Mismatched subsets arise because of the black-box approach used thus far to implement data exchange mechanisms. A preprocessor, about which little may be known, generates a dataset, which is then read by a postprocessor, also about which little may be known. There are several means to improve the match between the needs of a receiving application with the information provided by the sending one:

1) Improve the level of documentation for all exchange processes, especially by requiring mapping tables for all processors.
2) Define subsets that can be implemented and hold implementers to them. This reduces the problem, but still leaves the possibility of implementations of preprocessors that are subsets of the allowed subsets. Also unresolved are issues of exchange between pre- and postprocessors that address different subsets.
3) Introduce user control to the mapping process, allowing conversions to be partially controlled by the end-user.
4) Provide much better tools for defining and implementing mappings, so that the addition of a missing type is easily provided, either by a vendor or end-user.

3.7 SUMMARY

Today DXF and IGES, even with their shortcomings, are the most widely used methods for data exchange in the AEC fields. They are widely supported by various CAD and AEC applications, allowing transfer of geometry from one CAD system to another. They support, for example, the transfer of geometry from a CAD system to a rendering package, or extracting geometry for use in a brochure and report. It is in these areas that these two exchange formats are most effective. In addition, these formats serve well in making batch translations between one main CAD application and other subsidiary applications, where the subsidiary applications do not attempt to update the CAD data. Conversely, the limitations of these two formats can be viewed as the main reason that more use is not made of higher-level applications, dealing with building components, automatic detailing, or embedding more significant amounts of design knowledge.

The open issues listed above address the technical issues of carrying out data exchange. But there are other, more instance-specific issues in file conversion. These issues are made clearer by considering the various tasks a person might do in transferring data between two applications:

- Select the desired entities, omitting those that will not be utilized in the receiving application, possibly selecting them manually by location, by attributes or by external knowledge. In DXF and IGES, this issue must be dealt with by the person creating the neutral file. It thus requires the receiver to communicate with the sender about which entity instances are wanted. Robust remote communication—like email—was not widely available in the 1980s when neutral file transfers were first developed.
- Adjust attributes, sometimes deriving new attributes from existing ones, sometimes by doing table lookups, to generate the needed entities for the receiving application. These issues deal

with changing the type of the entities being exchanged. They were beyond the scope of both IGES and DXF.

- Adjust values to be closer to those desired, other times adjust them to simplify the dataset, to gain speed and/or accuracy. These issues deal with user judgment and expertise in the preparation of datasets for analysis or production. Practically all datasets incorporate such simplifications. Such issues require either human intervention in the mapping process, or the addition of the associated expertise to the mapping processor.

- Make new groupings or substructures, as needed for energy, some structural and other analyses. Aggregation or disaggregation is often required in order to go from one application to another—for example, from a construction drawing description of a wall to an energy analysis description.

- Reformat the data so that they are appropriate for the new application.

The last task has been the focus of most data exchange efforts, with only limited consideration of the other aspects. The other issues are not addressed by DXF or by IGES. Only a subset of the overall translation problem has been addressed.

3.8 NOTES AND FURTHER READING

This and later chapters had to deal with the many uses of the word "entity". It is used informally to refer to some general "thing" related to data exchange. It is also used as the name of some formal element in many data exchange systems. When a formal term is given, entity is capitalized. In all other cases, it refers to multiples of some informally defined entities. In these and all other cases, it is presented in lower case.

The development of the ICES package of programs, including STRUDL and COGO, was a milestone in the development of computer applications for building. The main published reference on this work is Roos [1966]. A good review of this and other early work is given by Teague [1968].

The wall case example presented in Section 3.2 also has been used in other comparisons, for different modeling languages. See Eastman and Fereshatian [1994].

Documentation on older versions of DXF is available in AutoCad® Customization Manual [1992]. The current version of DXF is available from the ACAD Customization Manual or from the World Wide Web site: http://www.autodesk.com/support/filelib/acad14/dxf.htm. The proposal to make DXF an official (British) standard is NEDO [1989]. A helpful next step beyond DXF alone is the addition of layer standards. The different layer standards are those developed through the American Institute of Architects [Schley,1990], the British Standards Institution [1992] and the ISO standards [Björk, 1997]. The movement to make DWG an official standard is being carried forward by the OpenDWG Foundation. Their Web site is http://www.opendwg.org/.

A publication on IGES is Smith [1983]. The various IGES standards include Smith et al. [1988] and IGES [1991]. The 3D Piping Protocol is documented in Palmer and Reed [1992]. The methodology for testing candidate processors for IGES is presented in IATC [1993]. In Europe, an alternative set of exchange formats was developed, including CAD*I [Schlechtendahl, 1988] and SET [1989].

There have been several reviews of these first-generation product model standards. A major survey was made by Bloor and Owen [1995], Chapter 7. See also Wilson [1987]. A description for

Backus Naur Form is presented in the Appendix of most books on programming languages. A fuller description is in Pratt [1975, pp. 301-309].

3.9 STUDY QUESTIONS

1. Define a small drawing in a CAD system with DXF exchange procedures. Export the drawing in DXF format and then import it back to the same CAD system. Examine each of the entities using the "Inspect" or similar operation and note any differences. Examine text placement, line weights and dimensions in particular.

2. Define a small drawing, export it using DXF and then import it into another CAD system. Print the drawing in both CAD systems and note their differences. Examine the entities using "Inspect" and identify any tolerance issues that arise. Discuss why these differences arose.

3. Repeat Exercise #1, but use IGES to export and then import the drawing. Examine the data "before" and "after" translation, noting any differences. From your knowledge of IGES, why might these transitions have occurred?

4. Repeat Exercise #2, but use IGES to export and import the drawing. Print the drawing in both CAD systems and note their differences. Examine the entities using "Inspect" and identify any tolerance issues that arise. Discuss why these differences arose.

5. Take a small drawing and export it through DXF and also IGES. Import the drawing to another CAD system using both translators. What are the differences? Which seems to be the better translation format? Why? Examine the files used in each exchange, in terms of their size. Which is more compact?

6. Examine the use of IGES for translation of finite element data. Using such an application that has IGES export, transfer the analysis results run in one application to the neutral file, then into another structural application supporting IGES postprocessing. What aspects of the element dataset are translated and what are omitted? Compare the numerical properties of the elements; have they changed and if so, how much? Is the imported data adequate to rerun the analysis?

7. Do Exercise #1 or #3, but instead look at the differences in editing capabilities of the entities "before" and "after". Examine the effects of translation on dimensioning and identify any problems encountered. Examine crosshatching and report any problems. Examine the effects of translation on polylines and report any problems.

PART TWO:

Current Work in
Product and Building Models

INTRODUCTION

The main part of this volume is this one, Part Two. It consists of six chapters. This part reviews the concepts and technology currently used in product data exchange generally, and reviews the current systems and capabilities *vis-a-vis* the needs of the building industry.

We begin in Chapter Four with a review of modern programming language and information modeling concepts. From their introduction in the late 1970s and early 1980s, different ideas regarding object-oriented design and programming, conceptual modeling and abstractions have influenced the development of databases, programming methods and all aspects of product modeling. Understanding these concepts is necessary for understanding the intentions and some subtle aspects of current building modeling efforts.

Chapters Five and Six review the organization of ISO-STEP and the information technologies it uses. Chapter Five begins with the evolving motivation and goals of STEP and its basic architecture. It then reviews three of the languages used in developing STEP application models: NIAM, EXPRESS and EXPRESS-G. EXPRESS AND EXPRESS-G are emphasized because of the heavy reliance on these languages in recent product modeling efforts. Chapter Five also reviews how EXPRESS models are used to build instantiated product models on files or databases using SDAI toolkits. It also addresses how translators implemented using STEP technology are tested for conformance. Throughout the presentation, strengths and weaknesses of the tools are discussed.

Chapter Six reviews the main Integrated Resources in Parts 041, 042 and 043. These include the ownership, measurement and geometric modeling schemas that are common resources adopted by most recent efforts in product modeling, both within and outside of the STEP organization. Along with Part 042, aspects of geometric modeling are surveyed. The chapter ends with a discussion of the types of geometry that are needed for different types of building representations.

Chapter Seven reviews three partial or aspect models of buildings. These are CIMsteel, COMBINE and Building Elements Using Explicit Shape Representation (Part 225). CIMsteel is the most successful building product model developed to date; it is widely deployed in Europe and is starting to be adopted in the US. COMBINE was a large and important prototype model developed around energy applications. It began to address issues arising when multiple applications use a single repository. Part 225 focuses on the data exchange of geometry. It is the most significant building-related product model developed to date within the ISO-STEP organization. It has significant use in Germany and other parts of Europe. The status of some other aspect models is also reviewed. The three aspect product models in this chapter point out both the current capabilities of product models and also the still open problems facing the field.

Chapters Eight and Nine review two current models that attempt to deal with the whole structure of a building. These include RATAS, a prototype Finnish National building model, and Part 106, the Building Construction Core Model (BCCM). RATAS was a research effort laying out groundwork for the development of national standards. It provided the basis for current data exchange efforts in Finland. The BCCM was initiated as a major building industry initiative. Currently it is in a passive state, with many of its participants now directing their energies to the IAI effort.

Chapter Nine is devoted to the Industry Foundation Classes, undertaken by the International Alliance for Interoperability (IAI), the most recent and ambitious effort to date to develop a complete data model for representing buildings. It draws upon previous work and introduces important refinements to the existing repertoire of modeling techniques. Interfaces with it have been implemented for a number of AEC software applications.

All of these models are reviewed in significant detail. Even so, the chapters are surveys and cannot be complete. Readers interested in additional details should refer to the original sources.

At the end of Chapter Nine, a review of the current capabilities of building models is attempted, using the IFC as the prime example. From this review we gain an understanding of their current capabilities—what building models are able and not able to do. This then sets up the issues that need to be addressed in further research.

CHAPTER FOUR
Modeling Concepts

Something red can be destroyed, but red cannot be destroyed, and that is why the meaning of the word 'red' is independent of the existence of a red thing... . When we forget which color this is the name of, it loses it meaning for us; that is, we are no longer able to play a particular word game with it. And the situation then is comparable with that in which we have lost a paradigm which is an instrument of our language.

Ludwig Wittgenstein
#57
Philosophical Investigations
1953

4.1 INTRODUCTION

The first generation of data exchange and information technologies in building was developed in the context of computer hardware and software technologies that existed in the 1970s and middle 1980s. Typical of that period was time-shared access to small mini-computers for interactive, departmental applications and batch, non-interactive access to mainframe computers for enterprise-level activities. On the software side, most applications were developed in FORTRAN running on the Unix or a proprietary operating system. During this time, however, new ideas were introduced regarding the way computer languages and systems should be organized. These ideas had a major impact on all areas of computing, including the direction of work in data exchange in the building industry.

One line of work that generated new ideas proceeded in three parallel areas—database designs, artificial intelligence, and theorem proving. It dealt with the structure of information in general, and sought to identify some basic relations between entities that capture certain kinds of human knowledge. Addressing what is called the *knowledge representation problem*, this work was the modern extension of a long line of effort in mathematics and logic to define a formal language whose terms and relations could capture the semantics of all knowledge. Such a language would potentially allow all implications of known facts to be derived. It was this quest in the 19th and early 20th centuries that eventually led to Bertrand Russell's paradox, showing that some facts would not be provable in such a language. Later, Kurt Godel proved in the 1930s that no such language could ever be complete. Despite such conclusions, the payoffs of partial solutions to the knowledge representation problem helped spur significant efforts in artificial intelligence, logic and databases to develop languages for representing knowledge. One result was the development of expert systems; other ideas stemming from this line of research led to the development of data modeling.

Another group of intellectual endeavors dealt with the concepts embedded in computer languages. The software industry in the early 1970s was grappling with new problems regarding what is now called software engineering. Computer programs were getting very large— hundreds of thousands of lines of code. At the same time, it was recognized that the costs of software were not only in its initial design and implementation, but in its maintenance over a lifetime of evolutionary refinement and updating. Some responses were managerial—suggesting how to form more effective programming teams. Others had to do with the organization of the programs themselves. FORTRAN, the first widely used programming language, allowed blocks of programs to be organized any way the programmers desired. Usually the data structures for a program were defined together in one module, with various modules of subroutines organized separately. Several researchers recognized that data structures and the subroutines that operated on them should be grouped together to facilitate their maintenance. Later, this idea was formalized as an *abstract algebra,* in much the same way algebras are formalized in mathematics, or in computer terms, as a set of *abstract data types.*

In parallel with the development of concepts about program structure, concepts about programming languages were evolving. A simulation language, SIMULA, developed in 1964 introduced the concept of a program *object.* It defined an object as consisting of data and methods, i.e., procedures and introduced inheritance and polymorphism. The ideas of abstract data types and object-oriented programming led to major rethinking of how programs should be structured. In the succeeding ten years, these ideas were refined and embedded in a new generation of programming languages, including Smalltalk, ADA, Objective C, and C++ (Java came later). Object-oriented databases that supported these languages soon followed. Abstract data types and object-oriented programming are ideas still influencing the development of software in the building industry.

Also during this period came the third development of new diagramming methods, which entailed rigorous syntactic and semantic definitions for designing information systems. SADT, a graphical language developed by Douglas Ross was one, followed in the late 1960s by the entity-relationship (ER) model developed by Peter Chen. The ER model, developed to support database design, showed that a graphical language could be a logical description of data and could be mapped, using formal methods, into a database implementation. This was greatly facilitated by the strong, logical basis for relational databases developed by Edgar Codd in 1970.

These new lines of research and thinking pointed out the limitations of IGES and other file-based exchange formats and suggested new approaches, ultimately leading to a new generation of data exchange methods.

This chapter reviews most of the computer science concepts used in succeeding chapters. These include the concepts embedded in object-oriented systems, and some that are associated with knowledge representation and conceptual modeling. Together, they form one part of the conceptual base for modern computer languages and also data modeling. Some recent work extending these concepts is also reviewed.

4.2 OBJECT-ORIENTED PROGRAMMING

The fundamental idea in object-oriented programming deals with the extension of the concept of *type.* The development of types in early languages was intuitive, dealing with REAL, INTEGER, CHAR, and BOOLEAN data formats built into the hardware of a computer's central processor operating codes. Each type held data in a different format. A set of operators that read, wrote and—most importantly—manipulated each type was implemented in the computer hardware (sometimes simulated in low-level software). A capability built into almost all programming languages was the ability to do *type checking*—to verify that the operators applied to some data were consistent with the format of the data accessed. Checking was supposed to catch such errors as applying a real arithmetic operation, like ADD, to characters, or a pointer operation, like ADDRESS DEREFERENCE, to a Boolean value. Type definitions and checking were important aspects of computer language design and implementation. The new idea was to support user-defined, high-level types that worked in the same way as those defined within the hardware. (NOTE: There are multiple uses of the terms *type* and *class* in programming languages. Here, the normal conventions are followed: a *type* is an abstract entity used to define object semantics and a *class* refers to the set of object instances declared to be of some type.)

The notion of abstract data types (ADTs) formalized a solution. The basic idea is that whenever someone defines a new data structure—to represent a shape, a set of attributes or a physical object such as a wall—they must also define the operators that create, manipulate, and delete that data structure. We call the data structure an entity or object. Embedded within the operators are the rules that update the object correctly. Other operators outside of the object are likely to violate these rules and, if applied to the entity, would "corrupt" it. Thus outside operations cannot directly manipulate an object's data, but can only call the object's methods to access or update its values. This security concept is called *encapsulation*.

In addition to the issue of type checking before execution, ADTs also introduced new formal notions of abstraction and security. During the maintenance phase of a program's life, encapsulation also allows the internal structure of an object to be revised without side effects, as long as the operator interfaces remain the same. Encapsulation thus isolates the effects of changes, creating strong modularity. ADTs also support reuse of software. If developers want to use an existing ADT in a new application, they should be able to simply name the type and start using it and its operators, with all its semantics intact. These concepts were explored in a range of experimental programming languages in the 1970s and 1980s, where ADTs were called objects. In the late 1970s, Adele Goldberg and others at Xerox PARC developed SMALLTALK, the first commercially available object-oriented (OO) language since Simula. Since then, other commercial OO languages have been introduced, including Objective C, CommonLoops, C++ and Java. A secondary goal of OO systems was to move programming languages closer to the languages used by people in problem-solving, which ADTs seemed to accomplish.

The emergence of object-oriented languages was seen as a great advance in making programming more intuitive as well as more robust. Advocates of OO languages suggested that we could think of software objects as if they were physical objects and could write programs in the same way real objects interact. As we shall see, however, the definition and implementation of these concepts are subtle and not as intuitive as one might hope. Today the basic concepts of object-oriented systems seem to be well-known, though the subtle differences that have arisen among different systems are probably not. Throughout the chapter, these differences are highlighted so that readers can distinguish the different "flavors" of object-oriented systems. These issues have influenced how people have thought about the structure of objects representing buildings and their parts.

4.2.1 STATE AND BEHAVIOR

The data structures within an object carry data that describe an object instance. If the physical object or concept being represented is static, then data describing it will be fairly static also, changing only as the object is designed or fabricated. Dynamic objects, however, will change over time. The changes in an object's data values characterize the object's *behavior*. For example, an elevator object's description not only identifies the dimensions of the elevator car, door size, capacity and so forth, it also may need to define its state at some moment in time—for example, when doing an elevator simulation. In this case, we may need to define the elevator's location at some floor level, its speed and direction, the position of its doors (open or closed) and how many people it is carrying. In general, then, the data need to be able to include attributes sufficient to define all possible states for the object's possible uses. Time and location in space are important dimensions in the definition of an object's state. An object's behavior can be defined as the possible transitions allowed from one state to another.

One issue in the design of any building representation is the generality of the representation. Which kinds of building components can an object type represent and which can it not? What is a sufficient representation of a building space—or any other building entity? Such questions can be answered by considering the possible states allowed by an object type. Can the space representation deal with the temperature of the space? How about the temperature changes over time? What about the temperature variation within the space—say, along the floor or next to a window? In general, an object's possible states are the Cartesian product of the range of the possible values for each of its attributes.

The mathematical concept of Cartesian product is fundamental to modeling. It is the domain of each variable describing something multiplied by all the other domains. For example, the range of possible locations—assuming some finite grid—is the Cartesian product of the x, y, and z locations in space. Of course, the value of most Cartesian products is huge. We are usually more interested in the *dimensionality* of the state—the names and types of attributes used to describe the object's state.

We are also interested in the rules that the operators apply to disallow certain object states. The operators incorporate rules regarding the logical correctness of some states of an object. Most object states do not allow negative area or volume, or values for these that are inconsistent with their shape representation. Locations may be restricted by the value of "connected to" attributes. Some location operators may check for spatial overlap, further reducing the allowed object states. In this way, an object's states begin to define the semantics of the object. We are interested in representing all semantically meaningful object states.

A term used to refer to the possible states of an object is its *universe of discourse*, or UoD. We might say, "the UoD of a space does not include temperature variation within the space". At a higher level, we are interested in a UoD of a whole model, which is the Cartesian product of all possible objects that can be defined in the model times the possible states of each object. We are not interested in the numerical value of the UoD of some model, but rather the properties of the UoD, such as
- Does it allow spatial conflicts among building components?
- Does it allow walls to be made of more than one type of construction?
- Does it allow fabrication and placement schedules to be associated with building components?

Until recently, there has not been a practical way to deal with such concepts as the structure or complexity of designs or representations. The UoD offers a qualitative metric for assessing the size, complexity and structure of a building model, a scheduling model or any other type of model. It should be noted that two buildings of the same size, designed for the same use, can differ tremendously in dimensionality. Highly modular designs that use standard parts—possibly represented in CAD systems using symbols—have low dimensionality in comparison to a design where there are curvilinear walls, varied details and no repetition. We would like the UoD of a building model to allow all the possible states that are meaningful, but disallow all those that are impossible to realize.

4.2.2 INHERITANCE AND POLYMORPHISM

A basic notion in object-oriented systems that goes beyond ADTs is *inheritance*. The notion of inheritance is fairly simple. Inheritance allows new object definitions to be based on existing ones; a new object type can be defined by referring to an existing object type and adding the differences between the referenced type and the new one. In different systems, the referenced object may be called *immediate ancestor, parent* or *supertype*. Similarly, the new object type is called *immediate descendant, child* or *subtype*. The relation between the two object types is sometimes called an "is-a" relation. Inheritance relations are transitive, so that a parent may be a child of another type, with the effect that a child object inherits the attributes of all parents, the parents' parents and so forth.

Some languages restrict inheritance so a child can have only one parent (single inheritance); others have no such restriction and allow multiple parents (multiple inheritance).

Inheritance automatically declares all attributes and all operators of the parent type in the child type. In single inheritance, there is no chance that duplicate operators or attributes with the same name will be inherited, but in multiple inheritance such a condition can arise. Multiple inheritance must include some means for disambiguating or eliminating the duplicates. A variety of means exist, such as giving precedence to the first one declared, or naming each duplicate according to the parent it came from. In single inheritance, any child object can have at most one parent and the result is often called a tree or hierarchy, because of its one-to-many branching. (An example is shown in Figure 4.2.) Multiple inheritance allows a many-to-many relationship among object types, which is called a lattice or directed acyclic graph. (An example is shown in Figure 4.3.) These terms arise frequently in distinguishing the structures created by different kinds of product models.

The mechanism and syntax for inheritance appear obvious, but have subtle implications. It is often thought of as similar to classical taxonomies of life forms. Thus a parent type might be "mammal" and its child types might be "dog", "goat", "horse", and so forth. However, the semantic implications of inheritance vary from language to language. The varying interpretations dealing with inheritance involve two concepts: one is polymorphism, the other is the way in which subtypes can be organized.

An exotic sounding term *polymorphism* refers to the ability in object-oriented systems to interpret an instance of a child class as an instance of any of its parent classes. Polymorphism applies to both single and multiple inheritance languages. A child object inherits the attributes and operators of its parent types and thus should have equivalent semantics as any of its parents. Thus an object may be of multiple types and belong to multiple classes—its own and all of its parents'—hence the term polymorphism.

An effect of polymorphism is that a child object instance can be operated on by both its own operators and those of its parents. However, the rules of inheritance generally allow child objects to be imperfectly compatible with their parent object types. For example, many OO systems allow a child to delete or redefine attribute definitions. Four levels of compatibility have been identified:

cancellation compatibility, where operations may be freely modified or deleted;

name compatibility, where a child must carry all the named entities of parents, but with no limits on how they may be changed;

signature compatibility, where the interface of all operators must match—that is, the operator names and their parameters;

behavior compatibility, where the behavior of all child operations must be equivalent to those of the parent.

Figure 4.1 presents some alternative examples. It shows three, alternative ways of defining types for `polygon` and `hatched_polygon`. The semantics of the two types follow one's intuition: the polygon is a sequence of connected points, while the hatched polygon additionally has a hatch fill. The simplest and most widely implemented level of compatibility is *cancellation compatibility*. This allows inheritance lattices to be organized in almost any way, with child attributes and operators being added, modified or deleted, as shown in Figure 4.1(a). Here, the child object, `regular_polygon`, deletes the `linetype`, `spacing` and `angle` attributes of the parent, `hatched_polygon`, because these attributes are not used in the child. The result, however, is that the parent operators cannot be applied to the child without crashing. The child object has overwritten the other parent operators so that when called from the child, they will not fail. In this example, the compatibility of the child object is some arbitrary subset of the parent.

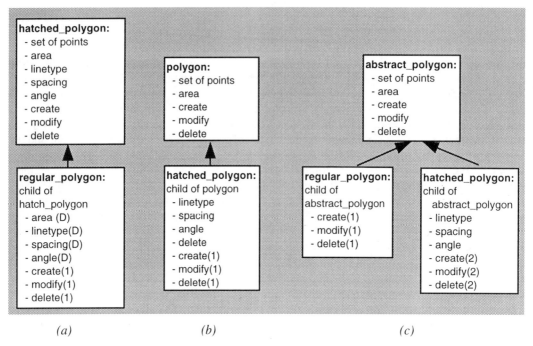

Figure 4.1: Three alternative interpretations of the class structure for polygons. The (D) denotes an attribute that has been deleted from the parent. A parenthesized number after an operator indicates it has been modified.

The second level of compatibility is *name compatibility*. Here, the language guarantees that any name introduced by the parent will also exist in the child, thus deletions are not allowed. Figure 4.1(b) shows an example of name compatibility. The simple polygon has been defined first and the crosshatch attributes are added in the child object type. *Signature compatibility* slightly improves upon name compatibility by guaranteeing that the operator arguments are also consistent. However, even with this level of compatibility, the detailed semantics of the polygons of a child may be restricted because there is no limitation on the internal modifications that can be made. For example, the parent polygon may allow polygons to be nested inside of other polygons, representing holes, with the direction of each polygon being used to identify whether it defines an inner or outer boundary—clockwise usually being inner. The overwritten operators of the child may not incorporate this convention or rule, leading to wrong results when an area calculation operation is applied.

C++, Smalltalk and Java support cancellation compatibility. No commercial languages enforce *behavioral compatibility*. Thus, there are limitations in all languages today in their use of inheritance. Objects are not as shareable and reusable as we might desire, with programmers frequently making errors because of their misinterpretation of an object's semantics. This is an open research issue in programming language design. These issues are potentially important because several proposals for distributing digital representations of building components and building information entail distributing software objects, for example, across the Web.

We turn now to how subtypes are organized. In our example of animal taxonomies, all child types are subsets of the parent type (all horses are mammals) and the subtypes are disjoint (no dog can also be a monkey). However, in Figure 4.1(b) it is unclear whether the polygon type is a superset of all polygons and includes the hatched_polygon type, or instead is meant to deal with non-hatched polygons and thus is actually disjoint from the hatched ones. Sometimes subtypes are disjoint and other times not. How can these cases be distinguished? Figure 4.1(c) shows a modified class hierarchy using the class `abstract_polygon`. Here, `abstract_polygon` is defined so that all subtypes are indeed disjoint; a `regular_polygon` without crosshatching is clearly distinguished from the more general class of all polygons. All operators are defined by name only in the parent type and all operator semantics are defined in the subtypes, so that revisions are not required. The concept of abstract type is supported in many object-oriented languages to facilitate this particular use of type hierarchies.

While behavioral compatibility seems to be the ideal, an unresolved issue is how to enforce it. One way to guarantee that the semantics of the child includes that of the parent(s) is to not allow overwriting of operators. However, this prohibits the name from possessing additional behaviors as an object type is refined and leads to many name variations of basic operators such as `create` and `delete`. There have been some efforts to add the essential conditions using rules, but these are cumbersome and lack intuitiveness. No adequate solution has been found. On the other hand, there have been good arguments for looser forms of inheritance and polymorphism. Despite these problems, these two concepts are extremely useful.

While most work dealing with inheritance considers it a means to specify new types, there are important conditions where its application to instances is also of concern. If an object model is populated with instances and a change is made to the definition of some types, then the matter naturally arises regarding what to do with the instances whose type definitions have changed. This sort of issue arises quite naturally when a building model needs to be extended in order to support an additional analysis application or to receive scheduling information. Most database systems also encounter the need to support extensibility and deal with that need in various ways. We may want to add new attributes to the building's walls, for example, to deal with analysis of the walls' smoke emission attributes. Some object-oriented databases support extensibility of this sort directly, adding the new attributes to all instances and initializing the new values to NULL. Other capabilities are needed also, such as specializing a current instance so that it includes new attributes, then migrating those instances in the original object class to the new, specialized class. These are still open research issues.

4.2.3 OPERATORS, METHODS AND SEMANTIC INTEGRITY

A weakness of preobject programming was that some external subroutine could operate on a data structure without a full understanding of the semantic rules that must be satisfied, possibly resulting in an illegal state. By associating a set of operators (called *methods* in some languages) with objects, abstract data types allow the rules to be embedded once into all operations associated with the object. All manipulations of the object must be done through these operators. This requires, however, that direct calls by external operators not be allowed. Thus all operations on some object ‵A´ that are required by other objects must be done by passing messages to object A's operators. *Messages* are operators on an object that are visible from outside. This is the encapsulation protection strategy introduced earlier.

Since the initial development of object-oriented systems, it has been found useful to distinguish different classes of operators beyond just those that are visible to the outside and those that are not. In various object-oriented systems, operators may be

public: accessible from any other object

private: accessible only from members of the type in which it is defined

protected: accessible from members of the class that defines it and its subclasses.

There are many variations and subsets of these three classes of operators, such as Java's *package* and C++'s *friend*. These variations have been defined in order to strengthen the type reliability of polymorphic objects, as described in the previous section. No definition of operator classes, however, has been found that resolves the ambiguity of polymorphic types.

A basic issue addressed by the abstract data type (ADT) concept and object-oriented programming is the specification of an object's semantics. Object semantics involve the object's behavior and states—that is, which behaviors and states are allowed and which are not. Can an elevator's dimensions be minus values? Can its dimensions be larger than the shaft it is in? Can its material not be fireproof? It also addresses the transitions between an object's states. For example: can an elevator be moving without closing its doors? Can an elevator go directly from floor 1 to floor 14? Such issues can be resolved by the proper definition of the operators and attributes of the elevator object.

Object semantics define rules that apply to an object and determine if it is meaningful or legal. These are called *semantic integrity rules*. There are at least three scopes for semantic integrity rules:

1. Some rules can be associated with attributes and can be embedded in the attribute type. Thus we can specify that all dimensions of objects be positive Real values. We can restrict the materials of an elevator to those that are fireproof. These rules were recognized before ADTs and embedded in some early programming languages.
2. Other semantic rules apply to relations between attributes: the dimension of the elevator car and its carrying capacity, or the state of the door and whether the elevator is moving or not. Relations between attributes can be managed only by the operators assigning values to them. Part of the attraction of ADTs was that operators could incorporate these rules, and encapsulation guarantees that they would be used. Note that these semantic rules apply to all instances of elevators and are defined for the elevator object class.
3. Still other semantic rules apply to relations between objects—for example, the relation between the shape of a space and the shape and locations of the walls bounding it; if two object instances are connected, this relation constrains their possible positions. These rules cannot be embedded in the objects that are related, because they do not have access to the other object(s) in the relation. In general, good object-oriented programming practice holds that such relations should be defined in operators that are defined in a higher-level object that composes the objects in the relation. For example, a wall and space relation should be defined by operators for room layout in a floorplan object that has wall and space as component objects. The connection relation should be embedded in operators that aggregate the object instances in the layout.

The concept of embedding rules in operators is fundamental to good object-oriented programming. It has direct application in the development of building applications. It suggests ways in which detailed knowledge and semantics can be embedded in a model of a building.

4.2.4 THE IMPACT OF OBJECT-ORIENTED SYSTEMS ON BUILDING MODELING

Object-oriented systems, even with the open issues and ambiguities reviewed here, have had a tremendous impact on software engineering and information technology in all fields. In the building industry, they have had an impact on at least two levels:

*   CAD systems and growing numbers of applications are implemented using object-oriented technology. Slowly, pre-object-oriented systems are disappearing. Current generation applications create and manipulate objects. The information that may need to be exchanged is embedded in objects.
*   The concepts of OO systems are applicable to data exchange and integration. They can be used to facilitate advanced uses of IT in the building industry. The full implications of some of these concepts for data exchange have yet to be fully explored.

From this review of the concepts associated with object-oriented systems, one can see that there are many variations and competing directions. These variations and the lack of a "correct" direction are why there has not been a standard model for object-oriented systems. At the same time, the concepts have had a major impact on the design of all modern languages and systems, including those for building modeling. The ideas of state and behavior are fundamental to the representations used in advanced building applications. They are important concepts useful in implementing data formats or databases that support multiple applications.

4.3 ABSTRACTIONS

Another important concept emerging in the late 1970s and early 1980s was that of abstraction. First articulated in papers by Smith and Smith in 1977, an *abstraction* of some representation is a second representation in which details of the first are purposely omitted. The omitted data reduce the complexity of the second representation and allow focusing on the information that remains. An abstraction can be applied to data that are already an abstraction (an abstraction of an abstraction) multiple times, resulting in an *abstraction hierarchy*. Abstraction hierarchies are important structures in both thinking about and organizing data within computers.

In order to examine more closely the concept of abstraction, an example of an abstraction hierarchy is shown in Figure 4.2. This is similar to but takes some liberties with the CAEADS building model reviewed in Section 2.5.3. As we traverse this hierarchy from top to bottom, the single term is replaced with a set of terms that the one word characterizes. Each of the terms may carry attribute and relation data as well as references to even more detailed terms. At the bottom level, a term refers to the set of attributes that describe it. We can assume that operators exist within each of these entities.

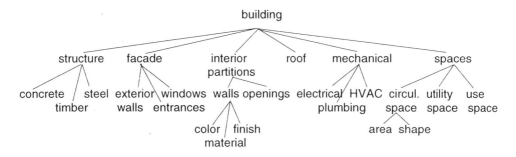

Figure 4.2: A high-level portion of a possible abstraction hierarchy of a building.

The notion of abstraction hierarchy was quickly recognized as useful for structuring a building description. The high-level terms and attributes roughly describe a building project in terms of the systems which comprise it, with the lower-level terms detailing the structure of each system. There may be multiple levels of abstraction. The growth of the hierarchy in a top-down sequence, corresponds roughly to the incremental refinement of a project, as carried out in a typical design process.

A closer look at Figure 4.2, however, identifies several different types of groupings. Abstraction is not a single concept, but rather a family of them. Four types of abstraction are now widely recognized. A few of these are intuitive and, in fact, have been used since the beginning of computing.

4.3.1 SPECIALIZATION

Specialization was one type of abstraction identified by Smith and Smith. Its inverse is *generalization*. Specialization is the conceptual term for "inheritance" or the "is-a" relation. Its semantics have been described in Section 4.2.2. Other terms often used are *subtype* and its inverse, *supertype.* Specialization is a relation between two entity classes and also applies to their instances. In one direction the relation is called a generalization; in the other it is called a specialization.

In Figure 4.2, the specializations in the building abstraction hierarchy seem to be structure, façade and spaces. We need a way to describe them textually. Below we define a relation, with the entities of the relation in parentheses. The supertype is named first, followed by subtypes:

```
specialization(structure: concrete, timber, steel)
specialization(facade: exterior_walls, windows, entrances)
specialization(spaces: circulation spaces, utility spaces,
          use spaces)
```

We assume that the `structure` type defines general static properties in its attributes, such as maximum shear, deflection, bending moment, that are "inherited" into each of its "subtypes". Facade attributes might include thermal conduction, waterproof-ness, and possibly ultraviolet degradation properties, which are inherited into the `walls`, `windows` and `entrances` subtypes. Spaces might have geometry, floor area and volume as primary attributes that are inherited into the three kinds of space subtypes.

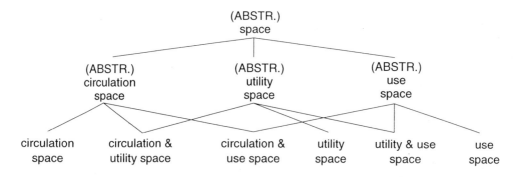

Figure 4.3: An abstraction hierarchy based on specialization that has two levels of abstract classes, before any instances may be defined.

In their original work, Smith and Smith proposed a particular semantic definition of specialization, different from most language implementations of inheritance and closer to the ideas used in conceptual modeling (see Section 4.3). They required that all specializations of a generalization class be both disjoint from each other and also subsets of the generalization class. In Figure 4.2, the specializations of `structure` are certainly disjoint. If `utility spaces` are always defined so they cannot also be `use spaces`, and `windows` are defined so they are never parts of a `exterior_wall`, then the Smith and Smith definition of specialization applies to them. However, it is often desirable to classify an object in more than one way; a room has both an assigned function (a use space) and is used for circulation. (This is an example of multiple inheritance.) If a space can have multiple classifications, then a specialization structure like that shown in Figure 4.3 is more appropriate. It defines two levels of abstract classes, then allows instances to be members of singular abstract classes or pairs of them. All the classes at the bottom level are again disjoint.

In Figure 4.3, all the possible specialization combinations are defined explicitly. Another way to define what combination of supertypes can be inherited into subtypes is to use *inheritance constraints*. In the example in Figure 4.3, we could impose a constraint that a space may be ANY combination of its specialized abstract classes. Alternatively, we could impose a constraint that a space can only be "ONEOF" its specialized abstract classes. Other constraints are possible. This approach is significant because it is used in the EXPRESS product modeling language, reviewed in detail in Chapter Five.

This discussion of specialization and the earlier discussion of inheritance point out the different strategies used to define specialization lattices and the different rules employed. It also suggests why translating a building based on a language using one type of inheritance conventions into a language with another type of inheritance conventions may be difficult.

4.3.2 AGGREGATION

The most basic type of abstraction involves grouping a set of data and giving the grouping a name. Smith and Smith called this form of abstraction *aggregation* and its opposite, *disaggregation*. Humans use the concept of aggregation extensively. For example, instead of saying "animal with short hair and small to medium size that barks" we say "dog". Instead of saying "a multiple story rectangular mountain cabin with steep roof and curlicue trim and bright paint" we say "chalet". Aggregation has been supported in many computer languages, in Pascal with the `Record` construct and in C with `struc`. It was also part of the earliest database systems, typically called a Record. However, these constructs do not define semantics and can be used to define several types of abstraction, in addition to aggregation.

Aggregation uses a single entity to describe a collection of entities. Sequentially, the single entity is typically named first, followed by the collection of entities, which are then assigned values. We normally call the collection entities attributes. In Figure 4.1, the following are aggregations:

```
aggregation(walls: color, material, finish)
aggregation(circul.space: area, shape)
```

The structure of an aggregation can be defined quite precisely. Before we can do so, however, we must define more precisely what is meant by *attribute*.

Attributes are entities that are used to describe the properties of another entity. The entity being described is the attribute's *reference*. The attributes can have no meaningful value without a referent. Because of this dependency, computer scientists often call attributes "second class" entities, while the referent entity is "first class". Attributes may be measures or properties of the referent entity. Together, the attributes are considered as defining the state of the referent entity.

In CAD system development, it has been recognized that a geometrical description is not a first class entity, but rather an attribute of another entity. Thus it should be possible to define some building component and even locate it, without defining a shape for it. At the same time, in design, we sometime define the attributes of an object before we decide what the object is. For example, we may create a shape as part of a building but not know what will go in it. Or we may define some thermal mass properties to capture, hold and later conduct heat to an inside space. In both of these examples we create and assign attributes to some object instance to be determined later. In object terms, however, the attributes must be aggregated into a named object, such as "building mass" or "space boundary", then later specialized into a more detailed version of the object.

Given an aggregation, all attribute entities refer to the single entity they describe. Their grouping is based on the fact that they all reference the same entity. Because the multiple attributes have the same referent entity, the semantics of the aggregation include a variety of integrity rules across the attributes. For example, all material properties must be realizable with a single material; all geometrical entities—like plan and elevation, for instance—must be consistent with a 3D model entity. The area and volume of the entity must be consistent with its geometry. When we say "area of room A", the "of" implies room A is the referent of area. We can say that this type of aggregation is based on *consistent self-reference*. These consistency rules typically are managed by the object's operators.

Aggregation is usually implemented as a class definition construct, defining one or more structures that are used for defining object instances. Inheritance, already described, provides means for abstracting the definitions of aggregations, resulting in a hierarchy or lattice.

Consistency and integrity are two terms used frequently in computing. *Consistency* in any type of data is important if the data are to be relied upon. Consistency, when applied to data, means two things: (1) duplicated data carry equivalent values and (2) data that can be derived from other data are also equivalent with what would be derived. Integrity is also used to deal with the quality of data, but it has a wider meaning. *Integrity* means that the data are correct, according to whatever relations or rules that are applied to it. Thus integrity is always relative to some definable set of semantic rules.

4.3.3 COMPOSITION

A second type of abstraction groups entities into *composite objects*, sometimes called assemblies. The entities that make up a composite object are related to it by the *part_of* relation. There has been extensive discussion of composition and the concept of composite object, mostly dealing with its logical definition. It is recognized as an important construct in many areas of engineering and design.

Semantically, composition is defined by a composite object class and a composed set of part object classes that define it. A part object may also be a composite object, resulting in a multi-level hierarchical composition. In composition, the attribute names of the parts and composite object may be independent of one another, in contrast to specialization, where they are inherited. For example, the attributes of a window or layer in a wall, in the core wall example presented in Chapter Three, are different from those of the wall itself. The parts are also not attributes, because the parts are first class objects. That is, they are independent of the assembly of which they are a part. Because a part is a first class object, it may be part of more than one composition. For example, a pump may be part of the electrical assembly and also part of the piping assembly; a wall may be both a shear element and also a thermal mass element (in some solar heating designs). This is not possible with even a complex attribute, such as shape. Thus composition can be distinguished from specialization and aggregation.

Because a composition has an organization, it also has rules that distinguish those organizations that are allowed and those that are not. If we are defining a pipe run, then there are a number of rules that define if the run is minimally correct. The object instances must be connected; their joining conditions must be satisfied regarding thread type and diameter, the flow directions of elements (pumps, pressure gauges) must be consistent, and so forth. Similarly, in the wall example introduced in Chapter Three, the wall segments and openings are composed into a wall and the wall composition must satisfy certain rules. The rules for walls include that the parts are tightly packed and contain no spatial overlaps. The rules of composition constitute simple but fundamental knowledge in most areas of design and construction, in terms of proper definition of complex object organizations, such as the pipe run and wall. Many rules are intuitive and so obvious that people take them for granted—a pipe run must have all its parts connected, that wall components cannot overlap spatially—but are not known to a computer. Others are subtle and reflect human concepts of expertise, for example, flow direction and flow rates in piping.

In the example in Figure 4.2, the hierarchy below `space` was earlier considered as an inheritance structure. However, it may also be considered as a composition of all the spaces within a building:

```
composition(spaces: circulation spaces,
                         utility spaces, use spaces)
```

These alternative interpretations have subtle but important differences in meaning. If space is linked to the different kinds of spaces through an inheritance structure, then space is a general concept whose attribute and relation definitions are inherited into the different specific kinds of spaces. If they are linked via a composition structure, however, then the utility spaces and the use spaces are parts of the composite object called space and there are rules of well-formedness regarding how they are composed together. In this case, the composition becomes useful as a means to define the necessary conditions of arrangement among spaces. These might include that space instances have only limited forms of overlap and they have specified kinds of adjacency relations. (Notice that these uses of the terms specialization, inheritance, aggregation and composition have precise meanings; intuition is a poor guide to their use.)

The recognition of composition as a distinct type of abstraction came only after the initial language developments of Smalltalk and C++. Even today, only a few commercial object-oriented languages support them in any explicit way. Thus, most object-oriented product and building models today treat them in *ad hoc* ways, with one object having references to other object instances, with no clear distinction whether the referenced object is an attribute or a part-of relation. However, if not built into a language, composition can still be defined using a low-level structural object, then embedded into higher-level structures in a semantically clear way.

4.4 RELATIONS

A relation is the generalized association of one object instance with one or more other instances. All the abstraction relations reviewed above—specialization, aggregation, and composition—are very general relations that apply across most domains of application. However, many other types of relations are domain or application specific and have to be specified within the types defined in a building model. Examples of relations needed within a building model include *placed within* (relation between some furniture or other located objects and some space), *electrical connection* and *structural connection*. Examples from the construction standpoint include *precedes* (relation between two tasks), *required resource* (relation between some material or equipment and some task) and *delivers* (relation between an event and product material). Thus, there is a need for a variety of special-purpose relations.

A partial response for the need to define special-purpose relations has been to provide ways to specify the properties of a relation. One property of a relation is its *cardinality* or *arity*. That is, does the relation exist only between one object and another single object? Or does it exist between one object and a fixed or arbitrary number of other objects? Simply put, is the relation 1-to-1, 1-to-2, 1-to-M or M-to-N?

The reason that this aspect of a relation has been made explicit is largely historical. The proper definition of relational database schemas has been dependent upon the definition of the arity. If it was a one-to-one relation or one-to-M, this relation could be implemented by merging the two objects in the relation into a single table, while M-to-N relations required a separate table for depicting the relation. Thus, defining the arity was an important issue in defining the proper schema for a relational database. All data modeling languages developed to support relational database design explicitly define the arity of a relationship.

In object-oriented databases, the issue is still relevant but somewhat different. In relational databases, all relations are automatically bi-directional. That is, if a relation is defined between A and B, then operations can go from either A to B or B to A. However, in object-oriented systems, relations are not bi-directional and the directionality of relations must be made explicit. In practice, if a relation is defined to be bi-directional, one must define two relations: from A to B and from B to A. If the relation is greater than one on either side, it means that a collection structure (lists, arrays or sets are examples of collections) should be used in defining that relation.

The treatment of bi-directional relations in object-oriented systems raises another problem, however. It is implied that the two relations are the inverse of each other. That is, whenever an instance of an a carries a relation to an instance of a b, the instance of b also has a relation to the same instance of a. This defines a constraint between the two relations a-to-b and b-to-a. Each must be the inverse of the other at the instance level. Several database and data modeling languages—for example, EXPRESS and UML—support an INVERSE constraint between two relations.

Other semantic properties of relations have not been specified in modern programming languages—for example, those defined in discrete mathematics:

Symmetric relation: a relation is symmetric if Relation(A,B) $\Leftrightarrow$ Relation(B,A). For example, *adjacent* is a symmetric relation.

Antisymmetric relation: a relation is antisymmetric if Relation(A,B) $\wedge$ Relation(B,A) $\Leftrightarrow$ B = A. An example is the size relation among areas: if *size-equal-or-greater(A,B)* and *size-equal-or-greater(B,A)*, then *size(A)=size(B)*.

Reflexive relation: a relation is reflexive if R(A,A) is True for all As. An example of a reflexive relation is *electrical_connection* when applied to a conductor object.

Transitive relation: a relation is transitive if Relation(A,B) $\wedge$ Relation(B,C) $\Rightarrow$ Relation(A,C). *Generalization* is an example of a transitive relation. *Adjacent* is an example of a relation that is not transitive.

(The above properties of relations are defined using logic notation, where

$A \Leftrightarrow B \equiv$ A implies B and B implies A,

$A \Rightarrow B \equiv$ A implies B,

$A \wedge B \equiv$ A and B, and

$A \vee B \equiv$ A or B.)

Future developments in programming languages may provide means for defining and checking such constraints on relations. Today, if a modeler wishes to use these kinds of constraints, they must be defined and managed by the modeler.

It should be noted that from a conceptual modeling viewpoint, object-oriented systems do an exceptional job in defining objects. They do not provide helpful functionality for defining relations, however. Since there are no facilities for defining relations, software developers must do it themselves. In the worst case, this results in relations being defined implicitly, using *ad hoc* methods. In the best case, developers will predefine a set of structures corresponding to relations and use them consistently in their product modeling.

4.5 CONCEPTUAL AND DATA MODELING

The work on abstractions, initiated by the efforts of Smith and Smith, resulted in a flurry of work addressing different ways to model information processes in support of automation. One of the impetuses for these new efforts was the previous development of useful *graphical information modeling languages*. The first of these, gaining early widespread use in the 1970s was SADT, developed by Douglas Ross. SADT and its more recent emulator, IDEF0, are process design languages that were developed to be interpretable by people. The potential for linking graphical information modeling with implementation emerged with the work of Peter Chen, who, for his Ph.D., developed the Entity-Relationship model.

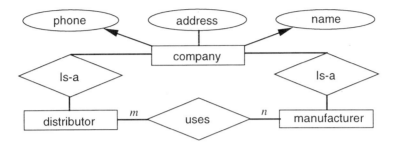

Figure 4.4: A small E-R model of a manufacturer and distributor and their relation.

The Entity-Relationship (E-R) Model was conceptually simple, being defined in terms of
   entities—diagrammed as boxes
   relations—diagrammed as diamonds
   and *attributes*—diagrammed as denoted by ovals

Entities may be people, parts, conceptual entities, packages of qualities or properties, or other nouns. Attributes are as they were defined earlier—simple properties characterizing entities. Relations identify conditions between entities. These may be a fixed condition, such as a connection; they may be a temporary relation, like the room some furniture is in; or they may be an abstract relation, such as between a company and the parts it supplies for a construction project. All entities, relations and attributes have their identifying name inside their corresponding shape. Attributes are dependent upon the entity they belong to. The arity of relations is defined. Inheritance is a special kind of relation, called *is-a*, where the attributes and relations of a parent entity are inherited into all child entities. The arity of *is-a* relations is not relevant. In Figure 4.4, the uses relation is many-to-many, with one distributor possibly representing multiple manufacturers, and a manufacturer possibly having multiple distributors.

The utility of the E-R model was in the development of methods for automatically translating an E-R diagram into a relational database schema. These methods evolved over time and have been formalized. They greatly simplified database schema design and took it beyond computer programmers into the hands of knowledge workers and domain experts who had only limited computer knowledge. In practice, most E-R models were edited and revised by programmers, but the gap between user and their tool definition significantly narrowed. The original E-R modeling approach was limited and has been expanded to include other types of relations. Soon after E-R, other graphical data modeling languages were developed, most notably NIAM, which is reviewed in Chapter 5.

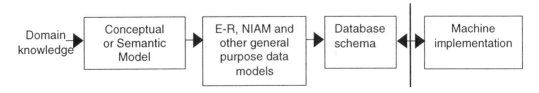

Figure 4.5: The sequence of models that have been proposed to translate from domain knowledge to a computer implementation.

E-R was partially motivated by earlier work in artificial intelligence to develop general models of knowledge. In parallel with the efforts to develop graphical data modeling languages was an effort to develop modeling languages that captured the reasoning of experts—what are now called expert systems. Some of these efforts were directed at making the task of programming or database design simpler. Others were addressed more at the conceptual level of defining requirements more precisely and formally, without a specific implementation strategy in mind. Thus the work covered a broad spectrum—ranging from domain-oriented conceptual modeling languages to implementation-specific data modeling languages. The range of different models and their sequence is characterized in Figure 4.5. Conceptual models are meant to capture the subtle knowledge in some area of expertise. They are often logic based. The general-purpose data models are somewhere in the middle between human and machine. Database schema languages, such as SQL or an object-oriented database language, must be inevitably oriented toward machine translation.

Automatic mappings between various language models have been a fundamental capability of programming languages since their origin. The long-term goal of all of these efforts was to span the knowledge gap between domain experts and computer scientists, allowing those with limited computer expertise to express naturally their conceptual knowledge in a form that could be automatically translated into a computer program or database schema.

4.6 SUMMARY

The concepts reviewed in this chapter had major impacts on all areas of computing including programming languages, databases and user interfaces. The development of a new generation of programming languages, incorporating ADTs, inheritance, polymorphism and encapsulation—called object-oriented systems—required new strategies if one hoped to translate objects from one engineering environment to another. It also suggested new techniques for programming data exchange and translation software. In the 1980s, the new conceptual modeling and abstraction concepts were having an impact on databases and data modeling. They seemed to allow for the definition of richer product representations, including those of buildings. Also, the automatic translation of a conceptual model into a database schema suggested that a parallel approach could be used in data exchange.

At the same time, the complexities encountered with IGES, when the development of domain-specific application interfaces were attempted, seemed huge. It was thought that the application of these new concepts to the problems of product data exchange might allow significant breakthroughs to the growing problems of data exchange in products and in building modeling. They held out the promise that the hard issues of modeling complex objects and their relations could be resolved.

4.7 NOTES AND FURTHER READING

For an excellent historical review of the knowledge representation problem, see Penrose [1989], Chapter 4. Another good piece on this topic is Hofstadter's *Godel, Esher, Bach* [1979].

ADTs can be formalized using the notion of sets that carry all possible legal values of a type. Using the set theory notion, operators within an algebra transform an input value from one member of the set to an output, which is also a member of the set (as in the set of real numbers in conventional high-school algebra). By only allowing manipulation by operators that are within the algebra, the integrity of the algebra can be guaranteed. To review the key concepts leading up to object-oriented systems, see Parnas [1973], Guttag [1977] and Dahl, Djikstra and Hoare [1972].

SIMULA was the first object-oriented language [Dahl and Nygaard, 1966]. This was followed (much later) by Smalltalk [Goldberg and Robeson, 1983].

The theory of types in computing languages is surveyed by Danforth and Tomlinson [1988]. Some different models of subtyping are presented in Wegner [1987] and America [1990]. The issue of inheritance, both single and multiple, is surveyed by Taivalsaari [1996]. Brachman has written an interesting article on the different meanings of inheritance [1983] and some of these issues are reviewed in more detail in Cook, Hill and Canning [1990]. Wegner and Zdonik [1988] address the different types of inheritance and their robustness in light of modifications allowed in child types. The issues of model evolution are reviewed from the database side in Kim [1992], Chapter 5, and from the building modeling side by Eastman [1992a]. Efforts to define a standard object model in the Unified Modeling Language (UML) faced the issues reviewed in this chapter and have generally left them to be resolved at the implementation level [Fowler, 1997].

Further reviews and surveys of object-oriented systems at the language level have been provided by Shriver and Wegner [1987], Danforth and Tomlinson [1988] and Meyer [1988], among others. Kim [1992], Bertino and Martino [1991] and Kemper and Moerkotte [1994] have written good surveys on object-oriented databases.

An important development in software engineering has been the development of software design theory. Among the object-oriented design methods are those for embedding rules into a complex object structure. Some good examples of such methods are presented in Meyer [1988], Coad, North and Mayfield [1995] and Coplien and Schmidt [1995].

The two seminal papers on abstraction by Smith and Smith are [1977a] and [1977b]. For an early application of the concepts of abstraction hierarchy to building models, see Eastman [1978]. An introduction to abstraction concepts in product modeling is in Shaw, Bloor and de Pennington [1989].

Some of the important papers in composition in the database area are Kim, Banerjee, et al. [1987], Kim, Bertino and Garza, [1989] and Kotz-Dittrich and Dittrich [1995]. Two attempts to include composition as a relation type in a language are in the Unified Modeling Language (UML) [Fowler, 1997] and in the EDM data modeling language [Eastman, Chase and Assal, 1993]. A good discussion of inverse and other constraints on relations is offered in Kemper and Moerkotte [1994], Chapter 9. Other abstraction relations have been proposed by Brodie [1981] and Navathe and Cornelio [1990]. There are many discrete mathematics texts that review the properties of relations, such as Stone [1973].

Conceptual modeling began with early efforts to model knowledge using semantic networks [Minsky, 1969]. See also Bobrow and Collins [1975] and Sowa [1984]. These later were picked up in the database field to model knowledge in databases [Peckham and Maryanski, 1988]. Some object-oriented languages include integrity rules as part of their structure [Meyer, 1990].

A description of SADT is presented in Ross [1977] and IDEF0 in [IDEF0,1981]. The first graphical data model used for databases was Peter Chen's thesis [Chen, 1976]. A recent presentation of the E-R Model with new extensions is given by Teorey [1990]. How an ER data model is formally translated into a relation database schema is presented in Ullman [1988, Chapter 2]. Two surveys of expert systems are [Dym and Leavitt, 1991] and Coyne, Rosenman et al. [1989]. A recent survey of graphical programming techniques and examples is presented in [Burnett, Goldberg, and Lewis, 1995]. For a discussion of the various roles of conceptual modeling, data modeling, semantic modeling and other types of modeling see Eastman [1993a].

4.8 STUDY QUESTIONS

1. For a specific type of object and an intended use, define the attributes of the object and the rules its operators should incorporate. Try to cover all the possible states for the intended use. Some example objects:
    a. an elevator as it is operating
    b. a delivery truck dropping off materials at a construction site
    c. a construction crane, as it moves material about the site
    d. a light fixture, door or other premade building component, as it is being delivered to the site, then installed

2. The following are some questions relevant to the universe of discourse of an object or set of objects comprising a design space:
    a. Are the UoDs for a building model the same throughout the building life cycle or do they vary with each phase of the cycle? If they change, how do they change?
    b. Can a building model fully represent an application if the resulting building design of the application has a larger UoD than the building model?
    c. Is there a relation between the UoD and the complexity of a building process used to create it? Does a large UoD during design make a complex design, or a design that is complex to build?

3. An issue in the design of object-oriented systems is where to define the operators and their semantic rules within an object hierarchy. Define an object model for a desk and chair. The desk has drawers. Define a simple set of attributes for each. Define the operators to create and edit each of the three parts. Discuss where the operators for manipulating each part of the ensemble should go and the rules they should incorporate.

4. Composition is an important relation in building models. Consider composition as an abstract relation that can be applicable to the following two examples: concrete slabs, made up with metal decking and reinforcing bars; and walls bounding spaces in a building. Outline a definition of composition that can be used in both cases.

5. For Question #4, consider whether the parts remain if the composite object is deleted or not. How would you define them so that the proper deletions were made if a floor slab is deleted?

CHAPTER FIVE
ISO-STEP

...the opportunity to develop EXPRESS was a chance to improve upon what seemed to be an inadequate way of thinking about and documenting what we know about information, which influences our lives so greatly. We are pushed into either of two corners: one that dealt with data and relationships only and another that entangled information with every conceivable computer application development detail. From my point of view, information is certainly more than the former and definitely should be kept apart from the latter.

Douglas Schenck and Peter Wilson
Preface
Information Modeling the EXPRESS Way
1994

5.1 INTRODUCTION

By 1984, the issues of CAD data translation, as summarized at the end of Chapter Three, suggested to a number of industry-based research groups in Europe and the US that the time was appropriate for a new generation of standards efforts. In the US, the new effort was centered around the Product Data Exchange Standard (PDES). About the same time, the International Standards Organization (ISO) in Geneva, Switzerland, initiated a Technical Committee, TC184, to initiate a subcommittee, SC4, to develop a standard called STEP (STandard for the Exchange of Product Model Data). The full title of the standard is *ISO1303 - Industrial Automation Systems - Product Data Representation and Exchange*. The STEP effort was initiated in part because different European countries were embarking on development of their own standards. Beside IGES and PDES, there was SET by the French, CAD*I by the Germans, and other efforts such as VDA-FS and EDIF. After initially operating as parallel but separate activities, the PDES and STEP efforts merged in 1991. Today, the international committees working on STEP meet quarterly, twice a year in the USA, once a year in Europe, and once a year in Asia.

This chapter reviews the overall structure of ISO-STEP and its approach to data exchange. It surveys the various languages—specifically NIAM, EXPRESS and EXPRESS-G—that are used in developing specifications within STEP and also some of the tools available for implementing STEP-based translators. In later chapters, we review additional facilities surrounding STEP and some of the exchange models developed with it. (The reader is warned that these international standards efforts use many acronyms and are a veritable "alphabet soup".)

The primary motivation for these new efforts in data exchange was the recognition that IGES and similar efforts had some basic weaknesses that were not easily corrected. The new effort had objectives to

- incorporate new programming language concepts, especially those dealing with object-oriented programming
- incorporate formal specifications of the structures defined, using the new recently developed data modeling languages
- separate the data model from the physical file format

- support subsets of a total model, allowing clusters of applications to be integrated without the overhead of having to deal with parts of a model irrelevant to a task
- support alternative physical level implementations, including files, databases and knowledge-based systems
- incorporate reference models that are common shared subsets of larger standard models

5.2 THE STRUCTURE OF ISO-STEP

An influence on STEP thinking was the work of the American National Standards Institute committee on database architectures and standards (ANSI-SPARC). This work distinguished between database definition and implementation, defining a layered architecture of abstractions and mappings. The layered architecture consisted of three levels:

Physical: the physical implementation of the conceptual schema on a file system; this level of definition defines the physical structure of data on disk or other media.

Logical: the information of interest, defined in a logical structure; this level of specification presents, in an implementation-independent manner, the logical structure of data and the relationships it holds; the logical level maps to the physical level.

Application: the information subset relevant to a particular application (later called a view); this level of definition specifies the information needed by a specific application and its format; the application level was implemented on top of and was derived from the logical level.

These levels distinguish the logical structure of the information from the format in which it is carried on some medium. This separation was a fundamental idea in the new STEP approach, as we shall see. It also separates an application level that is written in or on top of the logical level, suggesting a way to define subsets of a complete model.

The STEP Committee decided at the outset to use information-modeling methods to specify the required conceptual structure of the information to be represented. They accepted two information models as tools to formally specify the conceptual requirements—IDEF1x and NIAM. IDEF1x had been developed for the definition of defense planning in the US and NIAM was developed for business database schema design in Europe. The only popular information model not included was the original Entity-Relationship model (ER) and its extensions. Later, they also added EXPRESS-G. In addition, they approved development of an intermediate-level specification language for defining the logical structure of a model, separate from its physical implementation. For this, they commissioned development of the EXPRESS language.

Rather than define a complete information model and then take subsets of it, the STEP architecture started instead by defining various part models, called Application Protocols (APs), with the expectation that the APs would later be reconciled into larger domain-specific models. In this respect, the STEP approach has been very different from IGES. In the building context, this means that APs may be defined for structural steel, reinforced concrete, curtainwalls, mechanical systems and so forth. A complete building model would logically come only after the development of some sets of APs. This seems to have been based on the recognition that existing applications with rich semantic content are clustered into such specialized domains. Another motivation of the AP approach was the need to generate useful, incremental results quickly.

How are these various ideas organized into a data exchange system? The STEP system architecture identifies five classes of tools. They are diagrammed in Figure 5.1 with the information flows between various tools. The five classes of tools are

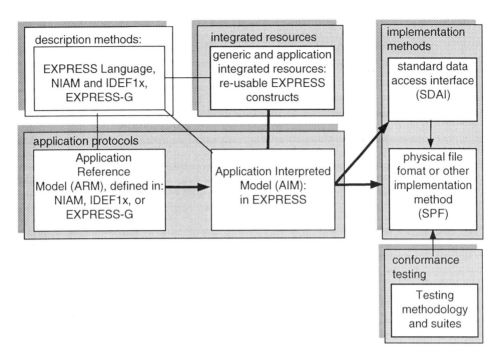

Figure 5.1: A diagrammatic representation of the different Parts of STEP, giving their names and how they are used. The thin lines designate language use, while the heavier arrows indicate a mapping realized by a translator. The one heavy line without an arrow indicates reuse of existing models.

1. A data exchange system utilizes various *description methods*, which are the information-modeling languages employed in specifying the information models used in the architecture, i.e., to define the Integrated Resources, Application Reference Models and Application Interpreted Models. The formal description methods include NIAM, IDEF1x, EXPRESS and EXPRESS-G.

2. *Integrated resources* are the common model subsets that get used repeatedly in the definition of application models. Models used in different domains are called *generic* integrated resources and include geometry, material properties and project classifications—that is, items that can be shared across multiple application domains. Model subsets that are industry specific are called *application* integrated resources. These include subsets for electronics, drafting, kinematics, finite elements and building. Presentation formats are called *constructs*.

3. *Application protocols* (APs) are developed for particular application contexts, using the Description Methods and Integrated Resources. An application protocol is partitioned into two aspects: an *Application Reference Model (ARM)* and an *Application Interpreted Model (AIM)*. An Application Reference Model represents requirements for an application in a form easily understood by knowledgeable users, for designing and assessing the information model. NIAM, IDEF1x and EXPRESS-G are initially used as languages for defining ARMs. An Application Reference Model is interpreted into an Application Interpreted Model, which is readable by both people and computers. EXPRESS is the language used to define AIMs. The AIM resolves all uses of the Generic Integrated Resources and integrates the model with Application Integrated Resources.

4. An application protocol is combined with an *implementation method* to form the basis for a STEP implementation. An implementation method typically includes multiple resources. The

STEP physical file (SPF) and the Standard Data Access Interface (SDAI) to the SPF are the Implementation Methods that have been developed thus far.

5.   Finally, each STEP application protocol and implementation requires *conformance testing*. A conformance test assesses the implementation in terms of its ARM and AIM and confirms that the STEP languages and tools have been properly used and interpreted. A Conformance Testing methodology is applied by accredited organizations. The testing includes application and interpretation of test suites.

The components of an AP include an Application Reference Model, an Application Interpreted Model and Conformance Testing requirements. Together, these provide the specific functionality for an application requirement in a self-contained form. The ARM states the needs of a particular application and a first order model definition. The AIM specifies a well-defined information exchange structure, organized for machine interpretation. The Conformance Testing Methodology specifies the process by which an AP will be accepted.

A major contribution to the field of product modeling has been the development of the EXPRESS language. EXPRESS was conceived as a means for specifying the structure of data, as an intermediary between an ARM and a Physical Implementation. It was conceived to support a variety of implementations, including flat file formats—either binary or ASCII—and also relational or object-oriented database formats. Originally, it was assumed that the translation from an ARM to an AIM would be done manually, and the ARM would serve to verify that the AIM was specified correctly. However, mapping tables were developed early that specified regular ways to translate the constructs of the IDEF1x and NIAM data modeling languages into an equivalent EXPRESS construct. Recently, translators or compilers that automatically generate an EXPRESS model from a NIAM or IDEF1x model have been developed. This has simplified the generation of Application Interpreted Models. Also, STEP activities have increasingly used EXPRESS—especially in its graphic format (EXPRESS-G)—for defining ARMs. Translation between EXPRESS-G and EXPRESS is widely supported.

Interpreted Resources provide a set of resource models that may be utilized within different Application Protocols. They are specified as an AIM, facilitating their assimilation into multiple APs. As ARMs are interpreted into AIMs, they are adjusted to allow integration of relevant Integrated Resources. Two levels of resources have been defined: Generic Resources, which have general use across applications, and Application Resources, which support one application or a cluster of similar applications. It has been assumed that Application Resources would be incrementally defined *post facto*, as parts of models developed in one application are found to be needed in others. More recently, this approach to Application Resources is being reconsidered.

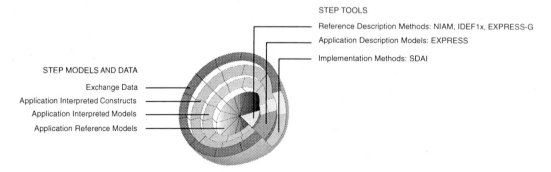

Figure 5.2: A layered representation of the ISO-STEP architecture.

Another way to visualize the STEP architecture considers STEP resources and models in terms of a layered system architecture, as shown in Figure 5.2. The base facilities are defined at the center, with the facilities that use them outside, in layers resembling an onion. The wedge in the lower right of the onion represents language resources. Thus information flows are between layers in Figure 5.2. The figure shows the three levels of the application model for generating exchange data, with reference and interpreted models being supported by Integrated Resources. The outer layers build upon the resources of the interior ones.

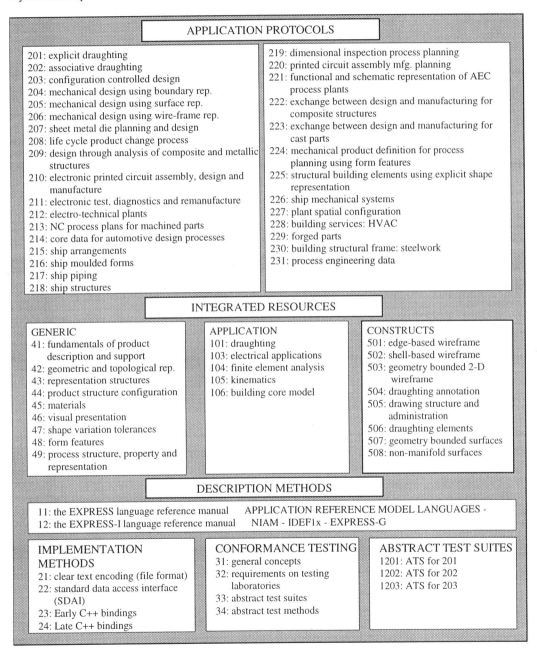

Figure 5.3: The numbers and titles of STEP Parts, as specified in early 1996.

The listing of the different STEP Part definitions that have been at least initiated, as of early 1998, is shown in Figure 5.3. Each ISO1303 series is published as a separate Part. Parts are grouped into one of the following classes: Application Protocols, Description Methods, Integrated Resources, Implementation Methods and Conformance Testing, as shown in the figure. Because the need for solutions for data interchange problems has been so great, pressures have been exerted to develop even limited solutions. As a result, draft forms of many Integrated Resources and Application Protocols have been developed in parallel, even with the Description Methods not being complete.

As can be seen, the ISO 10303 efforts are wide-ranging, and it is non-trivial to comprehend their full extent. The Part models initiated for the building industry thus far are 225 (structural building elements using explicit shapes), 228 (building services, HVAC), and 230 (building structural steel frame), and the Integrated Resource Part 106 (building construction core model). In addition, we must understand the Part models that the building-specific models use. These include Part 11 (EXPRESS), Parts 41-43, which provide the ownership, state, and geometry for building parts, and Part 45, which defines materials. For implementations, we need to understand Parts 21, 22, 23 and 24.

In the rest of this chapter, we focus on Part 11 (EXPRESS) and another frequently used ARM language, NIAM. We also will review Parts 21-24 to provide an overview of implementation tools. The next chapter, Chapter Six, surveys Parts 41, 42 and 43. Chapter Seven includes reviews of Part 230 and 228 and other work on aspect models, while Chapter Eight includes a survey of Part 106 (Building Construction Core Model) and other similar efforts.

5.3 IMPLEMENTATION CONCEPT

Most of the attention and people involved in STEP activities focus on the definition of various product models, both as Application Protocols and Integrated Resources. However, a product model is not of much use if it does not support implementation of a data exchange mechanism. That is its ultimate purpose. The purpose of a logical model, represented in EXPRESS, is to define the structure and format of data instances that correspond to that model, for exchange with another application.

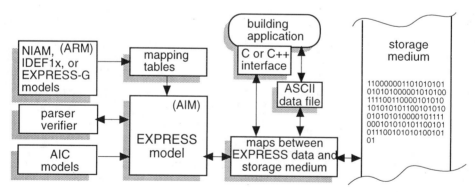

Figure 5.4: The general implementation approach used in STEP.

To gain an overall picture of the intended operation of STEP, it is useful to consider the process in three steps, which can be interpreted from Figure 5.4. The Application Protocol development process goes from the ARM (expressed in IDEF1x, NIAM or EXPRESS-G) to the AIM (represented in EXPRESS). It typically includes references to Application Integrated Constructs, for example, for defining standard geometries. The AIM is generated using mapping tables that define the correspondence between ARM constructs and EXPRESS. The AIM is verified, in part, using a parser/verifier.

In order to write STEP conformant data into a storage medium or read it, a *mapping structure*, written in C or C++, is defined that corresponds to the Entities in the EXPRESS model. Different methods can generate the mapping structure, which provides an in-between format between a building application and some storage medium. If the application can be extended to incorporate the mapping structure, it can copy its entities into the mapping structure's Entities. Alternatively, the application may write its entities to a file, which can be read by the mapping structure. The mapping structures have interfaces to write to various storage media, including a file and database formats. After one application writes to the storage medium, another can read it, using the same EXPRESS Entities that were used for writing. In this case, the mapping structures read the entities and either copy them to a second application's data structures, or to a file format readable by the second application. Later, in Section 5.8, we review this process in greater detail.

An important aspect of the STEP data exchange architecture is the use of mapping tables. Conceptually, a mapping table is a table of equivalencies, i.e.,

REAL in ASCII format	→	Real in IEEE double precision format
Double in EXPRESS	→	Double in C++
Cartesian point in Part 42	→	same on a file
Cartesian point on file in Part 42 format	→	Cartesian point in CAD application
Location point in CAD application	→	Location point in Part 42 format
Wall in CAD system format	→	Wall in Part 106 format

The function of the mapping table is to guide a program; when it encounters the entity on the left-hand side, the program is to write out the entity on the right-hand side. Clearly, the definition of such a mapping table ranges from being simple to extremely complex. All computer languages do the first mapping whenever they read or write a real number; the second mapping should not be any harder. A Cartesian point in Part 42 is three REAL numbers, so the second and third mappings should also not be hard. But what about the last one, regarding walls? How can we define the structure of a wall and then map it to some other format? The important point to remember is that all computer models are compositions of a few primitive data types and structures for grouping them. If the primitives and structures are defined in the mapping table, all higher-level entities defined as compositions can be mapped in terms of the primitives and structures. It is generally on this basis that mapping tables are generated, as we shall see. Mappings are required in many places for data exchange: between an external application and an EXPRESS model, between EXPRESS and a text file, or between EXPRESS and a database. Mapping and mapping tables are at the heart of data exchange and will be encountered many times throughout this and later chapters.

IGES and other early, neutral file formats had the problem that the data held in the model representation also included comments, scoping rules and other information structured for human, rather than computer, recognition. It was recognized that the specification for the information model should be separate from the data itself. This is realized in the above scenario, with EXPRESS providing the comments and format instructions for the data stored. The stored data can be very compact. The data could be written to a file, to a database or other information repository. In practice, the main work has relied on files.

The following sections review three of the tools used in STEP. We examine two ARM methods: NIAM and EXPRESS-G. Both are presented in detail, but serious use should be based on the official documentation. Examples are developed in both languages, facilitating comparison. EXPRESS is also presented, with sufficient detail for beginning use.

5.4 ARM DESCRIPTION METHODS

The STEP Description Methods address several different aspects of the specification of a data exchange process. The two most widely used aspects are the Description Methods used for defining ARMs and the Description Methods used for defining AIMs. ARMs are typically defined in either the NIAM information modeling language or EXPRESS-G. Another language used is IDEF1x. AIMs are uniformly defined in the EXPRESS schema definition language. (EXPRESS-G is a graphical version of EXPRESS, allowing translation from EXPRESS-G to EXPRESS to be almost transparent.) In this section, we provide an overview description of NIAM. Since the other language for defining Application Reference Models, EXPRESS-G, is a graphical implementation of EXPRESS, it will be reviewed after EXPRESS, which is discussed in the next section.

The development of an ARM begins with the definition of the Application Protocol's scope and requirements. What functional requirements should the AP fulfill? The STEP guidelines recommend the following be used to develop an ARM:

- the classes of external applications that the AP is to support
- example objects to be represented
- usage scenarios for the AP
- a definition of the scope of the AP, regarding both what is inside and what is outside the scope of the AP

These requirements are then defined more strongly as *Units of Functionality (UOF)*. UOF is a STEP term referring to the general entities, attributes and relations that convey the concepts within an AP. A critical membership criterion is applied: if any component of the UOF is removed, the concepts should be rendered incomplete or ambiguous. The UOF is usually defined in words; after agreement on it, the UOF is defined more rigorously, using an ARM Description Method. The purpose of ARM Description Methods is to specify at a detailed level the requirements for an Application Protocol or for a shared Integrated Resource. Thus, an ARM definition is the permanent, recorded basis for defining any later model.

5.4.1 NIAM

In 1977, Dr. G. M. Nijssen of CDC Europe developed one of the most widely used data modeling descriptions applied for the design of databases, the Nijssen's Information Analysis Method (NIAM). Dr. Nijssen developed NIAM as a database design methodology, supporting information exchange between a user and a computer, using elementary sentences. He had in mind relational databases and conceptualized NIAM as a means to translate written or verbal information into a database schema.

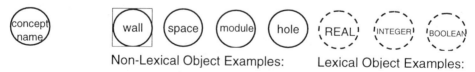

Non-Lexical Object Examples: Lexical Object Examples:

Figure 5.5: Objects in NIAM are of two types: Lexical Objects (LOTS) and Non-Lexical Objects (NOLOTS).

NIAM relies on an information-processing model of functional transformation—that is, it uses functions to transform information from one form to another. It provides a small set of constructs, which are the building blocks with which users can define higher-level constructs. There are two main primitives. One is the *object*, which is of two types: the *lexical object type* (LOT), and the *nonlexical object type* (NOLOT). The object in NIAM does not have any relation with object-oriented (OO) systems. A NOLOT corresponds closely to the concept of entity; it may depict conceptual classes of things or large aggregations of attributes. LOTS are those objects that are names or values of NOLOTS and have a type and coding rules. They are used for identifiers or attributes. LOTS are dependent upon the NOLOT they describe and are deleted if their NOLOT is deleted. Their graphical representation is shown in Figure 5.5. An Object with an enclosing box is a non-terminal, indicating that the details of the Object are defined elsewhere. Object is a class that has instances (called Entity in EXPRESS). Thus in Figure 5.5, there may be a variable number of `wall`, `space`, `module` and `hole` members. There may be a variable number of REAL, INTEGER and BOOLEAN LOTS.

The other main NIAM primitive is the *Role*, which corresponds to a relationship between Objects. Roles are usually binary and bi-directional—that is, they usually exist between two Objects and define a reciprocal relation. Occasionally, Roles have more than two Objects. The graphical notation for Roles is shown in Figure 5.6. Each Role has a box for each Object associated with it, with a name or attribute for the Object's part in the Role.

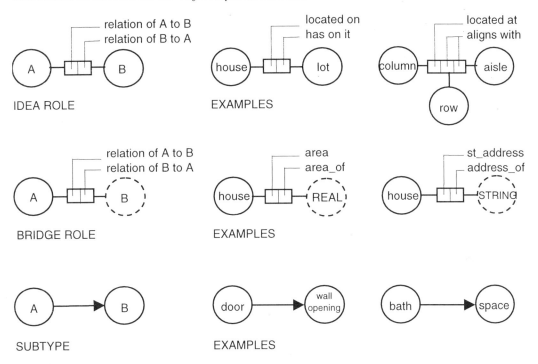

Figure 5.6: Roles within NIAM are usually binary. Roles are of two types: idea and bridge. A special type of Role is subtype, where the NOLOTs and Roles of one object are also those of another object.

Roles are of different types. Idea Roles are between NOLOTS—for example, between lot and house or between column, row and aisle. Bridge Roles are between a NOLOT and the LOTS that describe it. The box descriptors are the attribute names for the LOT. Subtype is an inheritance relation between two NOLOTS. These three Roles are the base structures within NIAM. The semantic richness of NIAM, however, arises from its many constraints, which are rules restricting the form of Role allowed between LOTs, NOLOTs, idea types, bridge types and subtypes.

One set of major constraints deals with *Uniqueness*. Some examples of Uniqueness constraints are shown in Figure 5.7. The bar on one side of a Role box indicates that that side has zero or one member in the Role and that no duplicates are allowed. Uniqueness allows one Object to identify the other. A continuous bar indicates that both Objects may be duplicated but the Role cannot. No bar indicates that the Role can be duplicated as well as the Objects. In the examples shown, a house may be on zero or one lot, but a lot may have any number of houses. A wheel is on at most one car, but a car may have multiple wheels. On the other hand, a door has only one threshold, and a threshold has only one door. Similarly, an auto has one engine, and an engine has one auto. A room is bounded by multiple walls and a wall bounds multiple rooms. Last, a room may be accessible by zero, one or more corridors, and a corridor may provide access to zero, one or more rooms. Thus a one-sided bar indicates a one-to-many relation, bars on both sides (not continuous) indicate a one-to-one relation and a continuous bar indicates many-to-many.

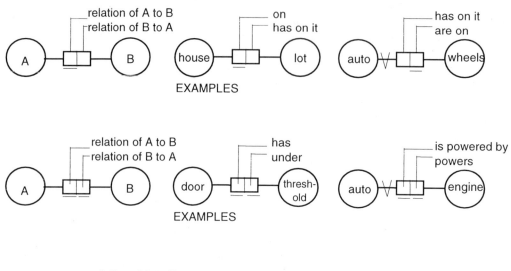

Figure 5.7: The uniqueness constraints delimit the arity of a Role and also the uniqueness of values. Uniqueness may be one-to-many, one-to-one, or many-to-many. The Total constraint is shown with a V intersecting the Role line.

Uniqueness is a fundamental relation in NIAM because of the major impact it has on the design of relational database schemas. Uniqueness provides a general definition of the arity of a Role. Sometime a specific number of Objects in a Role are required. Optionally, arity can be defined by noting the lower and upper bounds on the number of Objects that may participate in a Role. Examples are shown in Figure 5.8. This may be meaningful for Roles that are not one-to-one.

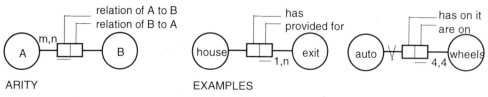

Figure 5.8: The arity constraints delimit the bounds on the number of objects that may be members of a Role.

Another important constraint is the Total constraint. It enforces the condition that at least one Object instance in a relation must exist. If it is deleted, the other Object in the relation is also deleted. Thus it defines dependency. In the examples in Figure 5.7, the wheels are dependent upon the auto and below it, the engine is dependent upon the auto. The total constraint is denoted by a V intersecting the Role line. In the two examples, if the auto is deleted, so are all its wheels and so is its engine.

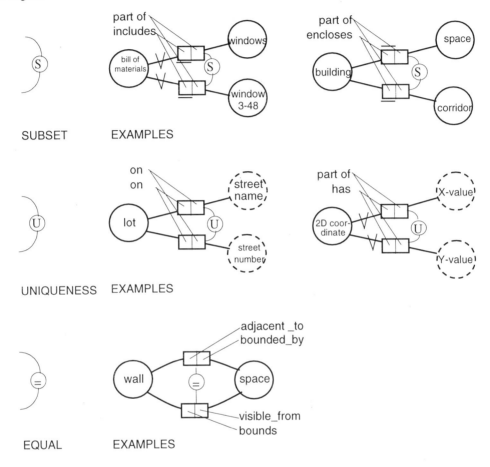

Figure 5.9: Constraints that apply to Roles are the subset, uniqueness and equality constraints.

Other constraints apply to multiple Roles. The subset, uniqueness and equality constraints are diagrammed in Figure 5.9. The subset constraint may apply to LOTs or NOLOTs. It specifies that an instance value for one attribute must also be an instance value of another attribute. In the examples of Figure 5.9, all specific window models are members of the set of all windows, and all corridors are members of the set of all spaces. The uniqueness constraint identifies a set of LOTs that uniquely define a NOLOT. The equality constraint requires that there are equivalent sets of Objects in the related Roles. In the example, one relation defines adjacency and the other defines visibility; both have the same membership.

Another set of constraints—inheritance constraints—applies to Subtypes. The mutual exclusion constraint imposes that no member of one subtype may also be a member of another subtype. As shown in Figure 5.10, an opening may be a door or a window, but no instance may be a member of both NOLOTS. (Sliding glass doors cannot be defined as members of both, for example.) Also, an instance of vertical circulation cannot be both a member of stair and elevator. Another Subtype constraint is Total, which means that there are no members of the supertype that are not members of the subtypes. For example, all pass-thrus are either flexible pass-thrus or rigid pass-thrus. Similarly, all members of vertical circulation are either stairs or elevators (dumbwaiters are excluded).

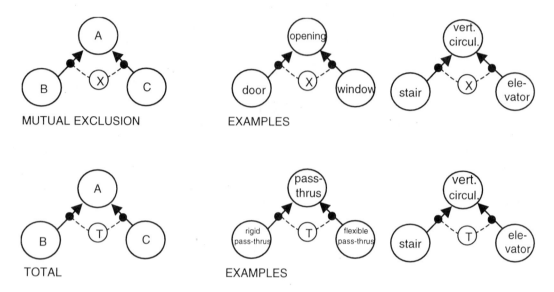

Figure 5.10: Mutual exclusion and total constraints apply to subtypes.

Supposedly, the conceptual basis of NIAM is natural language. It was conceived, however, to generate as output a relational database schema and is supported by various tools sold by Control Data Corporation and others. Its semantics were defined so as to have an implementation in relational databases. A variety of special constraint implementations have been developed for relational databases that correspond to the NIAM constraints defined above.

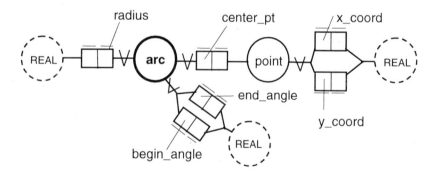

Figure 5.11: An arc modeled in NIAM.

5.4.2 NIAM EXAMPLES

In order to demonstrate the NIAM information-modeling language for product modeling, we use it to model the three examples introduced in Chapter Three: the arc, the bounded planar surface and the multi-view wall.

The arc is shown in Figure 5.11, defined as a point, two angles and a radius. All information is dependent upon the arc Object; if it is deleted, the other parts are also. All entities are unique, without duplicates. There is exactly one of each.

The bounded plane is shown in Figure 5.12, defined as a surface and a bounding polygon. In both cases, the definitions are simple: the surface is one of several subtypes, one of which is a plane. A surface can be of only one of these types. The plane is defined as a single point and single vector indicating its slope. The polygon is a sequence of points, where a point is defined as 0 to 3 coordinates. Again, all parts are dependent upon the bounded polygon; if it is deleted, the parts are also. All parts of the surface are unique to the surface and one-to-one. There may be multiple bounding polygons, however.

Aspects of the wall are defined separately, then combined later. The first part of the wall model defines the aggregate description of the overall wall and its top-level components (see Figure 5.13). The aggregate wall is defined here in terms of its geometry and thermal properties (other properties could easily be added). It is bounded by a number of boundary entities, of which wall is one subtype. The wall geometry has three views: plan, elevation, and BRep model. Each is defined by its component geometric entities. The geometry level carries a location for all the component geometric descriptions. The symbol has a many-to-one relation with geometry, while the location, solid and p-line have one-to-one relations. Multiple p-lines are defined supporting both elevation and plan. The thermal property has a single value, the U-value. All properties are dependent upon the wall, except pass-thru, which is defined independently of the wall and hence will not be deleted if the wall is.

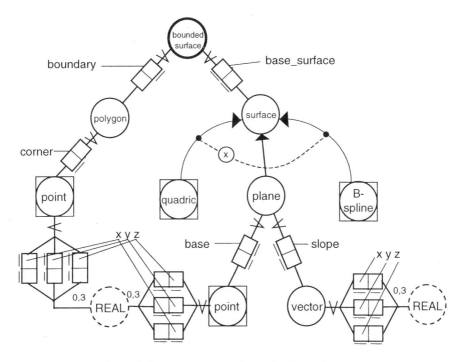

Figure 5.12: The bounded plane example, defined in NIAM.

As its parts, the wall has openings, pass-thrus, and segments. Segments and openings are dependent upon the wall, and have a many-to-one relation. Pass-thrus have a many-to-many relation and are not dependent upon a wall. Segments, openings and pass-thrus are all references to other definitions.

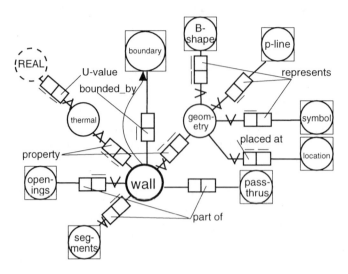

Figure 5.13: The top-level definition of the core wall, defined in NIAM.

The definition for openings is shown in Figure 5.14. Every opening has a thermal U-value attribute and a geometry Object carrying a location, an area, and a polyline outlining the opening. Filled openings are specialized from openings to carry the geometric definition of the filler. Two views are offered: a symbol for plan and elevation and a solid for 3D modeling. Windows and doors are specializations of filled openings.

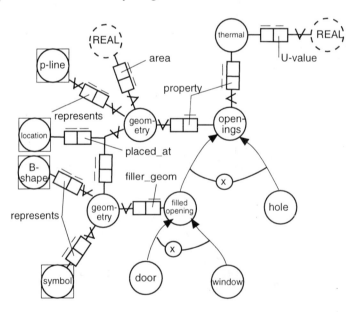

Figure 5.14: The NIAM definition of openings within a core wall.

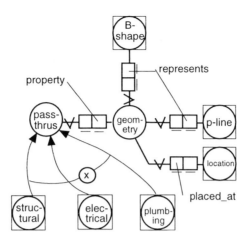

Figure 5.15: The NIAM definition of pass-thrus for the core wall definition.

A pass-thru is a generalization of any Object that passes through the wall using it as a chase (Figure 5.15). Pass-thrus are assumed to have one or two views: a centerline for rough layout (p-line) and possibly a solid model for detail layout. Geometry carries its location. Specializations of pass-thrus include structural, electrical and plumbing entities.

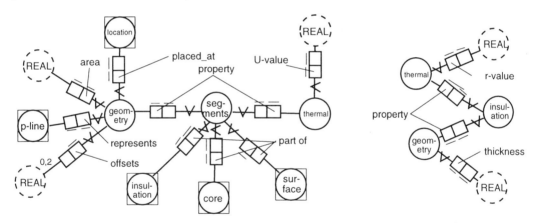

Figure 5.16: NIAM definition of a wall segment, its geometry and parts.

Figure 5.17: NIAM figure of wall insulation, as part of segment.

The major construct defining the solid part of a wall is the segment (Figure 5.16). A segment is a region of a wall with a consistently layered cross-section. At this level, a segment geometry specifies its location and carries a polyline defining its extent in elevation and the corresponding surface area. The segment geometry also has two offsets to each side of the wall, defining the thickness for this segment. It also has a U-value as the partial result from thermal analysis. The segment is further broken down into core, insulation and surface. All properties are dependent upon the segment, i.e., there is a total constraint. Most Objects have arity one. The coordinates are constrained to have up to two values.

Insulation has two properties: its resistance, or r-value, and its thickness, which often determines its r-value (see Figure 5.17). All properties hold one-to-one relations and depend upon the insulation Object.

The core of a wall segment defines its inner structure, which is assumed to provide its structural rigidity (see Figure 5.18). It is defined geometrically by a location, two offsets from a centerline, and a solid model (for exact shape definition). The core also has a thermal r-value. The core has a variety of specializations, some of which are included.

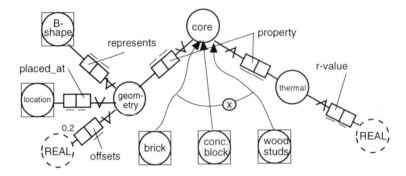

Figure5.18: NIAM definition of core of wall segment.

Surfaces are layered on both sides of the core to make up the finished segment (see Figure 5.19). Multiple surfaces may be laid on top of one another. The geometry of a surface is defined according to the side it is on, its order of placement and its thickness. An attribute r-value is included.

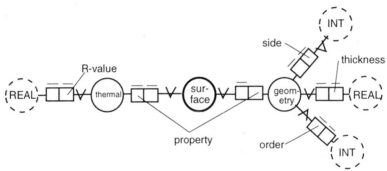

Figure 5.19: NIAM definition of the surface of a segment.

All of these aspects are combined into the overall NIAM core wall model, shown in Figure 5.20. It suggests the necessary complexity of a wall definition needed to support a range of applications. Some examples of core construction materials are offered as subtypes in the model.

5.4.3 SUMMARY

NIAM is an evolving language with alternative notations. The original documents present it as a conceptual-modeling language, meant to capture the semantics of a situation, independent of any database implementation. Later it began to acquire properties needed for automatic schema definition of relational database models. This evolution is evident from early publications describing the language to the most recent ones. We have presented a version that is an extension of the earlier versions, with an expanded set of attributes, which seems most commonly used within the product-modeling community. While this section presents most of the constraints developed or proposed for NIAM, new ones are added every few years.

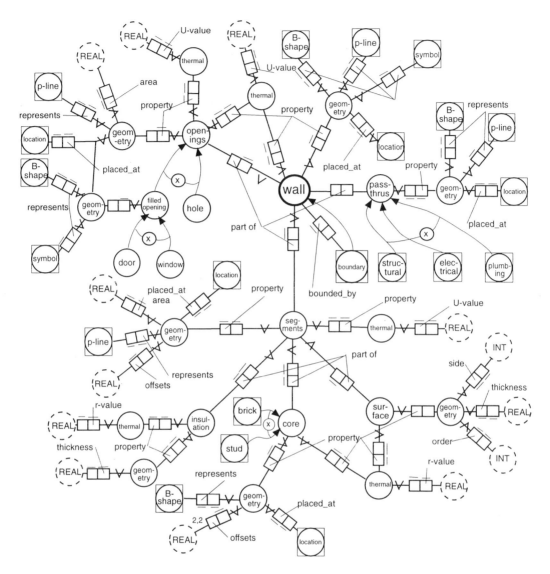

Figure 5.20: The various parts of the NIAM wall model combined.

NIAM's entity structure is extremely simple, defined in terms of Objects and Roles. It is elaborated by many kinds of constraints, some of which have subtle definitions. The texts written by the developers of NIAM emphasize the conceptual and intuitive aspects of modeling, but these seem ambiguous if there are not also clear operational distinctions. For data modeling, the operational distinctions are the specific forms of the data model required. A strength of NIAM is that it makes no distinction between a relation (defined as a Role) and an attribute. This distinction is imposed in some other models and, as such, is sometimes unclear. Also, NIAM provides a rich set of constraints or validation rules for which implementations have been defined in relational databases.

Yet for all this richness, several limitations are apparent for NIAM's use in product modeling. It has no means to represent several important constructs that are also missing from the relational model. For example, the rule that the order and side of a surface must be unique for a segment, and that the order must be a complete numbering, cannot be easily expressed. Composition rules that specify what is allowed in a particular composition cannot be specified. Also the consistency of geometrical relations, or the values carried at various levels of aggregation for thermal performance, cannot be expressed. As a graphical language, its use has been to define new schemas from scratch, not to revise existing schemas. Thus, it does not support model extension or evolution.

Though having a few shortcomings, NIAM has a significant audience and continues to be used to develop ARMs for various new application protocols within the STEP community.

5.5 EXPRESS - THE AIM DESCRIPTION METHOD

EXPRESS is the language developed within the ISO-STEP community for representing application interpreted models. It came into existence in a US Air Force project named Product Definition Data Interface and contracted to McDonnell Douglas. EXPRESS was developed by Douglas Schenck and has been under development since about 1986, when it received its current name. The purpose of EXPRESS is to represent a product model in an implementation independent manner, as interpreted from an Application Reference Model (ARM). That is, it is a representation of the knowledge to be embedded in a product model, interpreted to take advantage of shared Integrated Resources (reviewed in Chapter Six) and rationalized with other related application protocols.

EXPRESS's syntax is similar to that found in modern programming languages. It is used in the way one would define the data structures in a programming language. Its effect is also similar, in that it defines how instances of defined objects will be organized for use. It is different from a programming language in a few important aspects, however, and these differences must be kept in mind when examining or using it. This section presents its functionality along with its syntax. The following sections provide a working overview of EXPRESS Version 1.0.

5.5.1 SCHEMAS

The unit of definition in EXPRESS is a *schema*. A schema defines the *universe of discourse (UoD)* in which declared objects are given mutually dependent meanings and purposes. An EXPRESS schema is the normal representation of both AIMs and Application Interpreted Constructs (AICs). As defined in the EXPRESS manual, an EXPRESS information model consists of schemas that include *definitions of things* (entities, types, functions, procedures), rules defining *relationships between things*, and *rules on relationships*. EXPRESS includes a full, procedural language syntax for specifying rules. The rules defined in EXPRESS cannot be executed, however, but serve as guides for the implementation of translators. (Proposals exist to change this.)

EXPRESS is a block-structured language, like Pascal or C. All types have scopes, which are identified by the block in which they are defined. A block begins with the declaration of an Entity, Function, Procedure Rule or schema and ends at the end of Entity, Function, Procedure Rule or schema. When an identifier in one block is redefined in an inner block, the inner block declaration overrides the outer one, for the extent of the inner block. Any declarations outside a schema are within a global block referred to as the UoD.

An information model may be defined by more than one schema. There are special mechanisms to make cross references across schemas, but they require a different syntax in relation to references within the same schema. Thus, one schema can be linked to other schemas using the USE and REFERENCE specifications.

EXPRESS code may include comments and other non-interpreted text. These are bracketed with (* and *) . Comments may also be appended to a line with -- .

```
--Here is a comment:
(* Here is another comment. *)
```

5.5.2 IDENTIFIERS

The general convention that is followed is to define identifiers in lower case, and to capitalize all reserved words. Identifiers must begin with a character and may include other characters, digits and the underscore character. The characters for blankspace ( ), tab, arithmetic operations, period (.) and comma (,) are not allowed in identifiers.

5.5.3 BASIC TYPES

EXPRESS provides a set of basic types that are predefined and available to use in the definition of higher level types.

The Basic Types are

```
NUMBER, REAL, INTEGER, STRING, LOGICAL, BOOLEAN and
      BINARY.
```

INTEGER is an unconstrained whole number. REAL is an unconstrained rational, irrational or scientific number. Scientific numbers may be constrained to a specified precision. NUMBER is a supertype of INTEGER and REAL. STRING is a quoted list of characters, bounded by single quotes. LOGICAL types have the values (TRUE, FALSE, UNKNOWN) while BOOLEAN only has (TRUE, FALSE). BINARY is a vector of binary values, with user-defined encoding.

Both STRING and BINARY are variable length vectors. They may have a specified length. When a length is defined, up to that many may be assigned. When the length is specified and appended by FIXED, then all assignments must be of exactly that length. But, if it is not specified, the length is variable. For example,

```
s1: STRING;         (* variable length string *)
s2: STRING(10);     (* variable length up to 10 characters*)
b1: BINARY(10) FIXED;          (* exactly 10 bit flags *)
```

EXPRESS also has an enumeration type, where the possible values are defined explicitly. For example,

```
TYPE compass_direction =
  ENUMERATION OF (south, north, east, west);
END_TYPE;
```

5.5.4 CONSTRUCTORS

Variables and other structures can be aggregated into larger groupings, using *constructors*. Constructors group things of the same type; in EXPRESS they include the array, bag, list and set.

ARRAY is used to define an ordered list of elements of fixed size. Arrays may be concatenated, e.g.,

```
matrix : ARRAY[1:4] OF ARRAY[1:4] OF REAL;
```

In ARRAY, both the lower and upper bound must be defined (where lower_bound < upper_bound). There will be (upper_bound - lower_bound + 1) items in the array. Items are accessed by the array name and a subscript, e.g., matrix [3][1].

BAG is an unordered collection of like elements. Duplicates are allowed. Its lower bound and upper bound may or may not be specified. If the lower bound is specified, there must be at least the lower bound of things assigned to the bag. If the upper bound is assigned, it indicates the maximum number of members it can hold. If the bounds are not defined, it is assumed that the bounds are [0:?]. As an example,

```
bag_of_points : BAG[1:?] OF point;
```

means that there must be at least one point in bag_of_points.

LIST is an ordered collection of like elements, similar to the ARRAY, but LIST may have a variable length. Thus:

```
list_of_points : LIST[0:?] OF point;
```

means that there may be any number of points in the LIST, including zero. The LIST may grow or shrink but remains ordered, with the subscripts as assigned.

SET is an unordered collection of like elements. Duplicates are not allowed. SETs may be of fixed or variable size. An example might be

```
set_of_names : SET OF [1:100] OF name;
```

set_of_names must have a membership of at least 1 and not more than 100 members. If either or both the lower and upper bounds are left undefined, the default is that there may be zero or any number of members in the SET.

A generalization of all the constructor types is the AGGREGATE. AGGREGATE may be used in any declaration where any of the constructors may be utilized.

In general, access to items in any aggregate (ARRAY, BAG, LIST, SET) is by using subscripts ranging from the lower to the upper bounds.

Basic Types may be used to specify higher level user-defined Types, e.g.,

```
TYPE area = REAL;
END_TYPE;

TYPE name = STRING;
END_TYPE;
```

A type definition must end with END_TYPE.

Another type definition allows specification of a type that is a selection from among a set of types. EXPRESS calls this a SELECT TYPE. In other languages, such a construct is sometimes called a union type. A SELECT TYPE is also a generalization of other types. Some examples follow:

```
TYPE NUMBER = SELECT(REAL,INTEGER);
END_TYPE;

TYPE connection = SELECT(nail,screw,bolt,glue);
END_TYPE;
```

Constructors extend the means to specify Defined Types from Basic Types, and to specify other Defined Types from Basic or Defined Types. A limitation is that all elements of a constructor are of the same type.

```
TYPE column_row = LIST[1:?] OF line;
END_TYPE;

TYPE column_aisle = LIST[1:?] OF line;
END_TYPE;

TYPE color_value = INTEGER;
END_TYPE;

TYPE rgb_color = ARRAY[1:3] OF color_value;
END_TYPE;
```

The first two types listed above are defined for row and aisle. By starting with one, they are required to have at least one member in the list (not that the subscripts start at 1). If no subscript is defined for a constructor type, the default is [0:?]. Constructor types are only defined to provide high-level types to be used in attributes. That is, they are not Entities and cannot be instantiated.

5.5.5 ENTITIES

The general object type, from which instances of objects are made, is called the Entity. It supports definition of a wide range of complex elements, which we will consider incrementally below.

The basic Entity definition has the format:

```
ENTITY point;
  x,y,z : REAL;
END_ENTITY;
```

This point may have instances that may be independent of any line or arc. Instances of point may be defined, while a type may only be used to define an Entity or another type.

Entities can be inherited or subtyped into other Entities. For example,

```
ENTITY homogeneous_point
  SUBTYPE OF (point);
  w : REAL;
END_ENTITY;
```

An equivalent declaration would be for point to be declared as a SUPERTYPE of point1. An Entity may define both SUBTYPE and SUPERTYPE relations with other types. It may be a SUBTYPE to zero, one, or more than one other types. Similarly, an Entity may be a SUPERTYPE to any number of subtypes. Either the SUPERTYPE or the SUBTYPE may define a relation between the two Entity types. One or both Entity types may carry the relation declaration. However, all SUPERTYPE and SUBTYPE declarations, taken together, must be consistent with a directed acyclic graph. That is, an Entity cannot both a SUBTYPE and a SUPERTYPE of the same Entity, at any level of the relationship. There can be no cycles in the Entity subtype graph.

A SUBTYPE inherits all attributes of the SUPERTYPE, including DERIVED and INVERSE attributes. Domain constraints defined as WHERE clauses (discussed in the next section) are also inherited. Overwriting of rules, allowed in some languages, is not allowed in EXPRESS. This restriction addresses the consistency of polymorphic types (discussed in Chapter Four) in a strong manner.

5.5.6 ATTRIBUTES

Attributes in EXPRESS are of three general kinds:

Explicit: the values will be provided directly

Derived: the values can be calculated from other attributes

Inverse: captures the relationship between the Entity being declared and a named attribute

The attributes have a representation, which might be a simple data type (such as integer) or another Entity type. A relationship is established between the attribute being defined and the types or Entities that define it. In the default case, the schema instance is only correct if a value exists for the attribute, that is, if the relation is mandatory. If an empty value is allowed, an attribute is declared as OPTIONAL.

5.5.6.1 Explicit Attributes

Explicit attributes are the information units used to characterize the properties of some Entity. They may be classified as

Simple types: These cannot be further subdivided into elements; they are NUMBER, REAL, INTEGER, STRING, BOOLEAN, LOGICAL and BINARY

Aggregate types: These are groupings of the same type using one of the constructors

Entity type: These are object types declared by Entity declarations. Like other attributes, using an Entity as an attribute's data type establishes a relationship between two Entities

Qualified attributes are used to define a pathname from the current Entity to an attribute within an inherited Entity.

Inverse attributes capture the relationship between the Entity being declared and any named attribute that uses it as an attribute value. That is, inverse attributes define a backpointer referencing an Entity that uses the current one as an explicit attribute. Inverse attributes may be constructor types with cardinality or other constraints.

5.5.6.2 Derived Attributes

In addition to explicit attributes that are given assigned values, EXPRESS also supports derived attributes, which are not loaded as data, but derived from other values carried within an Entity. Derived attributes are identified by DERIVE:

```
ENTITY circle3d;
  center : point;
  radius : REAL;
  axis : vector;
DERIVE
  area : REAL := pi * radius ** 2;
END_ENTITY;
```

Thus the attribute area can be accessed just like other attributes, but is computed at the time it is accessed. The scope of attributes that can be accessed within a DERIVE expression is the Entity in which the DERIVE expression is located.

EXPRESS also provides a UNIQUE clause that can be applied to any attribute. It specifies that each instance of the Entity must have a unique value of that attribute that is not duplicated anywhere within the SCHEMA.

5.5.6.3 Inverse attributes

Often, an attribute defines a relation between the Entity carried by the attribute and some other Entity. For example,

```
ENTITY line;
   point_ref : ARRAY[1:2] OF point1;
END_ENTITY;
```

defines a relation between a line instance and two point instances. Sometimes we may require that the relationship is two-way. That is, a relation also exists from the point to the line it bounds. This may be defined as

```
ENTITY point1;
  x,y,z : REAL;
  line_ref : SET OF line;
END_ENTITY;
```

where only points that are part of a line may be of type point1. But now we have two relations that are related to each other. There are supposed to be two matching attributes; if line instance a has a point_ref attribute referencing point instance b, then point instance b should have a line_ref attribute referencing line a. Multiple lines may refer to a point1 instance. Any updates to a line or point instance would have to maintain this relation between the attributes.

The INVERSE attribute imposes a symmetry rule between an existing relation and itself. It automatically creates and maintains the relation between the attributes, always defining the inverse automatically from the relation. For the point1 Entity above, it would be defined as follows:

```
ENTITY point1;
  x,y,z : REAL;
INVERSE
  line_ref : SET OF line FOR point_ref;
END_ENTITY;
```

The line_ref attribute of point1 is the inverse of the point_ref attribute in Entity line. Any time that point_ref is updated, line_ref in the affected point1 will be also updated automatically.

5.5.7 RULES

EXPRESS supports the definition of a variety of rules that can implement many kinds of NIAM constraints, as well as other semantic conditions of importance to product modeling. It provides a variety of structures for embedding these rules. The derived attributes presented earlier describe one such structure. Domain rules describe another. They allow definition of restrictions on allowed values or combination of values in the attributes of an Entity. Domain rules are specified using a WHERE clause within an Entity specification.

```
ENTITY vector;
  a, b, c :          REAL;
WHERE
  length1 : a**2 + b**2 + c**2 = 1.0;
END_ENTITY;
```

The domain rule length1 is an integrity constraint of type LOGICAL (all WHERE clauses are of type LOGICAL). When accessed, it evaluates the expression and returns one of the values: TRUE, FALSE or UNKNOWN. UNKNOWN is used when some attributes are missing. If length1 is not TRUE, then an instance of vector does not conform to this specification.

In order to define complex rules, EXPRESS incorporates a fairly complete set of system functions. These include

ABS	–	absolute value	LOG10	–	base 10 log of value
ACOS	–	arc cosine	LOG2	–	base 2 log of number
ASIN	–	arc sine	LOINDEX	–	actual lower bound of values
ATAN	–	arc tangent	NVL	–	converts NULL to given
BLENGTH	–	number of bits in a binary			value
COS	–	cosine	ODD	–	TRUE if value is odd
EXISTS	–	TRUE if value exists	ROLESOF	–	returns set of strings of
EXP	–	e to power of value			references to value
FORMAT	–	string of a number	SIN	–	sine
HIBOUND	–	declared upper bound	SIZEOF	–	no elements in aggregation
		of constructor	SQRT	–	square root of value
HIINDEX	–	actual number of values	TAN	–	tangent of value
LENGTH	–	number characters in string	TYPEOF	–	set of strings of all types of a
LOBOUND	–	declared lower bound			value
		of constructor	USEDIN	–	set of strings used in given
LIKE	–	matches substring of string			Role
LOG	–	natural log of number	VALUES	–	numerical value of string

As can be seen, many of these are trigonometric and algebraic functions. The bounds checking, SIZEOF, USEDIN and other functions are provided to facilitate querying a product model. These can be used with the computational part of EXPRESS (Section 5.5.9) to define more complex and domain-specific functions for use in rules within a schema. General functions can be declared and used in defining DERIVE clauses or WHERE rules. An example is shown below. It computes the distance between two 3D points, where the points are as they were defined earlier.

```
FUNCTION distance (p1,p2 : point) : real;
(* computes the Euclidean distance between two points *)
  LOCAL  dist : REAL;
  dist := SQRT((p2.x - p1.x)**2 + (p2.y - p1.y)**2 +
                       (p2.z - p1.z)**2);
  RETURN (dist);
END_FUNCTION;
```

Access to a particular value is gained through a path accessed by a *pathname*. Pathnames are a sequence of attribute names connected by periods (.) traversing through an instance model. In the above example, p1 and p2 are the local names of parameters of type point within the function. Point has attributes x, y and z which are sequentially accessed from p1 and p2.

In writing a function to be embedded in a WHERE clause, there is no parameter that refers to the current instance. SELF is a primitive function that serves this purpose. It may be used in Entity and Type declaration or instance initializations. For example, the variables available in homogeneous_point above are

```
SELF\point.x;
SELF\point.y;
SELF\point.z;
w;
```

In the above pathnames, SELF refers to the current Entity. The \point refers to an Entity inherited into the current Entity. These pathnames are used to access these attributes from within homogeneous_point.

A well-known issue in object-oriented systems is the resolution of duplicate names inherited from multiple disjoint SUPERTYPEs. In EXPRESS, these conflicts are resolved by defining the pathname from which the attribute name came. Thus, if an Entity window is a SUBTYPE of both a wall_opening Entity with a location attribute, and a light_source Entity with a location attribute, each would be accessed using the following:

```
SELF\light_source.location
SELF\wall_opening.location
```

5.5.8 RELATIONS

Like object-oriented languages, EXPRESS does not have a structure for defining relations. Like most languages, they must be defined using attributes. On the other hand, the semantics defined in NIAM and other data modeling languages need to have a well-defined implementation.

5.5.8.1 Defining the Arity of Relations

Because of the prevalence of arity constraints in NIAM and other data modeling languages, EXPRESS had to have effective ways of representing them. While most languages allow a user to define a relation in one direction (either one-to-one or one-to-many), few allow many-to-many. That is, in the definition of a polygon made up of a list of edges, it is straightforward to define relations from polygon to edges. Using the line Entity defined earlier,

```
ENTITY polygon;
   edges : LIST [3 : ?] of line;
WHERE
   2_connected : (* check that the line's endpoints
                     are all two-connected *)
END_ENTITY;
```

Here, the polygon may point to three or more lines, indicating a three-to-many relation. This shows how a constraint on the arity can be defined in EXPRESS. The earlier line definition incorporates a two-to-many relation, where each line referenced two instances of point1, and each point1 could reference any number of lines. The WHERE statement checks that line endpoints connect to form a closed polygon. In many-to-many relations, an INVERSE clause is required so both sides of the many-to-many relation can be defined. In the example, a many-to-many relation occurs if a line may be part of multiple polygons, as is commonly the case in boundary representation solids.

EXPRESS solves the many-to-many problem by using the INVERSE relation. We redefine the previous line example and polygon above:

```
ENTITY line;
   point_ref : ARRAY[1:2] OF point1;
INVERSE
   loops : SET [0:2] OF polygon FOR edges;
END_ENTITY;

ENTITY polygon;
   edges : LIST [3:?] OF lines;
WHERE
   2_connected : (* check that the line's endpoints
                     are all two-connected *)
END_ENTITY;
```

The loops INVERSE attribute identifies a constrained SET of references, allowing a line to be composed into zero, one or two polygons. The polygon Entity specifies that it must have at least three lines within it. Thus this relation is many-to-many. In general, many-to-many relations are defined using the inverse clause, with both attributes in the inverse relation being constructors.

5.5.8.2 Supertype Constraints

In the general case, when SUBTYPEs are defined from a SUPERTYPE, relations may exist among the subtypes. A person, for example, may be specialized into male and female. It is not usually possible for an instance of person to be both of these types since the two categories are disjoint. On the other hand, a person may also be specialized into a parent and child, such that an instance may be both a parent and child. Supertype constraints allow the definition of rules on instances restricting their multi-type membership. These constraints apply to instances that are categorized as being of one or more SUBTYPEs of a SUPERTYPE.

EXPRESS provides constraints for the SUBTYPE clause to make these different cases explicit (the EXPRESS documentation calls them both constraints and operators). They are as follows:

  ONEOF—defines that the set of SUBTYPEs are mutually exclusive. An instance may be of only one SUBTYPE (as in the male and female subtypes).

  AND—used to compose both operands, by logical conjunction; an instance may be of both or all SUBTYPEs.

  ANDOR—defines that there is no rule and that an instance may belong to any subset of the SUBTYPEs. If no constraint is specified, ANDOR is assumed as the default. (This is in contrast to other object-oriented languages where ONEOF is the only option.)

These subtype constraints can be useful in product modeling because they allow definition of the different combinations in which Entities may be classified. Examples of ONEOF are spousal relations (husband, wife) and the types of materials used in construction (timber, steel, concrete). Examples of ANDOR might be type of mechanical equipment, where a set may be any mixture of electrical, piping, or air-handling. AND is used only in cases where there are multiple classifications of some high-level Entity type. An example of AND use follows:

```
ENTITY mechanical_part
SUPERTYPE OF (AND (power_part,handling_part));
. . .
END_ENTITY;

ENTITY power_part
SUPERTYPE OF (ONEOF(fluid_powered,electrical_powered,powerless));
. . .
END_ENTITY;

ENTITY handling_part;
SUPERTYPE OF (ONEOF(air_handling,liquid_handling,communication));
. . .
END_ENTITY;
```

In the above type structure, a mechanical_part always includes a power_part and a handling_part. This is a powerful feature, especially useful in product modeling classification. However, there is no direct implementation in existing programming languages, and developers implementing STEP interfaces must develop their own conventions for dealing with it.

5.5.8.3 Abstract Supertypes

A semantic constraint that can be expressed in NIAM and other data modeling languages is that a supertype can only consist of the members of its subtypes. That is, it can have no members of its own. In Figure 5.10, the TOTAL constraint expresses the fact that all pass-thru instances must be of type flexible-pass-thru or rigid-pass-thru. In the example of mechanical parts above, we may wish to require that all mechanical_parts have both a power_part and a handling_part, e.g.,

```
ENTITY mechanical_part
ABSTRACT SUPERTYPE OF (AND (power_part,handling_part));
```

```
FUNCTION distance(p1, p2 : point) : REAL;
(* compute the distance between two points *)
END_FUNCTION;

FUNCTION normal(p1, p2, p3 : point) : vector;
(* compute normal of a plane given three points on the plane *)
END_FUNCTION;

ENTITY circle;
   Center  : point;
   Radius  : REAL;
   Axis    : vector;
DERIVE
   Area    : REAL := pi * radius ** 2;
END_ENTITY

ENTITY circle_by_points;
SUBTYPE OF (circle);
   p1, p2, p3 : point;
DERIVE
   Radius     : REAL := distance( p1, SELF\circle.center);
   axis       : vector := normal(p1, p2, p3 );
WHERE
   Not_coincident : (p1 <> p2) AND (p2 <> p3) AND (p3 <> p1);
   Is_circle   : (distance(p2,SELF\circle.center) = radius) AND
                     (distance(p3,SLEF\circle.center) = radius);
END_ENTITY;
```

Figure 5.21: Example of code showing EXPRESS syntax as an algorithmic language.

5.5.9 PROGRAMMING CONSTRUCTS

EXPRESS incorporates a large set of language constructs to define the expressions for FUNCTIONs, WHERE clauses and DERIVE clauses. These are similar to those provided in standard programming languages, such as C. FUNCTIONs are used as repeatedly called routines in the definition of complex derivations or rules.

The whole syntax for EXPRESS is not provided here, but an example of a small program is shown in Figure 5.21 to indicate the style of the language. In addition, EXPRESS incorporates many useful functions that facilitate the easy definition of such rules presented in Section 5.5.7. Of particular significance are

In: an infix operator that tests membership in some aggregate. The right-
                 hand operand is checked to determine if the left-hand operand is a

member. Returns TRUE if it is a member, FALSE if it is not and NULL if either operand is missing

Like: a string matching operator that returns TRUE if the left operand is a substring of the right operand, FALSE if it is not and NULL if an operand is missing

Typeof: a function that returns a set of strings of all the types of which the parameter element instance is a member

UsedIn: a function that takes as an argument an Entity and a string containing an attribute name and returns a set of all the Entity instances that refer to the attribute

Figure 5.21 shows the outline definition of two FUNCTIONs (not fully defined) that are used to derive parameters from a circle that is defined by three points: a center and a begin- and end-point. From these, the WHERE clause derives radius and axis and checks that the points are not degenerate. EXPRESS also incorporates a query language, allowing data carried within an EXPRESS schema to be interrogated, compared and extracted.

5.5.10 COMBINING MODELS

EXPRESS facilitates the definition of abstract constructs that can be "imported" into other models for use. The two forms of importation are the USE FROM statement, which identifies a schema from which a list of Entities is to be imported into the current schema, and the REFERENCE FROM statement, which is used to reference Entities for use as attributes in the current schema.

```
USE FROM geometric_model
        (faceted_Brep, shell_based_wireframe);
REFERENCE FROM geometric model
        (manifold_surface, AISC_steel_spec)
```

The difference between these two forms of importation is that USE FROM allows subtypes and all other types to use the imported elements. This is done in a manner similar to the facilities in most programming languages, such as the include statement in C and C++. The REFERENCE FROM statement only allows use of the elements as attributes.

5.5.11 EXAMPLE

Figure 5.22 reproduces a standard example from the EXPRESS manual that shows a simple schema for defining a genealogical tree. It first defines three types: an array called date, a function for computing a person's age called years, and an enumerated attribute called hair_type. These are used to define person that has three variables for defining their name, a birthdate, a hair_color attribute and a relation to a set of the person's children. Person is subtyped into husband and wife. The recursive attribute children allows a person to refer to other persons that are their children. The INVERSE relation of children is named parents, so that any relation of a person to their children also carries the INVERSE relation of child to parent. This example utilizes the simpler concepts of EXPRESS.

5.5.12 BUILDING MODEL EXAMPLES

The STEP generic resources are available to represent standard geometric Entities. They will be reviewed in more detail in the next chapter, where we will again look at the arc and bounded plane examples.

This section presents the core wall example that was initially defined in Section 3.2. It is presented incrementally, so that various aspects of its definition can be discussed as we proceed. It is useful to compare this EXPRESS model with the roughly parallel NIAM model presented in Section 5.4.2, which has similar intended semantics. Since EXPRESS does not execute WHERE

```
SCHEMA example;

TYPE date = ARRAY [1:3] OF INTEGER;
END_TYPE;

FUNCTION years(d : date) : INTEGER;
(* computes an age to the current date from d *)
END_FUNCTION;

TYPE hair_type = ENUMERATION OF
    (brown,
    black,
    blonde,
    redhead,
    gray,
    white,
    bald);
END_TYPE;

ENTITY person
  SUPERTYPE OF (ONEOF(male, female));
  first_name    : STRING;
  last_name     : STRING;
  nickname      : OPTIONAL STRING;
   hair_color   : hair_type;
  birth_date    : date;
  children      : SET [0 : ?] OF person;
DERIVE
  age : INTEGER := years(birth_date);
INVERSE
  parents : SET [0 : 2] OF person FOR children;
END_ENTITY;

ENTITY female
  SUBTYPE OF (person);
  husband : OPTIONAL male;
   maiden_name : OPTIONAL STRING;
WHERE
 WI : (exists(maiden_name) AND EXISTS(husband)) OR
        NOT EXISTS(maiden_name);
END_ENTITY;

ENTITY male
  SUBTYPE OF (person);
  wife          : OPTIONAL female;
END_ENTITY;

END_SCHEMA;
```

Figure 5.22: The standard example of an EXPRESS model.

clauses and DERIVE computations, the EXPRESS model below conveys these features in words rather than pseudo-code.

```
SCHEMA core_wall_model;
USE FROM geometry (Brep,drawing,polyline,polygon);
USE FROM plumbing (plumb_entity);
USE FROM electrical(elect_entity);
USE FROM hvac(hvac_entity);
USE FROM floor_ceilings(floor_entity, ceiling_entity);
```

```
(* type definitions *)
TYPE resB = REAL ; (* resistance measured as reciprocal of
                    C, measured in BTU/sq.ft.*)
END_TYPE;

TYPE area_in = REAL; -- area in square inches
END_TYPE;

TYPE distin = REAL; -- linear distance unit in inches
END_TYPE;

TYPE UB = REAL; (*coefficient of transmission in
                    BTU/hr/sq.ft.   *)
END_TYPE;

TYPE boundary =
    SELECT(wall,floor_entity,ceiling_entity);
END_TYPE;
```

This model externally references various geometric models that are not included here, using the USE FROM syntax. It also refers to external mechanical systems elements that use the wall as a chase area. To deal with boundaries, it also uses floor and ceiling Entities. Some types are defined for some general measurements that in practice also have standard definitions within Part 41. However, we define them here explicitly and assume they will be resolved with supporting generic resources at a later "interpretation" stage of model development. A SELECT type is also defined, to group the different Entities that might bound a wall.

```
ENTITY wall
    SUPERTYPE OF (ONEOF ( core_wall );
    geom        : wall_geom;
    opening     : OPTIONAL SET [0:?] OF o_object;
    abuts       : LIST [0:?] OF boundary;
    DERIVE hs   : UB  := "derive hs for wall from hs
                            of segments and openings";
WHERE 3D_shape_consistent :="Brep consistent with
                            segments and openings";
INVERSE
    wall_abutted_by : LIST[0:?] OF boundary FOR
                                wall_abutted_by;
END_ENTITY;

TYPE wall_geom (ABSTRACT SUPERTYPE);
    3D_shape    : Brep;
    flr_plan    : drawing;
    elevation   : drawing;
    WHERE drawing_consistent :="plan and elevation are
                            consistent with Brep";
END_TYPE;
```

A generalized concept of wall is defined, which has a reference to a single ABSTRACT wall_geometry that includes three generic geometric views, for plan, elevation and solid model. The wall references a LIST of o_objects—openings in the wall. The wall has a DERIVED attribute for energy flow that is defined at this level so it can be applied consistently to any wall subtype. Wall also has references to a select type called boundary that defines the different Entities that bound this wall. It carries an INVERSE attribute backreferencing the other boundary Entities that may abut this one. Wall is then specialized into core_wall, as one general type of wall construction. Notice that a designed wall may be defined as wall or as core_wall. It may be defined first as wall, prior to determining its construction. Later, it may be redefined as a core_wall, when its construction type is decided.

```
ENTITY o_object
     ABSTRACT SUPERTYPE OF (ONEOF (filled_hole,hole));
     region      : polygon;
     r-value     : resB;
INVERSE
     part_of     : wall FOR opening;
DERIVE area : area_in: area_in := "compute area of polygon";
     hs  : UB     := "derive hs from r of filler and area";
END_ENTITY;

ENTITY filled_hole
   SUBTYPE OF o_object
     ABSTRACT SUPERTYPE OF (ONEOF (door,window));
     solid       : Brep;
     pattern     : symbol;
     r-value     : resB;
END_ENTITY;

ENTITY door;
END_ENTITY;

ENTITY window;
END_ENTITY;

ENTITY hole;
END_ENTITY;
```

The o_object is a generalized opening referenced by the supertype wall. It is ABSTRACT, meaning that only instances of its subtypes are allowed and ONEOF, which requires an instance to be of only one of the subtypes. It has an INVERSE attribute allowing any opening to reference the wall it is in. The filled_hole is an ABSTRACT subtype of the o_object; it includes geometric definitions of the filler and references the type of filler. It also carries a thermal resistance. Door and window are defined toward the bottom. Though they add no new attributes of their own, they inherit all the attributes and relations from opening and filled_hole, including the INVERSE attribute and thermal properties.

```
ENTITY core_wall;
     pass_thru   : OPTIONAL LIST [0:?] OF p_object;
     segment     : SET [1:?] OF s_object;
WHERE
     routing     := " pass_thrus only intersect core_walls";
     disjoint    := "all openings and segments are pairwise
                       spatially disjoint";
     coverage    := "all openings and segments cover the wall
                       elevation";
END_TYPE;
```

The core_wall is a subtype of wall. Thus it receives all the references to o_objects and geometry and references the bounding Entities. Core_wall references pass_thrus and segments. It has three constraints defined for it, restricting the possible combinations of data it can carry. They define the legal geometric definitions of a wall and constrain where pass-thrus may go.

```
ENTITY p_object
     pass_thru_entities : LIST OF pass_thru;
```

```
            center       : polyline;
            3D_shape     : Brep;
END_ENTITY;
```

The p_object is a generalized Entity that references other systems passing through the wall and represents their geometry in a way needed by the wall definition. It represents a single pass_thru and carries a polyline and a Brep representation of it.

```
ENTITY s_object;
            core          : core_type;
            insul         : OPTIONAL insulation;
            surf1         : OPTIONAL LIST OF surface;
            surf2         : OPTIONAL LIST OF surface;
            region        : polygon;
  INVERSE part_of   : core_wall FOR segment;
  DERIVE area       : area_in := "compute area of polygon";
            offset1       : distin := "derive thickness to side 1";
            offset2       : distin := "derive thickness to side 2";
            hs            : UB := "derive the thermal flows for this
                            segment from core, insulation and surface Rs";
END_ENTITY;
```

The s_object is a central construct. It defines the construction of a wall segment as a core (with optional insulation) with a sequence of surfaces on both sides. It carries an area and two offsets derived by summing the core offsets and surfaces on each side. It includes the polygonal region it occupies in the core_wall elevation. It computes an area and intermediate resistance for this region of the core_wall. The types of core construction and surfaces are not detailed; they would be subtypes of core_type and s_object, respectively.

```
ENTITY insulation;
            thickness     : distin;
            r_value       : resB;
  INVERSE part_of   : s_object FOR insul;
END_ENTITY;

ENTITY core_type;
            r             : resB;
            offset        : LIST [1:2] OF distin;
            3D_shape      : OPTIONAL Brep;
            INVERSE part_of   : s_object FOR core;
END_ENTITY;
```

The insulation is defined by a thickness and its resistance. It has an INVERSE attribute to the segment it is in. The core_type has a resistance, two offsets and an OPTIONAL Brep solid model defining it. It and other parts of the segment have been defined with INVERSE attributes, so that they access the wall instances to which they belong.

```
ENTITY surface;
            r_value      : resB;
            thickness    : distin;
            INVERSE part_of_1 : s_object FOR surf1;
            INVERSE part_of_2 : s_object FOR surf2;
END_ENTITY;

END_SCHEMA;
```

The `surface` carries a `thickness` and `r_value`, with `INVERSE` attributes back to the `s_object` it is in. It must carry two such attributes because of the two slightly different Roles it may play.

This EXPRESS model is a direct translation of the semantics of the core wall defined in Chapter Three and elaborated in NIAM in Section 5.4.1. It has not been simplified, generalized, or otherwise "tuned" for long-term use. These issues are taken up in Chapter Six.

5.5.13 SUMMARY OF EXPRESS

From the core wall example, we can see that EXPRESS is a rich language generally capable of representing a broad range of semantics associated with buildings or other products. It allows definition of special purpose types, Entities and relations needed for some area of product modeling. It supports sophisticated conceptual definitions of Entity supertype-subtype lattices whose semantics can be defined through SUBTYPE constraints. It also allows definition of complex attributes, including those that are functions of other attributes, making their derivation explicit. It allows definition of complex rules that specify the conditions required for a model to be consistent. It supports specification of two-way relations, allowing bi-directional access through the model structure and guaranteeing that these relations are consistent, using `INVERSE` attributes.

Some limitations also can be identified. Like most other object-modeling languages, EXPRESS does not distinguish relations between an Entity and its attributes and relations with other Entities. There is nothing equivalent to the TOTAL constraint in NIAM for identifying dependencies regarding potential deletions. This limitation points out the fact that EXPRESS models are static; there are no mechanisms to add or delete parts of a definition. Also, there is nothing equivalent to the UNIQUENESS constraint on sets of attributes for support of queries. There are numerous other differences. In general, however, EXPRESS is able to depict almost all of the semantic conditions defined for the core wall model in Chapter Three.

Given this introduction to EXPRESS, we turn to EXPRESS-G, which is the graphical subset of EXPRESS. We assess both in more detail at the end of this chapter.

5.6 EXPRESS-G

An integral part of the EXPRESS language definition is a graphical notation called EXPRESS-G. While it was defined as a means to depict EXRESS models, it is recognized as a STEP description method in its own right and is becoming frequently used for defining Application Reference Models.

EXPRESS-G allows easy definition of a major subset of the EXPRESS language. It defines a schema in terms of attributes, types, Entities of various type, and also references elements outside the current schema. It provides means to define relations, including Subtype, Derived and Inverse relations. Constrained elements can be identified, but the constraint rule or clause is not specified graphically. There are several slightly different flavors of EXPRESS-G. Figure 5.23 shows the graphical notation corresponding to one of these flavors. All Types but the Basic Types (Real, Integer, Boolean, String, Logical) are represented as a box of dashed lines. Basic Types are solid. Bars on the right or left denote different types. A box with an enclosed circle indicates a reference to a Type of Entity within another schema or diagram. Relations are represented by edges connecting boxes; Supertype/Subtype relations are shown with thick lines, while other relations are shown with thin lines.

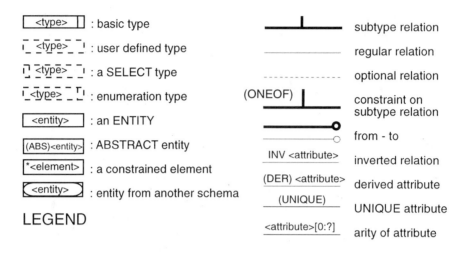

Figure 5.23: The notation of EXPRESS-G. Names of elements are within angle brackets (< >).

In this book, Subtype constraints are shown explicitly. In most other documentation, only the ONEOF constraint is represented, using a '1´ to denote it. We denote all three—ONEOF, ANDOR or AND—placing them in parentheses. All relations have a direction denoted by a small circle rather than an arrowhead. OPTIONAL relations are shown with dashed lines (subtype relations cannot be optional). Relations may be qualified with an INVERT constraint or a derivation. INVERTed relations are represented by a single line, with one attribute name shown above the line and the other below it. The arity of relations is shown as subscripts on the relation. Constraints on Attributes or relations are noted with an asterisk (*).

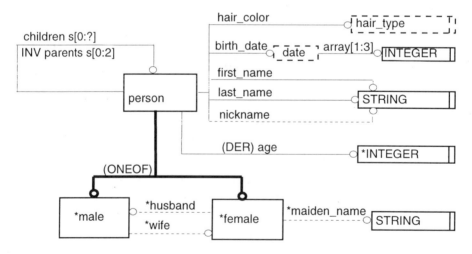

Figure 5.24: The EXPRESS-G representation of the hereditary model presented in EXPRESS in Figure 5.22.

By implementing EXPRESS-G within a graphical drag-and-drop environment, it becomes possible for a user to construct diagrams that, when interpreted by the computer, can be used to automatically generate most of an EXPRESS schema. It becomes a means to graphically define a schema to be used for data exchange. Figure 5.24 presents the example of the EXPRESS schema defined textually in Figure 5.22, in EXPRESS-G. By checking this figure's correspondence with the earlier example, the semantics of the notation are easily studied. The graphical notation facilitates understanding the overall structure of an EXPRESS model.

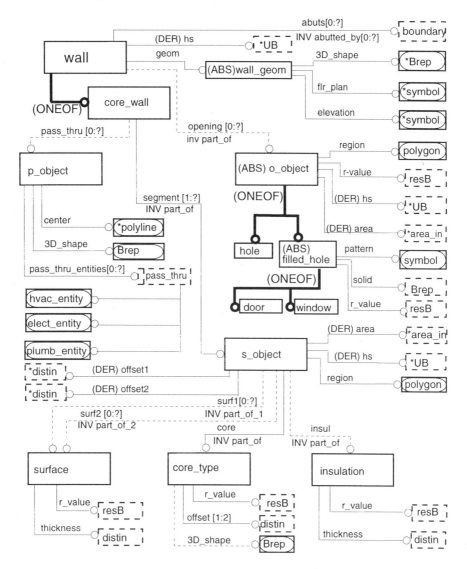

Figure 5.25: The core wall example defined in EXPRESS-G.

Figure 5.25 shows the larger example of the core wall that was developed in the previous section. Presenting an EXPRESS model in this way facilitates its understanding, and allows tracing and debugging. For most people, it is much easier to understand the graphic example rather than its textual version. Again, comparison of the EXPRESS model in Section 5.5.12 with the EXPRESS-G model indicates the semantic expressiveness of EXPRESS-G in relation to EXPRESS.

Once a model has been defined, it must be validated. That is, is the model complete, in the sense that there are no dangling definitions that refer to some Entity or type that is not defined? Are the Entity and attribute definitions that refer to external schemas correct in how the external schemas are referenced? Is the syntax of the model correct? These issues are aided by a variety of software tools.

A small software development community has grown up around the data exchange field. Originally, developers provided IGES translators and IGES file conformance testing software. Software firms, most of which grew out of university or industrial research labs, support the STEP enterprise with a variety of tools. A list of some of the vendors is provided in the Appendix.

The tools include
- EXPRESS compilers that check the syntactic correctness of an EXPRESS model
- cross-references tables that show where each schema, type and Entity is defined and where they are used
- tools that support the automatic generation of EXPRESS-G diagrams with associated browsers, supporting conceptual review
- tools that check if two EXPRESS models are the same or if they differ and how, so that different versions of an evolving model can be carefully compared
- code libraries that can access an EXPRESS schema and Read or Write data in the prescribed format

These tools are described in the Appendix.

5.7 ARM TO AIM INTERPRETATION

The translation of a reference model (ARM) to an interpreted model (AIM) involves rationalizing the ARM specification within the larger scope of STEP Part models. That includes determining the appropriate use of Generic Resources in the model and coordinating the model with the relevant Application Resources. For example, various APs developed in the AEC area should incorporate sufficient overlap to deal with such issues as spatial conflict testing. While initially the emphasis was on specifying ARMs in NIAM or EXPRESS-G, recently some ARMs have also been developed in EXPRESS. In this situation, interpretation means adjusting the already developed EXPRESS model to conform with the Integrated Resources and with the larger STEP environment, defined in other APs.

5.8 PHYSICAL IMPLEMENTATION OF AN EXPRESS REPOSITORY

An EXPRESS schema provides a template for generating a physical implementation of a product model. It defines a structure for storing data describing instances of a building part, assembly or a whole building. While there has been discussion of a range of possible physical format implementations, in practice there have been two types: a computer file that stores the data describing some product or a database that holds the data. Both of these approaches are described below.

The general facilities for developing physical level implementations of an EXPRESS model are defined in STEP Part 22, Standard Data Access Interface (SDAI). Part 22 defines consistent data storing and access mechanisms in terms of their functionality and their programming language interfaces. The interfaces are defined as a library of methods or procedure calls—generally called *language bindings*.

An EXPRESS Part model is a complex structure. One need only review an example model to grasp its complexity. In the core wall model presented in Figure 5.25, for example, what are all the attributes of a window? Looking at window alone, there are none, but in fact, a significant set has been inherited into the window through supertypes. Any interpretation of an EXPRESS model requires scanning up and down the subtype-supertype lattice to put together the structure of an

Entity. In addition, a model utilizes a variety of shared library integrated resources that are in different schemas, requiring going back and forth among multiple schemas.

In order to facilitate easier interpretation of EXPRESS models, a preliminary processing step is to transform the EXPRESS schema into a more easily read dictionary. The dictionary is a compiled form of an EXPRESS model. The dictionary is itself defined in EXPRESS. To give an idea of the dictionary's structure, the Entity and Attribute specifications are given below.

```
ENTITY entity_definition
    SUBTYPE OF (named_type);
    supertypes        : LIST OF UNIQUE entity_definition;
    attributes        : LIST OF UNIQUE attribute;
    uniqueness_rules  : SET OF uniqueness_rule;
    where_rules       : SET OF where_rule;
    complex           : BOOLEAN;
    instantiable      : BOOLEAN;
    independent       : BOOLEAN;
INVERSE
    parent_schema     : schema_definition for entities;
UNIQUE
    UR1: name, parent_schema;
END_ENTITY;

ENTITY attribute
    ABSTRACT SUPERTYPE OF
        (ONE OF (derived_attribute, explicit_attribute,
                                        inverse_attribute));
    name              : STRING;
    domain            : base_type;
INVERSE
    Parent_entity     : entity_definition FOR attributes;
UNIQUE
    UR1:  name, parent_entity;
END_ENTITY;
```

A structure in this format is defined for each Entity in a schema. For an Entity, the structure is fairly clear. Entity_definition is the name of the Entity. The subtype carries the name of the subtypes. The supertype carries all the supertypes of this Entity class. The attributes carries a list of attributes, as defined below it. Uniqueness_rules defines a set of constraints that apply to the Entity. Where_rules is a set of domain rules for the Entity. Complex indicates whether the Entity is the result of an ANDOR or AND subtype constraint. If TRUE, the Entity is the result of mapping multiple inherited Entities, else it is defined explicitly. Instantiable indicates whether Entity is an ABSTRACT SUPERTYPE. Independent indicates whether the Entity is independently instantiable or whether it relies on a REFERENCE clause. Parent_schema references the schema in which this one is declared. UR1 checks that the name_type is unique within the schema.

The attribute is a supertype of derived_attribute, explicit_attribute, or inverse_attribute. It carries a name and a domain. An inverse relation references the Entities that reference it. Its name also must be unique. Each of the EXPRESS constructs, including the schema, all types, all rules, constructor types of set, list, bag and array are specified in a manner similar to the Entity and attribute above.

Any approach to storing EXPRESS-formatted data must define the equivalency between each construct in EXPRESS and the format in the medium being used. For example, how is an attribute that uses a set constructor to be stored, or how is an INVERSE relation to be stored? These equivalencies can be represented in a mapping table, as described in Section 5.3. An example of a mapping table is shown in Figure 5.28. EXPRESS vendors have defined and implemented functions that realize such a mapping table for each repository medium to which they have developed interfaces.

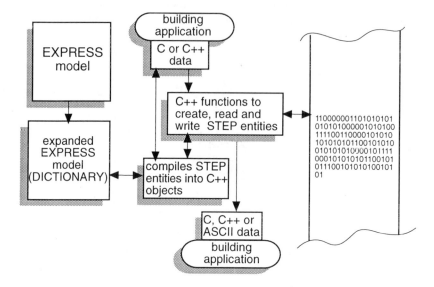

Figure 5.26: The early binding approach to developing a STEP interface.

There are two general strategies for implementing an EXPRESS data repository. These are called *early-binding* and *late-binding* strategies. In an early binding strategy, shown diagrammatically in Figure 5.26, each Entity in the EXPRESS schema for some application protocol is parsed and defined as a corresponding structure in a programming language such as C or C++. For example, an Entity made up of three attributes, a list and a set will be parsed and defined as a C struct or C++ object that carries three fields for attributes, a structure for a list and another structure for a set. These structures are created as a preprocessor to compilation, allowing code to be written that both reads or writes the Entities to and from the application and also to and from the storage medium. The read and write code related to the application entity instances is custom to the application. The read and write to the storage medium is provided by language bindings that read or write an Entity instance to or from the storage medium. Since the structures are defined and compiled with integrated reading and writing, they operate quite efficiently.

In the late-binding strategy, shown diagrammatically in Figure 5.27, the EXPRESS Entities and relations making up a model are identified during the execution of the exchange process. Instead of compiling the elements of the schema, the elements are identified and looked up in the data dictionary for each read and write. For example, using the early-binding example above, the late-binding interface would involve a function call to create an Entity, followed by three function calls to make three attributes, followed by a call to create a list structure, followed by a call to create a set structure. This sequence must be executed for each instance encountered. The advantage of late binding is that the functions used to read or write to/from a format can be defined once and used for any model. The model Entities can be easily revised. With early binding, custom functions must be written for each model Entity.

A particular implementation issue is the supertype constraints on inheritance. These identify a range of Entity types that may be defined as combinations of subtypes—using ONEOF, ANDOR, and AND. At the implementation level, each possible combination of subtypes is explicitly defined. For example, if a supertype 'A´ has subtypes "ANDOR(B, C, D)", then the following types are allowed: "A+B", "A+C", "A+D", "A+B+C", "A+B+D", "A+C+D", "A+B+C+D". "A" is inherited into all subtypes and all combinations of "B", "C" and "D" must be generated. There is no way to automatically generate these types. They must be defined by a programmer.

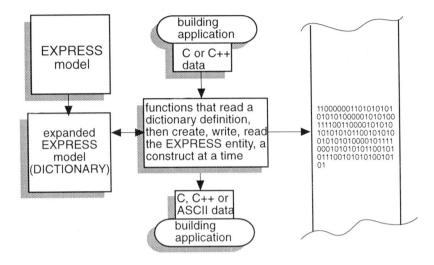

Figure 5.27: The structure of a late binding SDAI implementation.

5.8.1 TEXT FILE IMPLEMENTATION

The most prevalent implementation—and the one for which most tools have been developed—is using a text file. A file-based repository may be developed using either early- or late-binding strategies. Part 21 of the STEP standard fully specifies how an instance in any EXPRESS format should be organized in a physical file. Thus it defines how a dataset of Entity instances in EXPRESS format should be laid out. It uses a free format not based on columns and specifies a file composed of two sections, a header section and a data section.

The header section has three parts: (1) a file description part that provides an informal description of the file contents; (2) file registration information consisting of the filename, timestamp, author, organization, processor that generated the data, release status, and related information; and (3) the file schema format, defined for the EXPRESS AIM used, and any Application Interpreted Constructs used with the AIM. The second section consists of Entity occurrences, formatted according to a syntax uniquely mapping the EXPRESS schema into locations within the dataset.

Comments may be inserted into the physical file data section, delineated by /* and */. The file may include print format statements and "white-space characters" for enhanced readability. These are ignored when the file is parsed for translation.

Each construct and construct composition used in EXPRESS can be defined to specify a format for the EXPRESS schema on a file. Only some of the EXPRESS constructs are needed, and others are mapped onto a common format. The table in Figure 5.28 defines how each construct is mapped. The right side of the table shows the formats used in the physical file.

These equivalencies are built into functions that are then embedded into application programs that READ, WRITE, QUERY and CHECK the data stored on a file, using the conversions shown. The CHECK functions are used to validate the data carried within a model. Typical CHECK functions include validating that the attributes of an instance are of the correct types, validating that the instance carries values for all attributes that are not OPTIONAL, and validating that there is an appropriate pairing of INVERSE attributes.

EXPRESS CONSTRUCT	PHYSICAL FILE FORMAT
schema	---ignore---
array	list
bag	list
list	list
set	list
select	---ignore---
Entity	Entity
Entity as attribute	Entity name
Entity as supertype (no instance)	---ignore---
Entity as supertype	Entity
integer	integer
real	real
string	string
binary	binary
Boolean	enumeration
logical	enumeration
enumeration	enumeration
derived attribute	---ignore---
inverse	---ignore---
procedure	---ignore---
function	---ignore---
constant	---ignore---
remark	---ignore---
rule	---ignore---
where rule	---ignore---

Figure 5.28: The mapping table for EXPRESS constructs mapped to physical file constructs.

File level implementations are sometimes called clear text encoding. They are appropriate for writing out a model from one application program and reading all or part of it into another application program. Modifications of the data within the same file are not possible, because of the lack of a way to adjust the file's storage allocation.

Recently, a new non-standard text-encoding scheme has been developed using SGML, the Web-based data exchange language.

5.8.2 DATABASE IMPLEMENTATION

There have been a few database implementations that support storing and retrieving data in an EXPRESS-consistent format. Various implementations are possible. Most rely on an encoding and database representation of each EXPRESS language element as defined in the EXPRESS dictionary. These can be concatenated into high-level database calls to Create, Write or Read schema Entity instances.

Databases provide a higher level of functionality to the physical implementation of an EXPRESS model, which is reflected in the SDAI interface facilities. These include

- *Session operations:* start event recording, stop event recording, open session, close session
- *Transaction facilities:* start transaction in read-write access, start transaction in read-only access, promote to read-write access (from read-only access), commit, abort
- *Model access and management:* open repository, close repository, rename model

Database interface libraries have been developed and are marketed for a variety of commercial databases, including the Oracle relational database, and the ObjectStore, Versant, and ROSE object-oriented databases.

5.9 CONFORMANCE TESTING

Conformance testing is the evaluation of an implementation to determine whether it conforms to the standard. It begins with the development of a detailed set of tests and test purposes, while the AIM is still under development. Each implementation of an AP is required to have associated conformance criteria that can be used to validate any implementation of the AP. The test criteria include the options and variations that are to be exercised during conformance testing of an AP implementation.

In general, conformance requires satisfaction of the following criteria:
1. The information requirements of the ARM are preserved in the implementation. This includes all combinations of Entities and their attributes.
2. All AIM Entities, types and associated constraints are supported. Treatment of options shall conform to the AIM.
3. Only the constructs defined in the AIM are recognized and included in their implementation.

The implementation must also specify whether the the AP is required to be complete or whether subsets are allowed. If subsets are allowed, they must be defined.

These criteria are used to create a test suite for any implementation of the AP. This test suite is to allow testing laboratories to define and apply executable tests during conformance evaluation. The emphasis is on making the tests auditable and easily replicated. All controls are provided and standardized by STEP (with the exception of PIXIT, for which guidance in its definition is given), thus assuring regularity to the highest standards possible.

Guidelines for development of test suites include
1. Identify all Entities that can exist in the ARM without being a child of another Entity; these become test Entities.
2. Identify the correspondence between ARM and AIM Entities and verify that the requirements of all ARM Entities are satisfied.
3. If attributes have enumerated values, there should be tests for each possible value.
4. If attributes are of SELECT type, then each of the possible types allowed should be checked in the tests.
5. If an attribute if OPTIONAL, then the test suite should include cases that both include and omit the optional attribute.
6. Test cases should cover all combinations of types allowed by the supertype constraints.

There is a notion that an AP may be divided into *levels*, where a level is a subset of the overall AP. This provides the option to define partial implementations of the complete protocol. Each conformance level can be considered a miniature AP. This allows each AP to identify subsets of the protocol that can be considered "complete" for some purposes.

5.10 THE EVOLUTION OF STEP

ISO-STEP is one of the largest standards efforts ever undertaken, encompassing many areas of industry with worldwide scope. This is only practical if many different STEP activities are undertaken in parallel. A majority of the Parts identified in Figure 5.3 have on-going committees, which are either in the initial process of generating a version 1.0 AP, or are revising an existing one.

Across such endeavors, many aspects of STEP are evolving in parallel. At the base facilities level, the procedures involved in developing and gaining approval of APs are being refined to make them simpler wherever possible. At the same time, improvements in conformance testing are being sought. With over five years of practical experience using EXPRESS and EXPRESS-G, there are plans to extend and revise it to respond to known shortcomings. In this respect, STEP is different from other standards efforts. For example, in the early days of computing, there was no agreement on the coding scheme for characters. Specifically, there was the EBCDIC character coding used by IBM competing with the now standard ASCII coding. Well-developed working solutions existed and the standards issue was which alternative to select. All character-based devices now rely on ASCII coding as a fixed non-changing standard. STEP is not a fixed, non-changing standard. In this regard, STEP is more like a very large, related set of development efforts.

The ISO 10303 activities making up the STEP enterprise are largely funded through support of various interested organizations. Some are private, such as Boeing, General Motors, Bechtel and the large Japanese construction firms. Others are public organizations, such as the National Institute of Standards and Technology in the US and the various European Union organizations. Many small organizations, including civil engineers, software houses, and universities are also members. These organizations support the activities by providing staff time to develop, review and test proposed APs throughout their development process. Additionally, the standards are under public control and are not proprietary. In some ways, it can be viewed as the "United Nations" of industry standards.

We have covered much intellectual territory in the presentation of STEP, and there are other aspects that we will analyze in more detail in future chapters. It is important to gain a broad picture of all these dispersed activities, because effective contributions—either within the STEP framework or outside it—are difficult to undertake when such large but often poorly understood efforts in data exchange are taking place.

5.11 REVIEW OF EXPRESS AND EXPRESS-G

Here, we return to consider the role and functionality of EXPRESS as a data modeling language for product data exchange, especially for the building industry.

Product models are influenced by various interests. One intellectual goal of product modeling is the representation of all the information used in the conceptualization, design, construction or maintenance of a building—in other words, the creation of a semantically complete model. This goal is one of conceptual modeling, capturing the knowledge of domain experts. People holding this view accept that no application can yet use or manipulate all of the information they want to represent, but they contend that eventually applications will be developed to cover these aspects.

By developing a rich model, new applications may be realized more quickly. We might call this intention the "idealistic view". A different influence emphasizes that the goal of a product model is to provide data exchange between applications that exist today. As new applications are developed, the building model can be adapted to deal with them. Besides, as new applications are developed, they are likely to take unexpected forms and structures that may invalidate any effort to anticipate them. This view takes as its goal the easy mapping of data between applications. We might call this pull the "pragmatic view".

If we take the goal of building product modeling to be a practical one—that of facilitating the use of software applications in support of building activities throughout the building's lifetime—then the goal needs to be closer to the pragmatic intention. The basic reason for data exchange methods is to simplify the integration of computer applications into the workflow associated with a building task. Elaborating the definition of a building model increases its complexity; it may require a number of Entities that are derived or otherwise part of the model, but that are not now used. Thus the idealistic view adds issues and complexity to the pragmatic view of integration.

The design of EXPRESS was based mostly on pragmatic intentions. Its members are mostly practical engineers interested in exchanging what can be represented in current procedural programming languages, especially object-oriented languages such as C++ and Java. It has not emphasized conceptual modeling concepts (such as relations) or logic programming (and inference making). Thus it supports well the programming concepts now used to produce commercial application software.

The one area where EXPRESS includes some notion of conceptual modeling is also the most problematic in its application. The supertype constraints lead to both semantic errors in ARMs and also implementation headaches. Meaningless combinations of types are sometimes inadvertently allowed. At other times, implementations are limited and omit types that should be allowed, because they rely completely on the skill of a software programmer to implement the possible types correctly.

EXPRESS has been used by different Part Committees to define and implement a large number of complex Application Models, all listed in Figure 5.3. As these models have been developed, the inevitable restrictions, awkward aspects and limitations of EXPRESS have become visible. One issue is proper use of the rich semantics. For example, under what conditions should a Set constructor be used and when should a List be used? What is good practice for defining Basic Types in the definition of attributes? Guidelines for good practice have emerged as more models have been developed and the strengths and limitations of the language, as demonstrated in use, have become apparent.

Some areas of global functionality are missing. EXPRESS has been conceived as a format for exchanging data between two applications, with an implicit focus of using a file repository. However, there are many cases where a more permanent repository is desired, allowing multiple updates and incremental changes to the model data. Because physical implementations of databases have not been fully addressed, some issues associated with database implementations have not been adequately considered. How to communicate deletions and/or additions are two specific examples. A related difficulty is that EXPRESS was conceived as a format for exchanging data between two closely related applications using a single schema. It was not defined to support translation between unlike applications that need to utilize different schemas. Recent efforts to add mapping facilities, reviewed in Chapter Eleven, are aimed both at alleviating the static structure of EXPRESS and at allowing translation between heterogeneous types of applications.

A related difficulty is that EXPRESS has been conceived as a format for exchanging data between two applications with the same functionality as a file translator. Because physical implementations of databases have not been fully addressed, some issues associated with this kind of implementation have not been adequately covered. Some of the missing functionality include

- the ability of the user to select subsets of a dataset to send to an application
- the ability to check the consistency of the model over time as updates are made
- the ability to deal with concurrency control when multiple people are updating the model at the same time, and
- the ability to implement larger models encompassing multiple applications, such as those needed for buildings

Many of these issues go beyond the development of a building model *per se*, and suggest the need to address a broader set of issues—what might be called a modeling framework. Some of the issues pertaining to a modeling framework are addressed in later chapters.

Another limitation of EXPRESS is that the functions, rules and other relations that are defined within an EXPRESS model are not executable. The rules are currently used for specifying to "the translation writers" the relations that data is supposed to reflect. Some SDAI libraries support execution of Functions, Rules and WHERE and DERIVE clauses, but, in general, rules or clauses must be checked by the test suites developed for translators to and from the model.

Overall, the issues raised here are detailed ones, made apparent by the success and wide use of EXPRESS as a product model specification language. The evolving challenge will be to maintain its simplicity and clarity, while extending its functionality.

Currently, EXPRESS-G is used as a beginning ARM language, prior to developing more detailed models in EXPRESS. Throughout, it is used as an abstraction of EXPRESS, facilitating high-level organizational review and comprehension (which is how it will be used in the later chapters of this book). In these roles it works quite well. However, if EXPRESS-G is to be relied on for developing Application Reference Models by domain experts—not software engineers—it needs to offer better ways to express constraints and relations. These aspects of a model are not just technical issues best handled by implementers, but a basic part of defining the domain knowledge embedded in an application. As a result, less complete ARMs are defined. New data modeling languages have been developed and are in widespread use in other fields. Advances are continuously being made to NIAM and ER. Especially noteworthy in this regard is the Unified Modeling Language (UML), which is also designed as a pragmatic, implementation-oriented modeling language. It, however, has several semantic constructs not directly represented in EXPRESS-G.

5.12 THE STEP SYSTEM ARCHITECTURE

Before looking at examples of STEP models using this technology, it is worthwhile to assess the technology on its own and in terms of its own goals, as well as those we have defined earlier.

The notion of an implementation-independent standard, with alternative physical implementations, is based on the analogy that there may be different language compilers that allow implementation of a process on different machines. In the analogy, the semantics of the language are used to specify a process having multiple implementations. Similarly, EXPRESS allows the definition of a data representation that is then implemented in different media and storage environments.

STEP has made a conscious decision to partition the UoD of all data exchange into much smaller areas, which are oriented around clusters of software applications, using the Application Protocol concept. However, as Application Protocols grow in number and their overlaps and interrelations become apparent, several product areas have called for another kind of integrated resource, one that provides a framework for how different APs should be organized. So far, such framework models have been proposed for shipbuilding, process plants and for buildings. There are many issues regarding this new addition to the STEP system architecture that remain to be resolved. Some of them will be discussed in Chapter Eight.

5.13 NOTES AND FURTHER READING

The European efforts that were initiated parallel to or after IGES include SET [1989], CAD*I [Schlechtendahl, 1988], VDA-FS [1987] and EDIF [1985]. STEP was viewed as a pan-European effort to subsume these individual efforts. A review of these efforts is provided in Chapter Four of Bloor and Owen [1995].

The ANSI-SPARC database effort heavily influenced the STEP architecture. It is presented in Tzichritzis and Klug [1978]. The STEP system architecture and development methods are presented in two reports by Danner and Yang [1992a, 1992b], Burkett and Yang [1995] and also Bloor and Owen [1995]. The data modeling languages used in STEP include IDEF1x [1985] and NIAM. An early presentation of NIAM is [Verheijen and Van Bekkum, 1982] and another is Halpin and Nijssen [1989]. A much revised recent presentation of NIAM is Halpin [1995]. Translators that map an ARM language into EXPRESS are described in Chandhry [1992] and Poyet [1993]. An example of a tool that translates NIAM into a relational database schema is presented in De Troyer [1989].

The official EXPRESS document is ISO DIS10303 Part 11 [1991]. Another excellent presentation of EXPRESS is Schenck and Wilson [1994]. Its genesis can be traced from PDDI [1984]. STEP development methods are presented in Danner and Yang [1992a] and [1992b]. Details of the STEP implementation procedures have been omitted. The document specifying these procedures is ISO TC184/SC4/WG4 N34 [1992]. The Standard Data Access Interface for STEP is Part 22 [ISO/WD 10303-Part 22, 1993]. In the US, the National Institute of Standards and Technology maintains a Web site for coordinating ISO-STEP activities:

   http://www.nist.gov/sc4/www/stepdocs.htm

There have been many independent efforts to articulate the semantics embedded in data, for use in computer languages and databases—as well as product models. Good surveys of the general area are offered by Hull and King [1987] and by Peckham and Maryanski [1988]. The recent development of Unified Modeling Language (UML) provides another important conceptual model. It is easily accessible in Muller [1997]. Work assessing data models for use in product modeling include Hardwick and Spooner [1987], Fulton and Yeh [1988], Shaw et al. [1989], MacKeller and Peckham [1992], Eastman, Bond and Chase [1991], Eastman and Fereshatian [1994].

5.14 STUDY QUESTIONS

1. Select a catalog that presents information about some building product. Examples might be a catalog for windows, doors, window awnings, exterior material panels. Focus on products whose geometry is defined by just a few dimensions. Also, identify two or more computer applications that might use the information in the catalog. For one line of products within the catalog, develop your own NIAM, EXPRESS-G or EXPRESS model of the product line. The emphasis is to represent the information about the product needed by the applications.

2. Consider the information needed to model a process in a construction schedule. Consider the data in the context of current construction scheduling applications. A reference is Fischer, Luiten and Aalami [1995]. Define a NIAM, EXPRESS-G or EXPRESS model of a general

construction process. What kinds of information needs should be included in all such processes?

3. Given a general construction process model, as defined in Question 2, consider the specializations needs for particular classes of construction process. Consider, for example, concrete pouring, painting and excavation work. Define these specializations in an EXPRESS model. Again, consider this question in the context of current scheduling programs.

4. There have been few data models developed to define a space within a building. Consider such a model from an architectural design and building code perspective. Building codes in most countries have safety and habitability requirements for residential spaces. Two research efforts related to such an effort are Eckholm and Fridquist [1996] and Eastman and Siabiris [1995]. Develop a NIAM, EXPRESS-G or EXPRESS model for a general residential building space that could be use for checking against building code requirements. Ignore the relations of the space to its surrounding boundaries and focus on the attributes of the space and relations between spaces.

5. Given the space model defined in 4 above, extend it to deal with public spaces in different classes of building types, such as commercial buildings, auditoriums, gymnasiums and restaurants.

6. Compare the constructs in the major modeling language now used in business for data modeling, the Unified Modeling Language (UML), with the constructs in EXPRESS-G. Identify their similarities and differences. Present an argument whether or not UML should be added to the set of "approved" ISO-STEP ARM languages.

CHAPTER SIX
STEP Integrated Resources

6.1 INTRODUCTION

A fundamental component of the STEP information architecture includes a set of widely used generic structures that are grouped together and defined once, allowing reuse. Grouping expedites the definition of new models, because only the structures unique to the new model need to be defined. Grouping and reuse also facilitate data exchange, because the reused elements are frequently the shared aspects of different models that must be exchanged between applications. In this chapter, we review the elements that are incorporated in Parts 041, 042 and 043, the STEP Application Integrated Resources (AIRs). The STEP proposals for AEC-based models rely heavily on these AIRs. Part 041 deals with many of the contextual issues pertaining to part models. Part 043 deals with the origin, version and status of models, especially when embedded in other models. Historically, the basic information carried in all CAD systems has been geometry; this is the third area of focus of the integrated resources, defined in Part 042. Because of its importance, we review Part 042 last and in some detail, as it is the most important resource for building modeling use. Other AIRs have been defined, but are still in the early stages of development.

This chapter presents a set of EXPRESS definitions whose use is strategic in the part models presented in succeeding chapters. It gives a general overview of the Entity structures presented in EXPRESS-G. Significant constraints, defined in DERIVE or WHERE clauses, are noted. In general, STEP definitions are typically hierarchical, making maximum use of specialization to define common specifications only once, then subtyping them into the Entity classes which carry instance data.

6.2 APPLICATION CONTEXT (PART 041)

Most models have a context defined by the organization, management structure, ownership, security, certification, standard units of measurement used in that region and industry, and other aspects of the social and organizational environment in which the model was populated. These aspects are covered by Part 041. It has 18 different schemas:

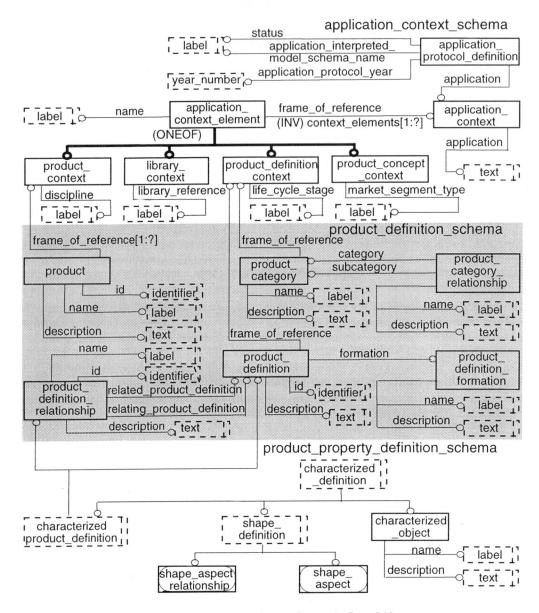

Figure 6.1: The first three schemas in Part 041.

1. application_context_schema: provides entities that identify the discipline, life cycle stage, market segment or library use of a product specification (see Figure 6.1).
2. product_definition_schema: product identification, naming and description, relationships and categories.
3. product_property_definition_schema: classifies properties according to whether they reference shape, shape aspect, an object or product definition (see Figure 6.1).
4. product_property_representation_schema: provides entities that can indicate the properties of one object are based on relations with another object.
5. management_resources_schema: provides a set of abstract supertypes that may be used in the definition of documents, certification, action requests, contract or organization assignments.

6. `document_schema`: five entities for defining documents, in terms of type, identifier, relationships, description.

7. `action_schema`: classifies actions—whether directed or not—and their properties—status, resources, methods, relationships.

8. `certification_schema`: two entities for defining certification type.

9. `approval_schema`: six entities defining approval status, role, providing organization.

10. `security_classification_schema`: two entities for security classification.

11. `contract_schema`: two entities describing contract types.

12. `person_organization_schema`: entities for identifying persons, organizations, their addresses and communication information. Relations of persons to organizations and between organizations can also be defined.

13. `date_time_schema`: the types and entities needed to define time and temporal relationships.

14. `group_schema`: two entities for defining group relationships.

15. `effectivity_schema`: four entities defining effectivity of objects, with a measure and date.

16. `external_reference_schema`: four entities that define external references and their relationship.

17. `support_resource_schema`: three types: identifier, label, text.

18. `measure_schema`.

The first three schemas of Part 041 are shown in Figure 6.1. They provide high-level classifications for all products. All of these schemas are rather short, with five to ten entities each, except for the last one, the `measurement_schema`, which defines in one place all the measures and units to be used in *all* STEP models. These measures and units are summarized in Figure 6.2. The table shows the measure value, the unit and the type of measurement. Only SI (metric) units are included in Part 041. The measurement Entities include constraints to check the units against the measurement type to see that they are consistent. The `measurement_schema` includes definitions for derived, count, descriptive and ratio measures.

MEASURE VALUE	UNITS	UNIT TYPE
length_measure	meter	length_unit
mass_measure	gram	mass_unit
time_measure	second, hertz, becquerel	time_unit
electric_current_measure	ampere, volt, ohm, farad, henry, siemans, weber, tesla, henry, coulomb	electric_current_unit
thermodynamic_temperature_measure	degree celsius, kelvin	thermodynamic_temperature_unit
amount_of_substance_measure	mole	amount_of_substance_unit
luminance_intensity_measure	candala, lumen, lux	luminance_intensity_unit
plane_angle_measure	radian	plane_angle_unit
solid_angle_measure	steradian	solid_angle_unit
area_measure	2 length units	area_unit
volume_measure	3 length units	volume_unit
ratio_measure	2 NUMBERS	ratio_unit
parameter_value		parameter_unit
numeric_measure	weber, tesla	numeric_unit
context-dependent measure		REAL
descriptive_measure		STRING
positive_length_measure	length unit	positive_length_unit
positive_plane_angle_measure	solid angle unit	positive_plane_angle_unit
positive_ratio_measure	2 NUMBERS	positive_ratio_unit
count_measure		NUMBER

Figure 6.2: Measures, units of measurement and unit definitions specified in the measure_schema.

6.3 REPRESENTATION STRUCTURES (PART 043)

In situations where multiple models are to be used together, the relationships among the models must be defined and managed. Part 043 defines high-level abstractions for grouping representations and provides general representational structures. Here, a representation corresponds to a STEP part model. Part 043 is used by and references other Integrated Resources.

The primary contribution of Part 043 is its definition of a framework for representations and several kinds of Rules that are applied to the more specialized entities below it in the inheritance lattice. The overall structure of Part 043 is shown in Figure 6.3. The main Entity is the `representation_item`. It is used throughout the geometry and topology schemas presented later and carries rules that determine the size of its more specialized representations.

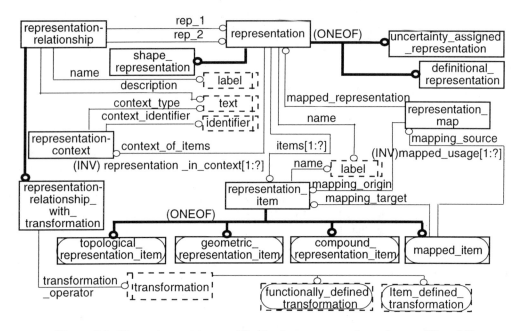

Figure 6.3: The major entities specified in the representation schema of Part 043.

`Representation` allows for a grouping of single `representation_items`, for example, multiple geometric representations of the same object. `Representation_context` reflects the fact that multiple representations may be used in a single context and provides an identifier and description of the context. `Representation_map` provides one way to identify the relation between mapped items. It identifies a `mapped_representation`; the target items are defined in a `mapped_item`, which identifies a set of targets and a representation that carries the origin to be used in the mapped representation. Maps are data exchange processes between two Part schemas. They will be discussed in Chapter 11. `Representation_relationship` provides a general way to define relations between `representations`. `Transformation` has the potential for describing how one representation is mapped into another. `Shape_representation` is a general handle for domain-specific shape representations.

Not visible in the EXPRESS-G diagrams, several functions, RULES and WHERE clauses are included in these entities. Two are worth noting. The `mapped_item` carries a WHERE clause checking that no mapped item is cyclically linked to itself, and a RULE applies to `representation_context` that the dimensions of these entities are internally compatible.

The structures defined in Part 043 provide a global set of handles and references for dealing with representations, especially where more than one representation is used in a model. The entities are used extensively in the geometry and topology schemas described below. The relationship and mapping between models is an issue increasingly recognized as a significant issue in integrating heterogeneous applications. The initial notions in Part 043 suggest that this part may become increasingly important in the future.

6.4 GEOMETRIC AND TOPOLOGICAL REPRESENTATIONS (PART 042)

Since their beginning, CAD systems have been geometric modelers, as described in Chapter Two. Geometric modeling became specialized so that today, the geometric modeling used in electronics, aircraft and automobile manufacturing, civil engineering and architecture share a common subset, but many aspects of the newer modeling techniques are domain specific. That common subset is the target of Part 042. It is based on several generations of research, refining and generalizing geometric models.

Because geometry is a foundation for so much of traditional CAD systems, and because of its central use in building models, we use this review of Part 042 as an opportunity to quickly outline and survey the organization of geometrical representations useful in building modeling. For those knowledgeable in geometrical modeling, this section will be familiar.

6.4.1 APPROACHES TO SHAPE REPRESENTATION

Common to all approaches is that shapes are represented in a space, typically defined as an infinite set of points. The most common space metric is a Cartesian coordinate system. The Cartesian coordinate system may be of one dimension (a vector), two orthogonal dimensions (a plane) or three orthogonal dimensions (a volume). A point is a zero-dimension entity that may be in a one-, two- or three-dimensional space. A line is a one-dimensional entity whose extent is designated by *length* that may exist in a one-, two-, or three-dimensional space. A surface is a two-dimensional entity whose extent is an *area* that may reside in a two- or three-dimensional space. A solid is a three-dimensional entity whose extent is a *volume* that may reside in a three-dimensional space.

Some simple geometrical shapes may be defined directly, but to be useful, most geometrical shapes must be defined indirectly using lower dimensional entities. These lower-level entities are combined in a structure called a *topology*. Thus shape representation involves two classes of definition: *geometry* represents single spatial entities and *topology* deals with the structure of lower-level entities into more complex higher-level ones. In order to distinguish them, topological entities are often given different names than geometrical entities. The geometric entities in Part 042 include *point, vector, curve* and *surface*, for zero-, one-, and two-dimensional entities, respectively. The corresponding topological entities include *vertex, edge, loop* and *face*.

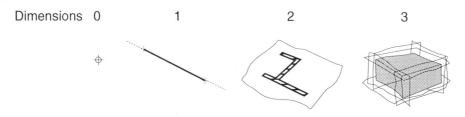

Figure 6.4: The 0-, 1-, 2-, and 3-dimensional shapes defined using boundary representation.

A useful way to conceptualize geometric modeling is to consider three types of geometric generators. All work across multiple dimensions, in one-, two-, and three-dimensional spaces, and are extensively used in various CAD systems.

Probably the most common and general class of shape representations is the class of *boundary representation*. Different boundary representations are characterized in Figure 6.4. A vertex (topological Entity) is defined by a point (geometrical Entity), which being of zero-dimension, is unbounded. An edge is closed or infinite curve bounded by two points. For example, an arc can be defined as a directional circle with a starting point and ending point. The two points partition off the segment of the curve of interest. A face is a surface bounded by a sequence of closed, connected lines. A solid is a three-dimensional space bounded by a set of surfaces. Because of the detailed control allowed in defining the bounding entities, the boundary representations can be used in a wide number of form-defining contexts. These include cases when a line may be partitioned into multiple (disjoint) segments or a space may be partitioned into a hollow solid or even multiple disjoint solids.

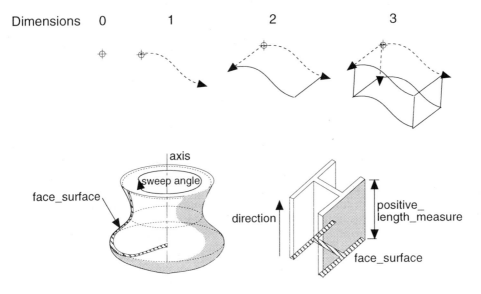

Figure 6.5: The sweep representations, generated by sweeping an entity along a bounded path. The entity may be zero-, one-, or two-dimensional.

A second, easily defined class of shape representations is the class of *sweep representations*. Sweeps are defined by an entity and a bounded path. These are characterized in Figure 6.5, both abstractly (above) and with common examples (below). Again, a zero-dimensional point has no sweep. A line is a sweep of a point and a surface is a sweep of an edge. A solid is a sweep of a bounded surface or face. Notice that all the entities must be bounded. Sweeps are specialized into two subtypes: the *revolved bounded_surface* and the *extruded bounded_surface*. The revolved shape is created by rotating the face about an axis, from zero to 360 degrees. The extruded shape sweeps the bounded_surface in a direction for some length. Sweeps are among the most frequently used general shape generators.

The third class of shapes is the class of *constructed binary shapes*. These shapes are shown in Figure 6.6. They are defined by combining simple geometric entities using the spatial *union*, *intersection* or *difference* operators, called the Boolean operations. Cases for the different spatial dimensions are shown. The zero-dimensional shapes simply add or subtract points. When combined, the one-dimensional edges are merged or trimmed. The faces and solids are similarly combined and trimmed. The three-dimensional case is called *constructive solid geometry* (CSG).

The main use of constructed binary shapes is for defining and editing shapes in CSG. CSG can be utilized in two alternative ways. Given a set of primitive shapes, such as box, cylinder, wedge and sphere, and the Boolean operations, CSG supports definition of arbitrarily boundary representation shapes that greatly extends the range of easily generated shapes. That is, CSG is used to sculpt shapes by union, intersection and difference. The shape is the set of faces that define a volume enclosing solid. This is the most common use of CSG.

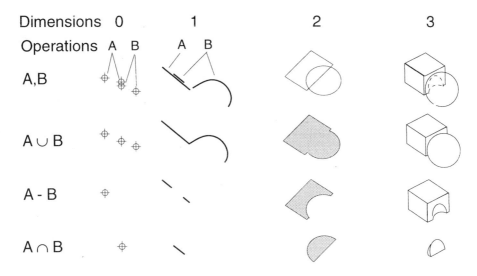

Figure 6.6: The constructive shapes, defined by union, difference and intersection.

The other way to use CSG is to represent the base shapes, the primitives or sweeps, parametrically as a creation operation with parameters, transformed to their relative locations. The base shapes are combined with the CSG operations, but rather than executing the Boolean operations, they are carried as an algebraic expression. An example was presented in Figures 2.6 and 2.7. The algebraic representation is very compact, taking much less space than the full boundary representation constructed from it. It can be executed at any time to generate a full boundary representation. The advantage of the unexecuted CSG representation is that it allows easy editing of the shapes. For example, a shape corresponding to the hole in a wall may be moved, changing the location of the window cut into a wall. A shape's bounded surface may be edited, changing the extent or width of a wall. There are great advantages of carrying *both* CSG approaches. They are, respectively, called (1) the *unevaluated* or *implicit* form, and (2) the *evaluated* boundary representation form. In three dimensions, the two combined CSG representations provide powerful sculpting operations that can be combined in shape-defining algorithms. While CSG modeling is covered in Part 042, recent extensions, called parametric modeling, are not.

Boundary, sweep and CSG representations are often combined to produce comprehensive shape modeling systems in CAD systems and other application programs. Given this quick review, we turn to the geometrical and topological representations provided in Part 042.

6.4.2 TOP LEVEL PART 042 GROUPING OF REPRESENTATIONS

Part 042 groups entities into three related but distinct schemas: geometry, topology and geometric models, as shown in Figure 6.7. *Geometry* defines points, curves and surfaces. Topology is the branch of mathematics that deals with connections and relations. It grew out of the study of graphs, knots and the structure of certain geometric shapes. A *geometric model* is a composition of both geometry and topology. It consists of a set of geometric representations bounded using topological representations and then possibly combined to produce complex shapes.

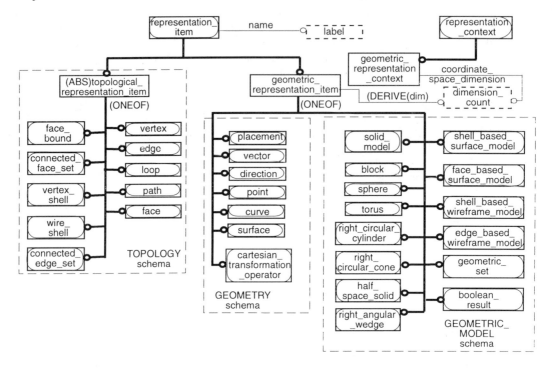

Figure 6.7: Top level structure of Part 042, geometric and topological representations. They are grouped into three schemas: geometry, topology and geometric models.

At the top level, Part 042 defines both geometrical and topological schemas, as `topological_representation_item` and `geometrical_representation_item`. Both are abstract types and subtypes of `representation_item`. `Geometrical_representation_item` includes a DERIVE clause that determines the dimensionality of the instance, carried in `dimension_count`.

The major entities included in Part 042 are presented and defined in the following figures. Some details, however, are omitted. Readers interested in the actual specifications should consult the complete specification.

6.4.3 GEOMETRY SCHEMA

The definitions of Part 042 have been derived from many years of experience implementing geometric modelers in CAD systems. In the presentation here, various low-level geometric entities are reviewed first and considered according to the classes to which they belong. These are then combined to define higher-level entities.

The geometry schema consists of the geometric properties of placement, vector and direction, the transform operator, and the geometric entities of point, curve and surface.

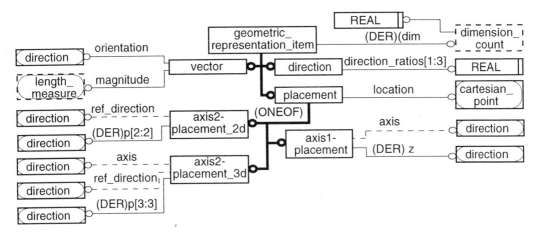

Figure 6.8: The inheritance lattice for geometric placement and orientation.

6.4.3.1 Geometric Properties

Before defining geometric entities, some general geometric properties are defined first, so as to allow their inherited use in the definitions that follow. These properties are diagrammed in EXPRESS-G in Figure 6.8 and shown graphically in Figure 6.9. The geometric properties include direction, represented as a *unit vector*. A unit vector is made up of the three cosines of a line in space from some origin, one unit in length in a particular direction. A general vector is also defined with an explicit magnitude, which may be other than unit length. A rule is enforced that the magnitude must be greater or equal to zero. Direction has no starting point. Placement consists of a cartesian_point defined by three coordinate values. Together, they define a location and direction, or alternatively a coordinate system origin.

Placement has three subtypes. Axis1_placement adds one optional direction that allows derivation of all three axes, and derives the z direction. A WHERE clause restricts the use of axis1_placement to three-dimensional space. Axis2_placement_2d consists of a reference direction and derives two axes in p defining the x- and y- directions (the notation [2:2] indicates a filled list of length 2). Axis2_placement_2d is two-dimensional. Axis2_placement_3d adds an optional reference direction and second axis from which are derived three axes in p. It is three-dimensional. All the specializations of placement have both a starting point and one or more directions.

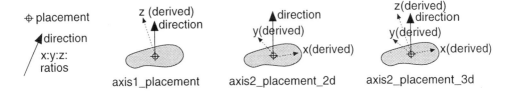

Figure 6.9: Characterization of the geometric placement and orientation structures. Derived axes are shown as dotted lines.

Through inheritance, each of these entities has a dimension_count attribute. These properties allow any piece of geometry to be placed in space and two or three axes to be defined for the local coordinate system in which the geometry is located.

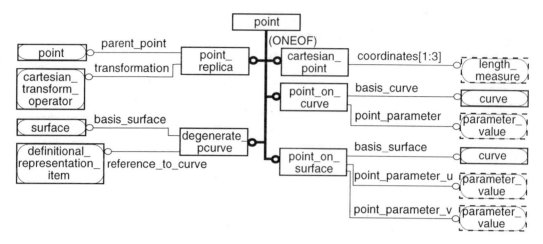

6.10: The definition of points in Part 042 and its five specializations.

6.4.3.2 Point

Point is a specific location in one-, two- or three-dimensional space. Its entities are shown in Figure 6.10. In Part 042, they are defined first as a general abstract point with five specialized subtypes, covering the subtypes needed within the Part 042 schema. Cartesian_point is defined as three orthogonal length measures, defining a point in 3D space. As specified, at least one length must have a value. Point_on_curve is defined as a reference to any type of curve and a parameter specifying its length position along the curve. Point_on_surface references a surface and u and v parameters defining a two-dimensional placement on the surface. Point_replica is a reference to a cartesian_transform_operator that transforms another point to some location; the reference point may be any subtype of point. Degenerate_pcurve is any geometric Entity that degenerates into a point; it references a surface and a definitional_representation.

6.4.3.3 Cartesian Transform Operator

The cartesian_transform_operator is a subtype of geometric_represent-ation_item, inheriting its dimension_count. It is defined by a local origin, specified by a cartesian_point, and optionally two direction axes, as shown in Figure 6.11. There is also an optional scale and derived scale.

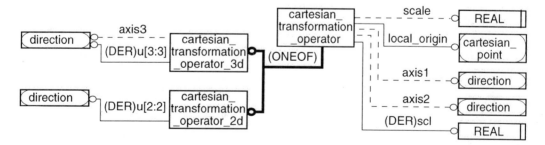

Figure 6.11: The Cartesian transform and its 2D and 3D specializations.

The two specializations carry a DERIVED u unit vector that orders the axes in 2D or 3D. The cartesian_transformation_operator_3d carries a third axis for three-dimensional space. These definitions correspond to one of several ways used in current CAD systems to represent location, orientation, and, optionally, scale. Notice that all the transformation entities are global. At this level, they are not able to refer to another transform to which they are relative. This can be represented at a higher level.

6.4.3.4 Curves

Curve is a generalization of all one-dimensional continuous entities. A straight line is a special case of curve. There are many forms of "curved" curves, which are important in many areas of product design. Curves are used not only to make one-dimensional forms, but also as parameters for many forms of surfaces. With curves, we begin to see how the lower-level entities are used to define higher-level ones. Curve includes all single direction entities in two- or three-dimensional space. It is an abstract type used to reference others and is specialized into eight subtypes, several of which have further subtypes. Curve and its eight subtypes are presented in Figure 6.12.

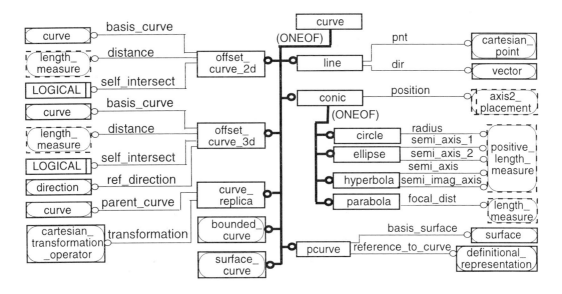

Figure 6.12: The EXPRESS_G definition of curve and its eight subtypes.

The simplest curve is a line. Line is defined as a cartesian_point and a vector. The vector passes through the point and the line is unbounded. A rule requires the point and vector to be of the same dimensionality. The set of conics includes circle, ellipse, hyperbola and parabola. The conic carries an initial definition of an axis2d_placement, which is a select type, defined as either axis2_placement_2d or axis2_placement_3d. From this a circle is defined as a single distance—its radius—and an ellipse is defined as the two distances, assumed to be the two intercepts along the two defined axes. The parabola is defined as a focal_dist measure along the axis that is the parabola's focus. The hyperbola is defined by two distances along the first axis, defining the two foci of the hyperbola. A pcurve (parametric curve) is defined as a surface and a definitional_ representation (see Figure 6.3).

Curve also includes an offset_curve-2d, defined as a basis_curve and offset distance; it also carries a flag if the offset curve includes self-intersections. An offset_ curve_3d is also defined, which carries the attributes of the 2d offset, plus a ref_ direction. There is also a curve_replica, similar to the point_replica, carrying a `parent_curve` and `cartesian_transformation_operator`. `Bounded_curve` and `surface_curve` are described below.

A major subclass of curves in Part 042 is the class of `bounded_curves`. Again, `bounded_curve` is an abstract type only used as a holder for its subtypes. It includes all classes of curve that are bounded (or trimmed) and also bicubic spline curves. `Bounded_curve` also includes `composite_curve`, as well as `polylines`, which are composed of straight lines. The various types are shown in Figure 6.13.

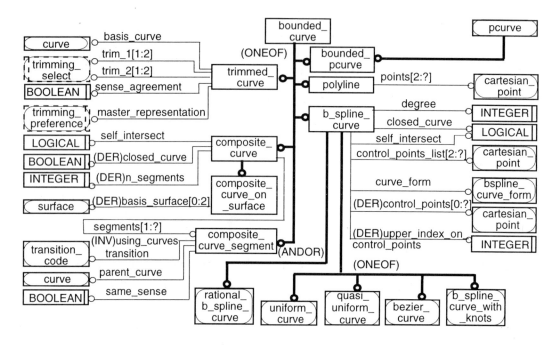

Figure 6.13: The definition of bounded curves in Part 042.

`Polyline` is a sequence of two or more connected points. `Trimmed_curve` is a curve with two sets of trimming parameters. `Trimming_preference` identifies the parameter type, which is carried in two pairs in `trimming_select`, which is either a `cartesian_point` or `parameter_value`. These are used to identify the boundaries of up to two segments of the trimmed curve. A `sense_agreement` flag identifies if the direction of the trim entities is in agreement with the direction of the `curve`. `Composite_curve` is a sequence of `composite _curve_segments`; it derives the number of segments and whether the segments are closed. Currently, the flag for self-intersections is not `DERIVED`. `Composite_curve_on_surface` is a subtype of `composite_curve` that derives its zero to two `basis_surfaces`. `Composite_curve_segment` is one element of a `composite_curve`. It has a WHERE clause checking that it may be any type of `bounded_curve` other than a `composite_curve`, `pcurve` or `surface_curve`. The `transition_code` attribute is a flag identifying whether the segment is `discontinuous` or `continuous`, and the degree of continuity.

`B_spline_curve` is the class of well-known blending functions of the same name that apply over a sequence of control points. Each b-spline function is based on a polynomial of a certain order m. The polynomial will be of degree $m-1$. Most common b-splines are of order 2 (quadratic) or 3 (cubic). The general form of blending function is

$$\mathbf{p}(t) \;=\; \sum_{k=0}^{L} \mathbf{p}_k N_{k,m}(t)$$

where:

$\mathbf{p}_n$ is the nth control point

m is the order of the B-spline functions

$\mathbf{T} = (t_0, t_1, t_2, \ldots)$ is the knot vector

The Part 042 representation of the b-spline curve is a sequence of control points, the degree, flags denoting whether the curve is closed or self-intersecting. It also carries a `curve_form` flag, denoting whether the curve is of any of the special types: (`polyline_form`, `circular_arc`, `elliptical_arc`, `parabolic_arc`, `hyperbolic_arc`, or `unspecified`). From these attributes are derived the `control_points` array and the `upper_index_on_control_points`. This general form is specialized into four subtypes, which may be rational or not (these are not detailed in Figure 6.13).

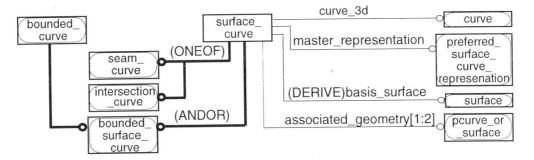

Figure 6.14: The surface curves in Part 042.

Another curve class is the `surface_curve`, shown in Figure 6.14. These are curves on a reference surface or curves defined by the intersection of two surfaces. The general `surface_curve` is defined as any type of `curve`, the one or more `surfaces` or `pcurves` it lies on and a DERIVED surface. Because such curves cannot always be analytically defined, the `preferred_surface_curve_representation` provides a choice of the approximate representation. Two specializations are defined, where `intersection_curve` checks that there are two different surfaces in `associated_geometry` and `seam_curve` checks that the two surfaces are the same, representing a self-intersection curve. Another form of `surface_curve` is `bounded_surface_curve`. It identifies one or more other surfaces that define the bounds of the initial `surface_curve`.

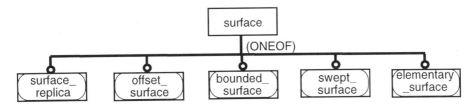

Figure 6.15: The Part 042 subschema for surfaces.

6.4.3.5 Surface

The last major class of purely geometric definitions comprises surfaces, which are major building blocks for defining three-dimensional shapes. The top-level class hierarchy for surfaces is shown in Figure 6.15. Surface is an abstract Entity class used for grouping the others.

Planes and simple self-closing surfaces are defined as elementary_surfaces, shown in Figure 6.16. Since elementary_surface is defined by an axis2_placement_3d, which defines all three axes, this is sufficient for defining a plane (assumed to be unbounded and in the X-Y plane). Similarly, an axis and a radius define a cylindrical_surface (unbounded) and a location and a distance define a spherical_surface. The toroidal_surface is defined on the X-Y plane about a location with a major_axis and minor_axis. A specialization of a toroidal_surface is used to represent self-intersecting toroids, e.g., where the major_axis ≤ minor_axis. Conical_surface is defined with a radius and angle. By placing all elementary_surfaces the same way, it is easy to replace one with another.

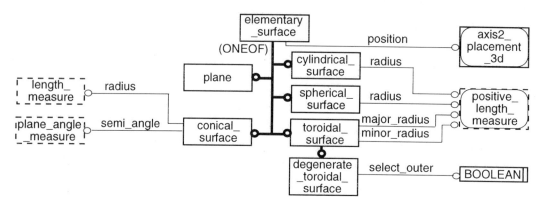

Figure 6.16: The subschema for elementary surfaces.

Part 042 provides two swept surfaces, shown in Figure 6.17. These are the surface_of_linear_extrusion and the surface_of_revolution. The general swept_surface is defined as a swept_curve, which is general and may be open or closed. The linear extrusion specifies the vector of the extrusion (again, unbounded). The surface_of_revolution relies on an axis1_placement, which defines an axis location. It also has a derived line that marks the axis. Also shown in Figure 6.17 is offset_surface. It constructs a surface at a distance from a reference basis_surface, where a positive value places the new surface above the basis surface and a negative value places it below.

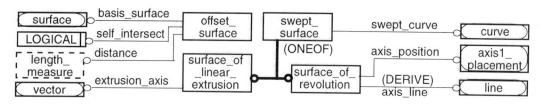

Figure 6.17 The subschema for swept and offset surfaces.

It should be noted that some forms of extrusion frequently used in the AEC area are not covered here: extrusion along a complex polyline composed of lines and arcs, such as needed for welded piping; extrusions between two cross-sections, such as needed for ductwork; and spiral extrusions, as encountered in the handrails of spiral stairs. These examples are depicted graphically in Figure 6.18.

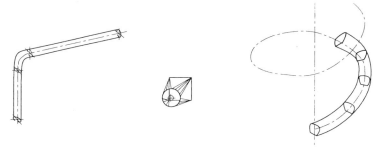

Figure 6.18: Some extruded shapes frequently encountered in AEC products that are not easily representable using the swept surfaces provided in Part 042.

Bounded_surface, shown in Figure 6.19, defines all the surfaces that are trimmed. There are four types. Also shown are the subtypes of b_spline_surface. The rectangular_trimmed_surface is defined as a basis_surface with two u and two v parameters bounding it on four sides. It has a rectangular boundary. Also defined is a rectangular_composite_surface, which is a composition of surface_patches, each with a sense and transition in both directions. Curve_bounded_surface is a basis_surface with one or more boundary_curves bounding it (multiple boundary_curves depict a bounded surface with holes).

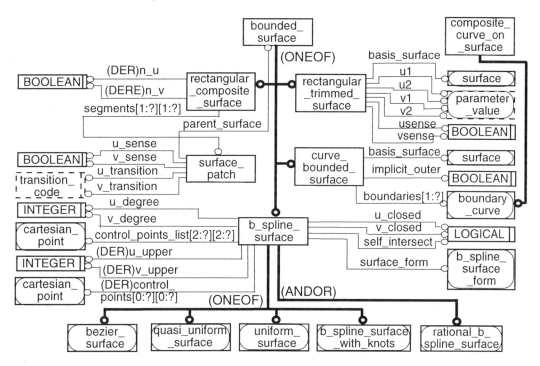

Figure 6.19: The Part 042 subschema for b-spline surfaces.

B_spline_surface is defined at the general level by a set of control_point_list, u and v degrees and whether the surface is closed along the u or v axes. From these are derived the upper u and v and a new set of control points. The type of b_spline `surface_form` is carried as an enumerated type. Four subtypes of `b_spline_surfaces` are defined (not detailed here).

6.4.3.6 Discussion

This is an elegantly defined set of Entity definitions that have taken into account many issues of shared high-level objects with specialized subtypes. Many of the definitional weaknesses encountered in IGES and other earlier standards efforts have been resolved.

The geometric entities have been developed for general use, with emphasis placed on certain kinds of curved surfaces, without close attention to the needs of the AEC industries. The limitations in the definition of `swept_surfaces` have already been noted. Another common use is the definition of centerlines made up of lines and arcs, which are used in numerical control bending and cutting of piping and tubing. Of course, these omissions can be handled in additional definitions in AEC-specific parts.

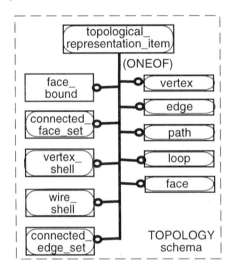

Figure 6.20: Overview of the topology schema.

6.4.4 TOPOLOGY

The topology schema provides structures for composing geometric entities into more complex shapes.

A portion of Figure 6.7 is reproduced in Figure 6.20 above, showing the component entities in the topology schema of Part 042. Some of the less intuitive entities are illustrated in Figure 6.21. *Vertices* are the topological equivalent of `cartesian_point`. They bound a curve to define a topological *edge*. Edges are sequentially connected and closed to form a *loop*. There are several special cases of loop. One or more loops bound a surface to define a topological *face*. A face may have an outer loop and multiple inner loops, as shown in the left part of Figure 6.21. Edges typically have a direction and this direction is often fixed with regard to the orientation of the face. Most often, the bounded part of the face will be on the right-hand side of the directed edge, when looked at from outside. A special case in the bounding of a face is when a loop is reduced to a single vertex, such as at the top of a cone, shown in the right of Figure 6.21.

A connected set of faces that enclose a volume forms a *shell*. One or more shells define a solid shape. Like loops on a face, a shape has one outer shell and possibly multiple inner shells, the inner shells forming hollows in the shape. Notice that two edges bounding adjacent faces coincide in forming the face connection. Also, a face may self-intersect, creating an edge seam, as shown at the right in Figure 6.21. Topological structures are combined with geometrical structures to define full solids in the next section. Here, we deal with the topological relations.

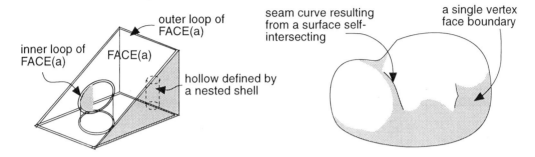

Figure 6.21: Some examples of special topological features: inner and outer loops, nested shells, faces bounded by an edge and a vertex.

Topological entities are needed to describe many of the same entities defined geometrically: topological vertices correspond to points, topological edges correspond to a bounded curve, topological faces correspond to bounded surfaces. Other topological entities are needed for sequences of edges and sets of faces. They are described below, in bottom-up order.

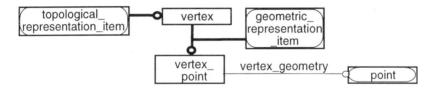

Figure 6.22: Model of the topological vertex.

Vertex is an abstract topological Entity. It is inherited, along with the general geometric_ representation_item, into vertex_point. Vertex_point also references a (geometrical) point. It has no additional structure. Later, we will see that many other topological entities reference vertex.

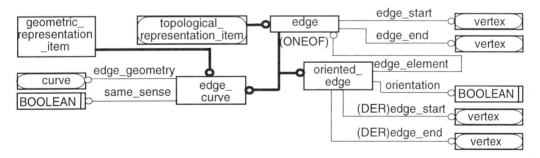

Figure 6.23: EXPRESS-G model of the topological edge and its specialization edge_curve.

Edge is also an abstract topological Entity, shown in Figure 6.23. It is defined as a curve (of any type) bounded by two vertexes. The two vertexes are specified at the general edge level and subtyped into two edges: edge_curve and oriented_edge. Edge_curve defines the curve (any subtype of curve) that the two vertexes bound. Several of the curve types in

Figure 6.12 have a direction. A same_sense flag tells whether or not the curve direction and the start and end vertexes are consistent directionally. An example of such a curve is shown in Figure 6.24. Edge_curve is a subtype of geometric_representation_item, inheriting a dimension_count attribute. Oriented_edge references an edge (certainly an edge_curve) and adds an orientation flag. Given the orientation, it derives which is the start and which is the end vertex.

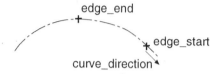

Figure 6.24: Geometric example of an oriented edge, with start and end vertexes ordered inconsistently with the edge direction.

The loop and path topological entities are similar to the geometrical curves that have a flag designating they may be a closed_curve (see Figure 6.13), but are used to bound a surface to define a face. They offer a limited number of representations. Loop is always closed and may be defined either in poly_loop by a sequence of cartesian_points or in loop_edge as a sequence of oriented_edges, inherited through path.

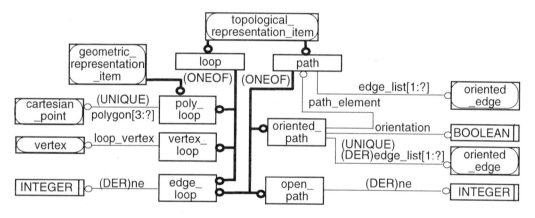

Figure 6.25: Model for the topological loop and path entities.

The derived value ne identifies the number of edges in the list. A special case is defined by a discontinuity within a face defined by a single vertex, as discussed earlier. (An example of such a case is a single face bounding a sphere, defined topologically as a surface bounded by a single vertex as shown on the right in Figure 6.26. See also Figure 6.21.) Path is a sequence of oriented_edges. The path may be closed (an edge_loop) or open. In both cases, the cardinality of the edge_list is DERIVED. The path also may be oriented.

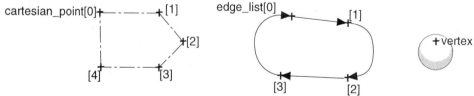

Figure 6.26: The three different types of topological loops. The first is defined by a sequence of Cartesian points, the second by a sequence of edges and the last by a single vertex.

A major topological structure is `face` and two associated structures, `face_bound` and `connected_face_set`. This structure is shown in EXPRESS-G format in Figure 6.27. An interpretation of these structures is presented in Figure 6.28. A `face` has an attribute `face_bound`, which provides one or more bounding loops. Face has three subtypes: `face_surface` references a `surface` and a flag that defines whether edge orientation and face orientation are consistent; `sub_face` allows one face to be made up of other (sub)faces; `oriented_face` adds an orientation flag to any of the face types. Face_bound has a subtype `face_outer_bound`, denoting that the `face_bound` is an outside boundary (see Figure 6.21); only one `face_outer_bound` is allowed per `face`. Face_bound is defined by a `loop` and an `orientation` flag.

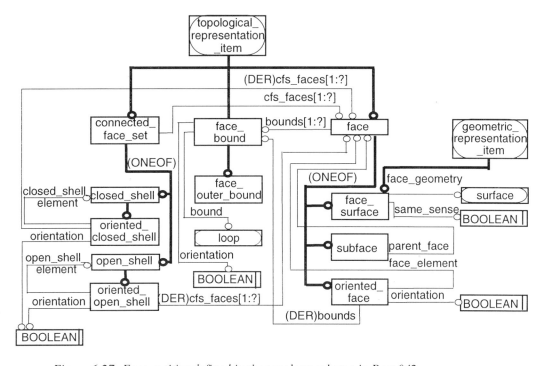

Figure 6.27: Face entities defined in the topology schema in Part 042.

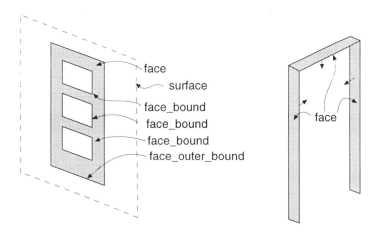

Figure 6.28: A face, with one face_outer_bound and three face_bounds, indicating inner bounds. An oriented_open_shell defining a door jamb is also shown.

`Connected_face_set` is a set of faces that are connected edgewise. When all faces in a set are connected pairwise to edges and have a consistent orientation, the set is considered *closed*, meaning that it completely partitions space into two disjoint regions. `Closed_shell` is the fundamental structure defining an evaluated solid model, also called a `boundary_` `representation`. `Connected_face_set` allows definition of a collection of not closed `faces`, used in mechanical engineering to identify a *feature*, a subset of geometry having specific function or semantics, such as a pocket or groove. `Connected_face_set` has two subtypes: `open_shell` and `closed_shell`. Either can be oriented.

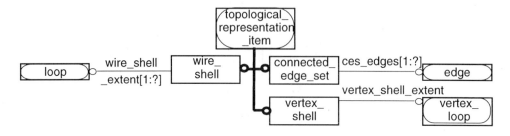

Figure 6.29: Additional topological entities defined in Part 042.

Some additional, special case topological structures are defined in Figure 6.29. These include a `vertex_shell`, where the complete topological structure of a shell is a single `vertex`. Also included is `wire_shell`, defined as a set of `loops`, and a `connected_edge_set`, corresponding to a set of `edges` that may not close into one or more `loops` (also used to define features).

6.4.5 GEOMETRIC MODELS

The Geometric Model Schema in Part 042 represents geometries connected by topologies, with emphasis on various forms of solid modeling—on shapes that enclose a volume. In order to understand these models, we present a bit more background.

In general, current generation CAD and engineering design systems rely on a variety of methods of solid shape form generation, which can often be mixed together. The constructive approach, introduced in section 6.4.1, relies on a predefined set of parametric primitive shapes such as those defined in Figure 6.30, which are combined using the Boolean operations of union, intersection and difference. The sweep approach relies on sweep operations—for example, sweeping a plane along a line or arc or circle as shown in Figures 6.5 and 6.33. The third way is to incrementally build up a shape by composing a set of bounded surfaces together until they are closed, suggested in the left portion of Figure 6.21. The Part 042 Geometric Model Schema fully supports the first two of these methods, with partial support for the third. It supports a fairly broad set of primitives and the two most widely implemented sweep shapes: linear extrusions and shape of rotation.

A fundamental capability of all solid modelers is the implementation of the Boolean operators of union, intersection and difference, taking two or more solid shapes as arguments and returning a solid that is the product of the operation. The result of most implementations is a *manifold* solid model. A manifold solid model is one that is well formed, without extra faces, edges or vertices. Such a model satisfies the extended version of Euler's Law:

$$F - E + V - L = 2(S - H)$$

    where:

         $F ::=$ *number of faces bounding the shape*
         $E ::=$ *number of edges bounding faces in the shape*

$V ::=$ *number of vertices bounding edges in the shape*
$L ::=$ *number of interior loops in the faces bounding the shape*
$S ::=$ *number of shells bounding the shape*
$H ::=$ *number of holes through the shape*

NOTE: The edges E in the above equation are counted as the connected edges in a polyhedron, not the edges bounding faces as described in Section 6.4.1. When considered according to Euler's Law, the edges bounding a face are considered "half-edges" that are matched with their coincident edge to define an Euler edge.

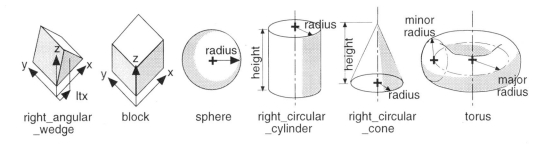

Figure 6.30: The six primitive solid shapes provided in Part 042.

There has been much research on geometric modelers and the Boolean operators. It has been found useful to extend the operators to deal with shapes that are not manifolds obeying Euler's Law, but that carry "extra" elements, such as centerlines, connection-points, construction lines, and so forth. Such shapes are called *non-manifold* shapes. These can be supported in Part 042 by composing solids with other geometric entities (though the Boolean operations do nothing with them).

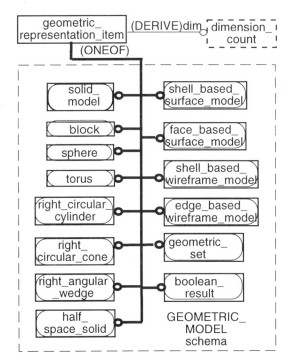

Figure 6.31: The top-level EXPRESS-G structure for a geometrical model.

The geometric model schema of Part 042 applies the distinction, introduced earlier, between an "evaluated" model and the "unevaluated" model. In the evaluated model, all the Boolean operations have been executed, resulting in zero or more closed shells that form the resulting shape. The evaluated shape is a boundary representation or *BRep*. The "unevaluated" model is defined as a sequence of Boolean operations on primitive or swept-generated shapes, specified by their input parameters and the Boolean operators that combine them, represented as an algebraic expression (or equivalently, an expression tree). With this background, we turn to the geometric model schema of Part 042.

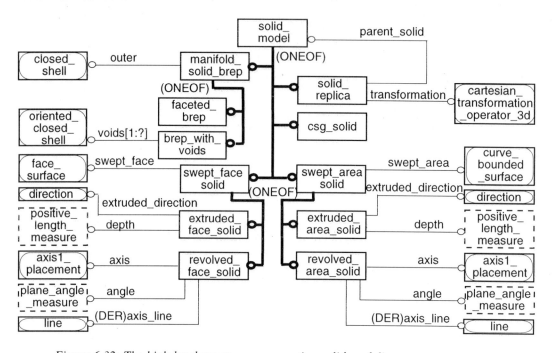

Figure 6.32: The high-level structure representing solid modeling.

The geometric modeling schema portion of Part 042, presented earlier in Figure 6.7, is repeated in Figure 6.31. The left side shows the basic `solid_model` type followed by a set of primitive shapes. The shapes are depicted in Figure 6.30. The bottom left `half_space_solid` and the right-hand side types are all various special conditions felt to be needed to support solid modeling and feature-based modeling. All the types are specializations of `geometric_representation`.

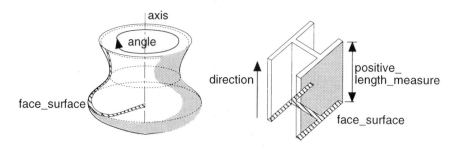

Figure 6.33: The revolved shape of rotation and linear extrusion solids and their corresponding parameters.

The solid modeling subschema is represented in Figure 6.32. It defines an abstract class `solid_model` and five subtypes below it. `Manifold_solid_brep` is the most common type of solid model, consisting of a `closed_shell` in its most general definition. (`Closed_shell` is described in Figure 6.27.) `Faceted_brep` has no additional definition, and is assumed to be restricted to planar faces. `Brep_with_voids` carries multiple shells representing hollows inside the one outer shell. `Solid_replica` is similar to a symbol, creating an instance of any type of solid by referencing the solid and associating it with a transform.

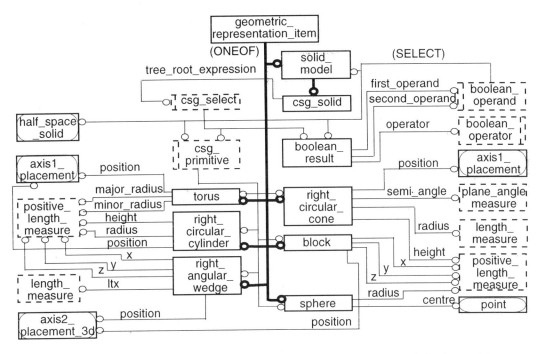

Figure 6.34: The structure for representing the unevaluated csg_solid, along with the various solid primitives.

The swept shapes are of two alternative types, based on what form of closed surface is swept: `curve_bounded_surface` (a geometrical Entity) or `face_surface` (a topological Entity). `Swept_face_solid` and `swept_area_solid` are the same except for the Entity being swept. Both may be extruded through a linear distance, defined by a direction and length, or swept through an arc or circle, defined by an axis and angle. Examples are shown in Figure 6.33. The revolved shapes may optionally derive a line denoting the axis of rotation.

The `csg_solid` model, shown in Figure 6.34, represents an unevaluated (implicit) solid model. The `csg_solid` type is abstract, only referencing the types below it. It has one attribute, `tree_root_expression`, whose value type is a SELECT type in EXPRESS, meaning it can be one of several types, in this case either a `csg_primitive` or a `boolean_result`. A `csg_primitive` is another union type that can be any of the CSG primitives below it. Each of these primitives has a position or placement plus the necessary parameters to size them. Most parameters are `positive_length_measures`. The `right_angular_wedge`, however, has as its `ltx` measure a `length_measure` (that can be positive or negative), so that the distance to the apex of the wedge may be positive or negative, allowing the wedge angle to be acute or obtuse. (Notice these are different Entity types from those similar surface entities described in Figure 6.16.) The `boolean_result` represents one Boolean operation and its two operands. `Boolean_operand` is another SELECT type, which may have another `boolean_result`, `csg_primitive`, `solid_model` or `half_space_solid` as the type of its value.

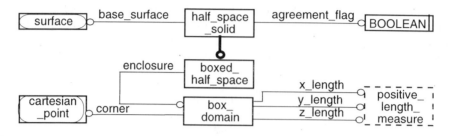

Figure 6.35: The half_space_solid representation, used to partition shapes by interposing a surface.

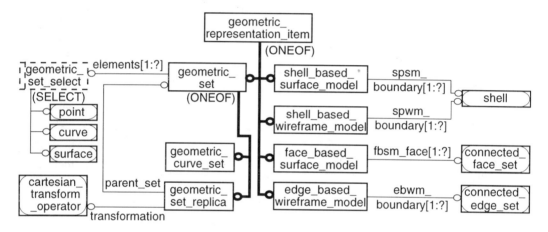

Figure 6.36: Representations used for special case solids.

The `half_space_solid`, presented in Figure 6.35, is used to cut away a solid by subtraction. It consists of a `surface` (which has an orientation), which partitions space into the side of the surface that is solid and the other that is "outside." An `agreement` flag identifies whether the orientation of the surface is consistent with the inside of the solid being defined. If not, the other half space is used. The `half_space_solid` can be specialized so that only part of the surface inside a box is applicable. Because the bounding surface need not be a plane, the half space may cut out cylinders, cones and spheres.

Last, we identify some "special" representations, added to support the needs of some particular applications. These are shown in Figure 6.36. They allow various collections of geometry, shells, faces or edges to be modeled. Together, the `manifold_solid_brep`, the `csg_solid`, the various primitives and the `half_space_solid` provide a basic set of solid modeling representations that cover a range of cases needed in building design and construction.

6.5 APPLICATION TO TEST EXAMPLES

The arc and plane surface example representations can be directly represented using Part 042. The arc is represented as a `trimmed_curve`, where the curve is a `circle`. The arc class would be defined as shown in Figure 6.37. Conic provides an `axis2_placement` (a type which may be either `axis2_placement_2d` or `axis2_placement_3d`). This defines a plane and location on it in a two- or three-dimensional space. `Circle` provides the radius, as a `positive_length_measure`. The `trimmed_curve` provides `trimming_select` values (one or two points or parameters) that define the boundaries of the arc. It also provides a `sense_agreement` flag, which identifies (in this case) whether the arc is from `trim_1` to `trim_2` or the reverse. This model would be mapped onto a file or database format, into which arc instances could be mapped. The arc also could be defined in terms of a b_spline curve.

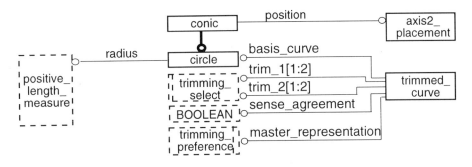

Figure 6.37: The data model for representing an arc, using Part 042 as a basis.

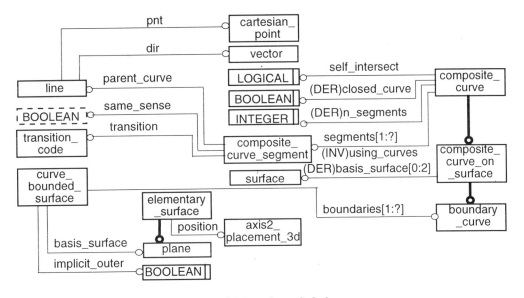

Figure 6.38: The EXPRESS-G model for a bounded plane.

The bounded plane example would be defined similarly. It is shown in EXPRESS-G in Figure 6.38 as a `curve_bounded_surface` (shown on the left of the figure). It references a `surface`, in this case an `elementary_surface` of the type `plane`. The `elementary_surface` provides an `axis2_placement_3d` Entity that is sufficient for locating the plane (a point and two reference vectors, see Figure 6.9). The surface is bounded by the sequence of `boundary` attributes in the `curve_bounded_surface`. `Boundary_curve` has inherited many attributes; the important ones for this example are the `composite_curve_segment`, `n_segments` (DERIVED) and `closed_curve` (DERIVED). The `composite_curve_segment`'s `parent_curve` attribute references a `curve`, in this case a `line`. Each line has a beginning `cartesian_point` and a direction `vector` (that explicitly defines any interpolated point along the line). In sequence, these will define the bounding of the surface. The number of segments in the curve is derived along with whether the curve is closed. Notice that the same_sense flag in the `composite_curve_segments` allows a line's vector to be in the reverse direction. The `transition_code` allows the intersections to be smoothed. Alternatively, the planar face could be represented as a single face in a `face_based_ surface_model`, depicted in Figure 6.36.

6.6 DISCUSSION

Part 042 provides a relatively complete set of the basic geometric representations found in most current CAD systems. The representations are generally sufficient for defining complex entities needed for modeling buildings and construction. A few omissions have been noted. In general, well-developed data exchange facilities for Part 042 alone could replace those that have been developed for IGES. However, Part 042 is simpler, cleaner, and the actual exchange files would take much less space

The style of EXPRESS models is complex, based on their heavy reliance on inheritance within hierarchies of Entity classes. In IGES, either a type was found that matched (or could be used to approximate) the desired entity, or translation was not possible. In EXPRESS, there is a rich vocabulary of Entity types that can be composed for almost any purpose. The two relatively simple examples are complex to understand, and these are for simple objects. Because of this complexity, a variety of parsing, display and mapping tools have been developed, which developers must rely on for the effective use of STEP models. The reliability of these tools and the completeness of translators to and from applications have become important issues.

This part standard is relatively new and will receive incremental refinements over time. Its well-conceived structure should enable it to provide the base-level geometry for data exchange efforts for the foreseeable future.

6.7 GEOMETRY NEEDS FOR BUILDING MODELING

Given such a general "tool-kit" of geometrical entities, it becomes difficult to sort out which portions are needed for building modeling—that is, what representations should be carried in building and building component models. Some efforts have avoided making any selection of the Part 042 entities and left them for later work, while others have made selections. Here, a few basic guidelines are offered, based on common sense reasoning.

There are several different aspects of the building that need to be modeled, each requiring different geometries and thus requiring separate consideration. First and most obvious are the building components, of both major composed forms, such as walls and roofing systems, and of single fabricated elements, such as windows, cabinets, plumbing fixtures, wallboard and piping. Second are the more abstract geometrical concepts that designers use for developing early models of buildings, which specify the forms that the individual components eventually comprise. Last are building spaces, which are often derived shapes, defined by the components that bound them. We will consider them in this order.

6.7.1 BUILDING COMPONENTS

Two broad classes of materials are used in fabrication: preshaped sheets and cross-sections, and plastic material formed for the specific piece.

Wood and metal are generally preformed, consisting of sheets (metal sheets, plywood, chipboard panels, gypsum board) or stock made with a fixed cross-section (wood cut sizes, wide flange and other steel sections, piping). Elements cut from sheet material are generally defined as polygons (possibly nested to define hollow shapes) cut orthogonally to the surface of the sheet. Later, non-orthogonal or other cuts may be made at the detail level. Elements cut from stock of fixed cross section are typically rough cut to length, then later detailed by making additional cuts, notches and holes. In both of these cases, the material is well depicted using sweeps. For the sheet material, the sweep section is the polygon cut from the sheet, with an extrusion distance being the thickness of the material. For the constant section material, the section is the sweep section and the length corresponds to the cut length. Most detailing of these materials is not supported by a sweep representation. Instead, the sweep-generated solid is stored as a boundary representation and constructive operations are used to subtract pieces from the shape, or if welding or bonding is used, pieces are added to the shape using union operator. Thus as we move from raw materials to detailed construction elements, three different three-dimensional representations are needed: sweeps, unevaluated CSG models and last, evaluated CSG models. The base shape is a swept solid and constructive operations are used to detail it. Constructive operations are especially important at the fabrication or shop drawing level. This sequence is shown in Figure 6.39 for both sheet and constant cross-section shapes.

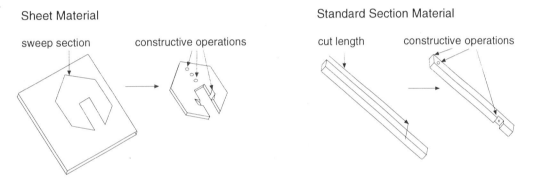

Figure 6.39: Examples of reshaping of material stock, first as rough cut, then as detailed elements.

Whether the CSG model needs to be evaluated or not depends on many issues, some external to the issues of geometric representation. Evaluation restricts changing of sections or other large-scale changes, thus providing a level of version control. It is also necessary for spatial conflict testing and a host of other tests. However, the evaluated representation can always be generated from the unevaluated description. Not all CAD systems or applications support unevaluated geometry, so the applications being used may impose other considerations.

A full boundary representation is very expansive in the computer space required for its definition, in that each surface and intersection is defined explicitly. On the other hand, the unevaluated `csg_solid` is very compact. For shop drawing representations, it also supports definition of families of shapes, defined as an unevaluated model and a table of parameters for individual shapes. An example of a single parametric shape is shown in Figure 6.40. The Tee is defined according to width, web center, web thickness, depth, flange depth, and length. From this shape up to three holes are subtracted—two cylinders and one box cutout. Two cylinders are first translated, a cylinder shape defined and subtracted. Last, a slot is cut from the web. In all cases the subtracting shapes are made one unit longer than the piece being cut. A hole can be defaulted to zero, allowing a variable number of holes (any number less than the number explicitly carried in the parametric model).

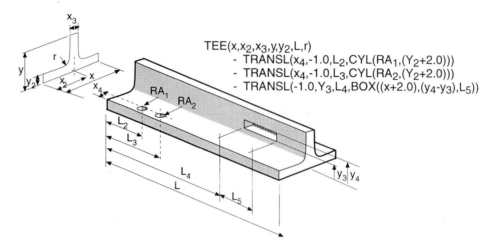

TEE(x,x$_2$,x$_3$,y,y$_2$,L,r)
- TRANSL(x$_4$,-1.0,L$_2$,CYL(RA$_1$,(Y$_2$+2.0)))
- TRANSL(x$_4$,-1.0,L$_3$,CYL(RA$_2$,(Y$_2$+2.0)))
- TRANSL(-1.0,Y$_3$,L$_4$,BOX((x+2.0),(y$_4$-y$_3$),L$_5$))

Figure 6.40: A structural tee with two flange holes and a web cutout (reinforcing not shown).

Most fabricated elements, such as doors, windows, wood and steel framing, are constructed as compositions of stock shapes and thus can be generally represented as a `csg_solid`. The example of core wall construction, introduced in Chapter Three, consists of a packed set of wall sections, defined in elevation as a polygon and with a thickness. If the thicknesses vary for different segments of a wall, then these wall segments define the geometry of the wall as a set of packed swept polygons. Of course, if all segments of a wall have the same two offsets from a centerline, then a single elevation polygonal section can be used to construct a swept shape to define the complete wall.

A detail issue is that the current implementation of EXPRESS does not support unevaluated descriptions, where the width or height are variables and not a numerical value. The current SDAI and other support tools allow representation of object instances according to some model. The tools do not support storing a custom model, only instances of a predefined model. Work is now proceeding to extend EXPRESS to support the carrying of variables and other needed capabilities to represent parametric models.

Freeform Wall   Formed Stairway

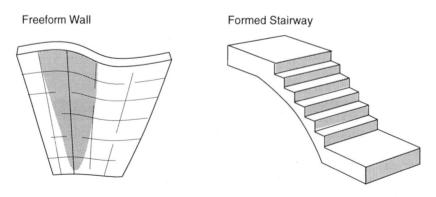

Figure 6.41: Two examples of cast concrete shapes, based on different shape models.

Plastic shapes, such as those made from plastics or concrete, may be formed in infinite ways, unconstrained by any original stock or standard shapes. Historically, carved stone was another plastic material. Depending upon the shape desired, any of the three, form-generating approaches may be most appropriate. Some examples are shown in Figure 6.41.

This quick overview indicates that constructed elements have a need for all three approaches to geometric modeling. No one method is appropriate for representing all types of constructed shapes. All three are often applied to the same element as it is refined through the design and fabrication process. The boundary representation, however, is the foundation representation used to store and possibly exchange shape information.

6.7.2 BUILDING CONCEPTUAL ELEMENTS

The conceptual development of a building design varies tremendously depending upon the design processes utilized. Because the function of the building often has more to do with the spaces that the building encloses or with its visual character than with its construction, innovative buildings are seldom conceptualized by laying out walls on a floor slab. Instead, spaces, masses, and forms are represented and manipulated at various levels of abstraction. Some aspects of the design process of architects are reviewed in Chapter One.

Computer-based design conceptualization and development processes involve a radically new medium in which to work, and new approaches are being actively explored, both in architectural schools as well as in professional practices. Each process tends to emphasize a different set of geometrical operations and their corresponding representations. Two different processes are described below, indicating how different processes require different shape representations.

One often articulated process has evolved from paper-based methods and relies on the feasibility study and its space program as the primary generators of a building form. In this process, polygons are used to represent the internal spaces, which are then arranged in a way that responds to both internal functions and relations, and also external relations with fixed features of the site. The boundaries of the polygons are then characterized as walls or other types of space boundaries, allowing generation of a floorplan. Later, the walls are given thickness and swept vertically to create 3D walls. Special operations both interpret walls and insert doors and windows within them to articulate the wall forms. Strong support for this type of process requires a mixture of evaluated and unevaluated 3D representations. An evaluated geometric model is needed to review the current design, while an unevaluated representation allows easy revision of wall elevations, door and window locations.

A different conceptual design process relies on solid modeling to define a 3D form of the building's exterior, possibly based on shapes corresponding to functional types (a school auditorium, a theater, a laboratory wing of a hospital). The 3D form is then sectioned or sliced to identify floor levels that are partitioned into spaces and different space functions. In this case, CSG-type solid modeling operations initially play a heavy role, from which floors are derived and partitioned into spaces. Here, 2D operations are required to partition the floor sections, which are later given width and extruded as walls. Of course, both of the processes described need to support going back and forth in each direction. Both rely on 2D and 3D representations. These are only two possible development processes; there are many others.

The general point is that conceptual design processes involve quite different geometric operations and representations, depending upon the specific processes being followed. Many different processes are used at various times during design, each potentially relying on a different geometric representation or different operations. No one representation is appropriate in all conditions. The range of representations and operations needed to support a range of processes grows quickly into a large set. Because of the variety of geometric representations needed and the range of operations now used in design, conceptual-level design is one of the most difficult to augment computationally. At the same time, many creative designers are now using 3D modeling for conceptual design development, often with innovative results.

6.7.3 BUILDING SPACES

In some design processes that start with spaces and use their edges as centerlines, an approximate shape is explicitly defined at the outset. However, wall thicknesses and various details impinge upon the space shape continuously. Depending upon the intended use of the space, various levels of accuracy regarding the shape of the space will be required. Occupancy studies require one level of accuracy, while airflow analysis using computational fluid dynamics requires another.

In most of these cases, the shape of the space will be derived from its bounding solids. The bounding shapes may be extrusions, but more likely will be detailed with cabinets, lighting fixtures and other constructions. The shape resulting from such a group of imposed elements can only be represented using the boundary-representations, using the constructive Boolean operations to subtract the bounding walls, floor, ceiling and fixtures from a space shape to derive the final space shape.

In summary, designers rely on many different form-generating solid representations. However, as the building is refined, both its construction elements and spaces are only generally representable as boundary representations. The boundary representation is the most general representation available, but is rather inflexible, because of its elimination of generating parameters. The extension of parametric modeling to allow further elaboration of 3D forms, without losing the form-generating parameters, is a major thrust of current research and product development in geometric modeling. At the fabricated product level, the components are often represented parametrically, in an extended form of CSG.

From this short discussion, it appears that the range of geometric representations defined in Part 042 covers the range of shapes needed for representing a final building, as designed and constructed. It is less clear that it adequately supports the process of designing. There are many issues pertaining to ease of use, not addressed here, that suggest the need for further evolution and refinement of the representations carried in Part 042.

6.8 NOTES AND FURTHER READING

Part 041 can be reviewed from ISO TC184 Part 041 [1997]. Part 042 can be reviewed from ISO TC184 Part 042 [1995]. Part 043 can be reviewed from ISO TC184 Part 043 [1997].

Geometry was the earliest area of study in CAD and computer graphics. A good review of computer-based geometric modeling is offered by Faux and Pratt [1979]. See also Hill [1990] and Foley, van Dam et al. [1991].

Topology, as applied to shapes is well reviewed in Giblin [1977] and also in Flegg [1974].

The development of solid modeling as a way to combine faces into volume-enclosing shapes was a major breakthrough in computer-aided design. It has led to major advances in CAD, in the ease of use of CAD systems, and in developing automated shape generation systems. This work can be directly traced to two Ph.D. theses that—almost in parallel—developed the basic concepts; these are presented in Braid [1973] and Baumgart [1974]. There is a continuing literature on solid modeling. Requicha [1980] deals with the formal properties of solids and provides a survey of systems in the early 1980s. Baer, Eastman and Henion [1979] provide a general survey. Two good texts are Mantyla [1988] and Hoffman [1989]. The use of solids in feature-based design, having some application to architecture at the detailed design stage, is presented by Rosen [1993].

The development of Euler's Law regarding shape topology, discussed in Section 6.4.5, is nicely reviewed from a philosophical perspective in Lakatos [1977]. A set of operators for creating and manipulating solids, based on Euler's Law, is presented in Eastman and Weiler [1979].

There are a growing number of programs that detail the construction of walls, floors and other pieces of a building. Good examples are Triforma® from Bentley Systems, Architectural Desktop® from Autodesk, Inc., and products from Graphisoft and Eagle Point Software.

Some quite different examples of methods for developing the conceptual design of buildings can be seen by reviewing the papers of the SEED project undertaken at Carnegie-Mellon University [Flemming et al., 1995]. An example of a set of software for converting two-dimensional DXF-type files to reliable surface models, with doors and windows and detailed stairways, is presented by Lewis and Sequin [1998].

6.9 STUDY QUESTIONS

1. Given the perspective regarding geometry needs presented in Section 6.7, select the subset of Part 042 that seems most critical for modeling building geometry. In particular, what might be omitted?

2. Look up the design of the Guggenheim Museum in Bilboa Spain, designed by Frank Gehry. It was described and presented in some detail in Menge [1998]. How might this building's forms be modeled in Part 042? Consider both the exterior and the structure.

3. New geometric representations are introduced frequently. A recent addition is Virtual Reality Modeling Language (VRML) [Ames et al., 1997]. It defines some general primitive 3D solids and also some general surface representations. Propose a means to represent *indexed_facesets* in VRML in Part 042 without loss of information.

4. For the EXPRESS wall model defined in Chapter Five, propose a geometric representation using Part 042.

5. Given the wall representation developed in EXPRESS-G in Chapter Five, outline the material representations needed for construction use. That is, what material property information should be associated with the entities and how? Consider the wall as a structural bearing wall. What other properties would have to be added to the wall?

CHAPTER SEVEN
Building Aspect Models

In hierarchic systems, we can distinguish between the interactions among subsystems, on the one hand, and the interactions within systems—that is, among the parts of those subsystems—on the other. The interactions at the different levels may be, and often will be, of different orders of magnitude.

Herbert A. Simon
Chapter 4: The Architecture of Complexity
The Sciences of the Artificial
1969

7.1 INTRODUCTION

The original intent of ISO-10303, STEP, was to define middle level "application domains" that addressed data exchange needs related to the organizational level of engineering departments. By addressing these middle-level integration issues, the STEP organization felt that progress in resolving immediate practical problems could be made in the course of refining and stabilizing the overall STEP methodology. In manufacturing, examples of middle-level domains were part casting, numerical control machining and sheet metal forming. Parallel application areas in building might involve processes associated with structural steel, structural concrete and wood framing. Also embedded in integrated resources associated with manufacturing were finite element modeling and kinematics. These correspond to particular kinds of analysis or simulation. Parallel kinds of applications in building modeling might be energy analysis, lighting analysis, or acoustic modeling. Adopting the terminology used by many, this book calls these middle-level information domains *aspect* models.

In this chapter, three major examples of traditional aspect models are reviewed: the CIMsteel project, part of the ESPIRIT program sponsored by the European Community; the COMBINE project in energy modeling, part of JOULE, another European Union project; and Part 225, Building Elements Using Explicit Shape Representation. Part 225 is the first complete Part model draft specification in the AEC area. These three efforts are considered the most influential efforts to date in the development of aspect product models.

7.2 CIMSTEEL

The CIMsteel project was a major initiative under the European Community's EUREKA umbrella (Project EU130). First instigated in 1987, the overall goal of this wide-ranging project was to improve the efficiency and effectiveness of the European constructional steelwork industry. The focus of CIMsteel was the application of Computer Integrated Manufacturing techniques to steel fabrication. A key strand, representing some 20% of the total project, was the development and deployment of standards for the digital representation of technical information for steel building-type framed structures. The goal of this activity was the establishment of an "open standard" that reflects the international market for structural steelwork and hence could be used globally, by any software company or end-user organization.

The CIMsteel project formally ended early in 1998, by which time it had involved some 70 collaborating organizations in nine European countries. The resulting data exchange standards, known as the CIS (CIMsteel Integration Standards), have already been widely implemented. The CIS/1 are the result of a major, long-term collaborative effort involving researchers, contractors, steel fabricators and a group of application vendors. Although the collaboration was based in Europe, significant inputs were received from groups in the USA and Japan. The digital standards strand of the CIMsteel project was led by Dr. Alastair Watson of the School of Civil Engineering, University of Leeds, UK. Post-CIMsteel work on the CIS continues at Leeds. A CIS User Group has been established, hosted by the Steel Construction Institute, UK.

7.2.1 THE FIRST RELEASE OF THE CIS

The formal CIS/1 specifications were first made available to software developers in late 1995. The description that follows is based on Part 1 of those specifications plus other near final working documents.

CIS/1 is based on Version 4 of the LPM (or logical product model), a model which was incrementally developed and tested by the CIMsteel project. The LPM is concerned with the engineering information that arises in the design, fabrication and erection of steel framing, as used in building construction. It is independent of application-specific software and may be considered to be a standardized container for carrying product information. The long-term intention was to address the building lifecycle, although initially the CIMsteel project focused on outline design through detailing. The current scope covers the main structural steel work as well as secondary work such as purlins, siderails and cleats. Parts can be formed in a variety of ways; they can be formed as prismatic (rolled) sections or cut from 2D rolled plates, supporting rolled, cast or cold-formed fabrication. Joints may be welded or bolted. Analysis results are carried in the model so as to allow detail design to be carried out in light of load requirements.

While CIMsteel was a major industrial effort, there were many areas where the "full information model" for steel frames was deliberately truncated in CIS/1. These restrictions on the scope of the data model were based on practical considerations including the significant learning curve and organizational re-engineering involved in adopting the model, and the limitations of current implementation tools dealing with large models and datasets. The goal was to reduce the overall complexity of the model and limit the use of some entities and attributes by the interfacing vendor applications. The restrictions on CIS/1 included

> *no detail design information or code references*
> *no second order elastic analysis*
> *bearing surfaces not separated out*
> *solid modeling of geometry not included*
> *holding down bolts are not distinguished*
> *no cambered beams*
> *no material specifications for bolts*
> *no movement joints*
> *no dynamic, moving or destabilizing loads*
> *no handling of bridges, offshore structures or transmission towers*
> *no cost information*
> *no control information for NC machinery*

Initial work on CIMsteel relied on IDEF1X (the third ARM modeling language) as a conceptual modeling language and on the use of STEP technology, where possible. The intent was to develop and test a series of prototypes, which would evolve into a STEP-compliant model. In practice, they used EXPRESS as their data modeling language and developed several tools to support their data exchange work, in particular some early PC-based tools.

The section below presents an overview of the LPM employed in CIS/1, based on an interpretation of the EXPRESS model defined in an early "research" release of the effort that does not incorporate recent production revisions. The full model is quite large, with over 120 entities. An overview and survey of some of the model's major aspects are presented here. Readers interested in fuller aspects of the model should contact the CAE Group at Leeds University.

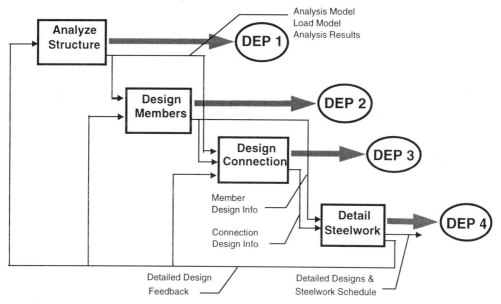

Figure 7.1: DEPs mapped onto a simplified activity model.

7.2.2 DATA EXCHANGE PROTOCOLS

Although the LPM is presented as a series of distinct schemas, a single, high-level schema serves to unify it. Particular data exchange translators are implemented to support one or more of four Data Exchange Protocols (DEPs), each addressing a part of the building lifecycle. A schematic representation of these different DEPs and their relation to the engineering process is shown in Figure 7.1. The initial stages are to support design, first dealing with the overall structure and then members. This is followed by connection design. The last stage deals with member fabrication. In order for the CIS to support the different tasks within the structural steel product area, it provides different views of a steel-framed structure. The views include centerline, rough-cut geometrical model and detailed geometrical model. A detailed representation of loads, as well as various levels of joint details and welding specifications, is also included.

7.2.3 DATA EXCHANGE MODEL

A high-level overview definition of the LPM is given in Figure 7.2. We will describe it first in order to establish a roadmap for the more detailed presentation of portions of the overall model that comes later. The model begins with a skeletal building model, shown at the top of the figure, consisting of site and project. Within the building model skeleton, the STRUCTURE is the only defined subsystem. The STRUCTURE is defined as a hierarchical organization of ASSEMBLY entities, which may be of two types: DESIGN_ASSEMBLY or MANUFACT(uring)_ASSEMBLY. These provide two main views of the structural system, two different ways of structuring the part information.

The elements that make up a steel framework consist of S_PARTS. There are three kinds: PRISMATIC_PART, SHEET_PART and COMPLEX_PART. Prismatic and sheet part geometries are defined implicitly, that is, parametrically. COMPLEX_PART geometry is defined explicitly. (Explicit geometry is not included within the model.) The DESIGN_ASSEMBLY and MANUFACT(uring) ASSEMBLY can be decomposed into smaller units of the same type. At the most detailed level, they are made up of LOCATED_PARTs, LOCATED_JOINT_SYSTEMs and AS(sem)BLY_CONNECTIONs.

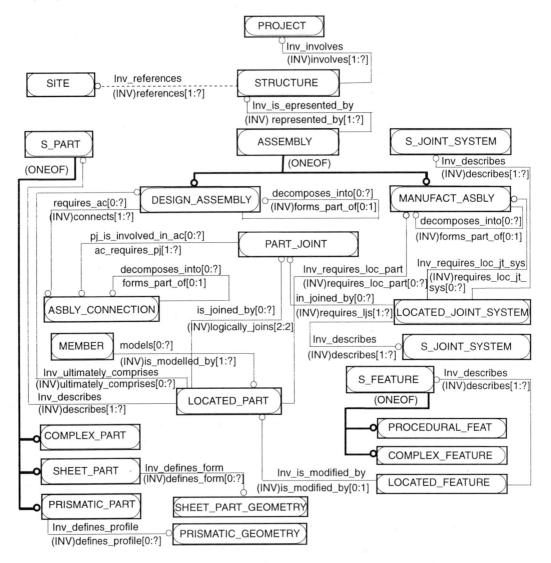

Figure 7.2: Overview of the CIMsteel model, indicating its high-level structure, focusing on the design and manufacturing aspects.

Joints are also parts, called PART_JOINTS. A PART_JOINT defines the connectivity between two LOCATED_PARTS. These are combined and the joint realized as specified in a LOCATED_JOINT_SYSTEM. Another type of entity is a feature, called an S_FEATURE. A feature is some aspect of a joint, such as bolt hole or weld. Features either remove material from a LOCATED_PART or define a surface treatment. Features may be procedural or complex. Each of these systems—for parts, joints and features—will be defined in more detail below.

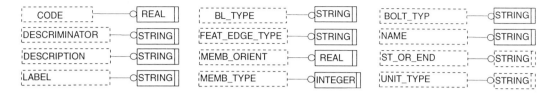

Figure 7.3: The base types used in defining the CIMsteel Model.

In order to review the various parts of the CIMsteel model in detail, we need to know the high-level Types that were predefined and used throughout the model. Some of these are shown in Figure 7.3. They define some basic types repeatedly used throughout the model. The last two are enumeration types, with the UNIT_TYPE being SI or Imperial.

We now turn to examining the model in more detail. Each project has a high-level description, as defined in Figure 7.4. A PROJECT is the top-level entity that involves one or more STRUCTUREs. A Project has a client name and address, a description and a method of payment. Each STRUCTURE is an independent steel framework. Each STRUCTURE has a UNIT_SYSTEM of measurement and a global COORDINATE_SYSTEM. The structure is related to and accessible from the site, which identifies both address information and also the exposure and altitude of the site. Two address(es) are provided: one for the site and another for the owner. This top level of the CIS allows one project to consist of multiple structures, possibly located on different sites. Each entity has attributes needed for identification and location.

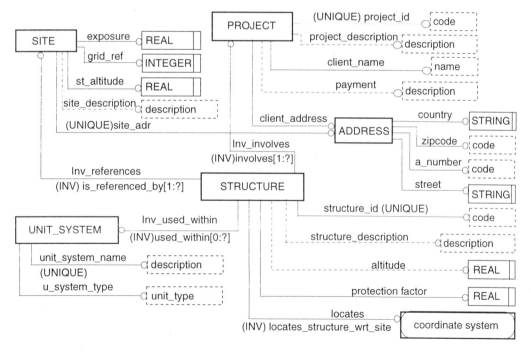

Figure 7.4: The top-level entities describing a CIMsteel project schema.

Figure 7.5 presents the information associated with a structure. Each STRUCTURE, along with its subelements, has a unique identifying CODE and DESCRIPTION. It has a location, defined by a COORD(inate)_SYSTEM, altitude and protection_factor (fire rating type). A structure is defined as a set of ASSEMBLY entities, subtyped as either DESIGN_ASSEMBLY entities or MANUFACT(uring)_AS(sem)BLY entities. Each ASSEMBLY may be decomposed into subassemblies of the same type. In fact, these two types of assemblies provide two different ways of grouping the same entities into an aggregation hierarchy, for operating on the entities for different purposes. A DESIGN_ASSEMBLY is made up of AS(sem)BLY_CONNECTIONs and LOCATED_PARTs. LOCATED_PARTs are joined in pairs as PART_JOINTs, which are grouped into AS(sem)BLY_CONNECTIONs. CIS/1 does not include bay or aisle groupings, evidently relying on a more general concept of DESIGN_ASSEMBLY to generalize this notion and be useful in other kinds of organizations. A MANUFACT(uring)_AS(em)BLY consists of LOCATED_PARTs and LOCATED_JOINT_SYSTEMs. MEMBER is an abstraction of a LOCATED_PART, used for representing parts in analyses, as we shall see. Though the structure introduces several levels of abstraction, when all is reduced to essentials, we see that a DESIGN_ASSEMBLY is a composition of S_PARTs and S_JOINTs, both of which have located versions. GROUPs may be used to collect various ASSEMBLYs for use in fabrication or other areas.

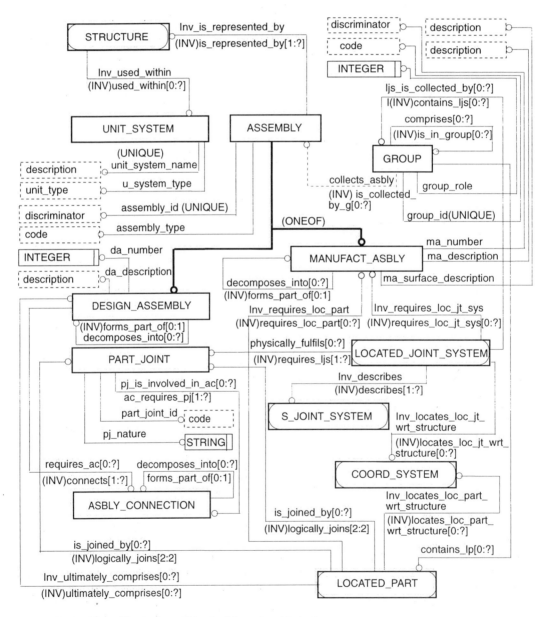

Figure 7.5: The main entities making up a structure.

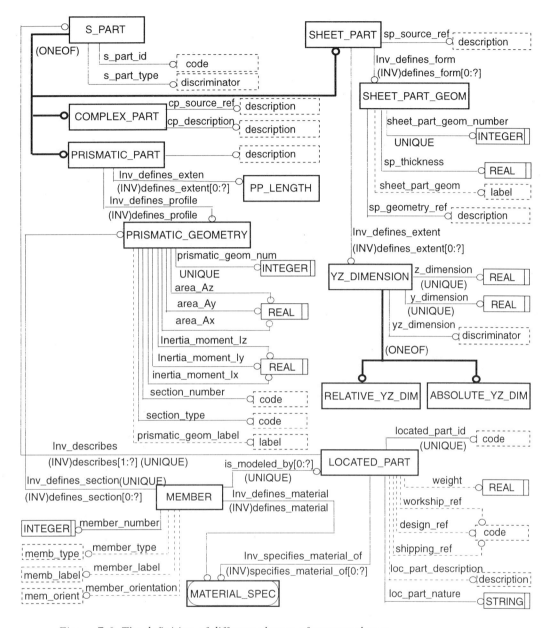

Figure 7.6: The definition of different classes of structural part.

The general parts of the structure are called S_PARTs. As shown in Figure 7.6, an S_PART is specialized into one of three particular classes of shaped part: SHEET_PART, PRISMATIC_PART or COMPLEX_PART.

SHEET_PARTs are assumed to be made from bent, corrugated or flat sheet material. The shape of the section of the sheet is defined by its sp_geometry_ref description. The constant thickness of the section is carried in sp_thickness. Its Y and Z dimensions (X is its thickness) are carried by the defines_extent attribute. Corrugations or bends are oriented in the Y-direction. The extent dimensions are specialized into a relative or absolute version. The absolute version fixes how the cut is to be made from stock and the placement of corrugation and bends within the part. The relative version allows the part to be cut anywhere within a sheet.

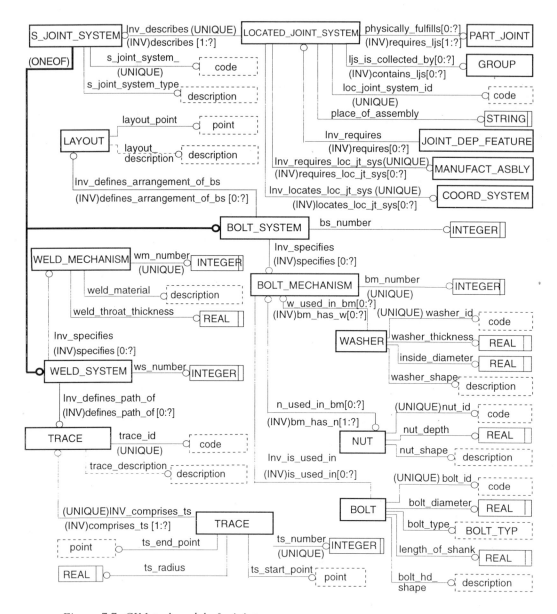

Figure 7.7: CIMsteel model of a joint.

PRISMATIC_PARTs are assumed to be extruded sections, such as wide-flange beams or L-shaped beams, and are defined by a constant section and a length. They are the main parts used in the model. Each PRISMATIC_PART has a length, carried in PP_LENGTH. The profile is carried in PRISMATIC_GEOMETRY, consisting of a precoded section_type and SECTION_NUMBER, and defined by three areas and three moments of inertia. The explicit geometry of the section is not available for direct calculation, but is defined by reference to an externally defined shape. However, the areas and moments are considered sufficient for analysis calculations, though not for others, such as spatial conflict testing. The COMPLEX_PART is used to define castings and fittings that require solid geometry to describe them. In CIS/1 they are only defined by reference to an external DESCRIPTION.

A structural part during design begins as a MEMBER. After analysis, based on loading assumptions that are used to derive moments and stresses, it becomes one of the three subtypes of S_PART and is fixed as a LOCATED_PART. It is assigned attributes associated with its manufacture: weight, shipping_ref(erence), workship_ref(erence) and more. Thus each S_PART is part of two assemblies, a DESIGN_ASSEMBLY and a MANUFACT_ASBLY. These compose the S_PARTS as needed for these two different views. The LOCATED_MEMBER and MEMBER also have a MATERIAL_SPEC(ification).

Joints are an important part of a structure. They are modeled within the LPM as shown in Figure 7.7. An S_JOINT_SYSTEM is specialized into either a WELD_SYSTEM or a BOLT_SYSTEM. The WELD_SYSTEM is specified by a TRACE, made up of TRACE_SEGMENTs. A BOLT_SYSTEM is made up of one or more BOLT_MECHANISMs, consisting of BOLTs, NUTs, and WASHERs, each with an identifier carrying the grade of the bolt and its critical geometry.

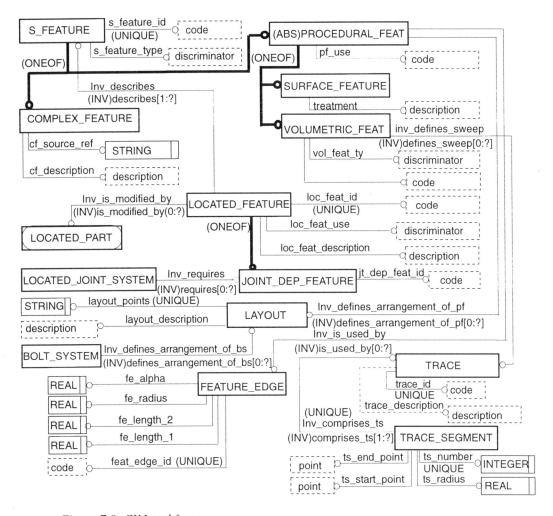

Figure 7.8: CIMsteel features.

Similar to parts, joints, when located as parts of a LOCATED_JOINT_SYSTEM, become part of a MANUFACT_ASBLY, which carries the attributes associated with fabrication. Remember from Figure 7.5 that a PART_JOINT joins two LOCATED_PARTs.

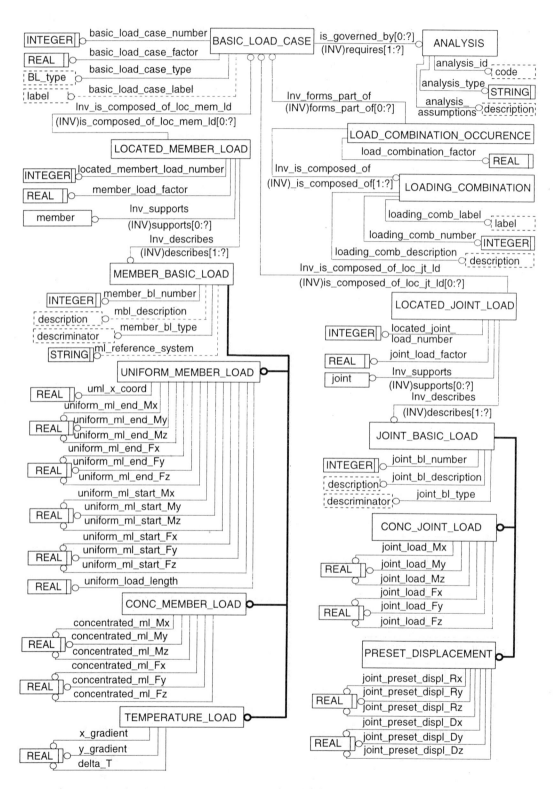

Figure 7.9: The loading conditions carried by CIMsteel.

Still another useful way of organizing information about a structural steel system is by its features. A feature may be a composition of holes, welding traces or other elements that define a pattern, that may be stored and assigned to a design condition. It may be a bracket for attaching a curtainwall to the structure, or a predefined joint system, or a notch for pass-thrus. A feature then is a collection of individual elements that modify a LOCATED_PART by removing material or adding a finish treatment. The model view of features is shown in Figure 7.8. A LOCATED_JOINT_SYSTEM is defined in terms of one or more JOINT_DEP(endent) _FEATURES. JOINT_DEP_FEATURES are defined from a LAYOUT, which is an arrangement of bolts or given TRACES that define weld paths.

An S_FEATURE is specialized into either a COMPLEX_FEATURE or a PROCEDURAL_ FEATURE. The PROCEDURAL_FEATURE consists of a LAYOUT (of bolts) and a FEATURE_ EDGE. These are inherited into either a SURFACE_FEATURE or a VOLUMETRIC_FEAT(ure).

7.2.4 MODEL OF LOAD CONDITIONS

Version CIS/1 of the CIMsteel model also carries the loading conditions used as input for structural analyses. This part of the model is shown in Figure 7.9. It involves applying loading conditions to members and joints. Member loads are associated with a MEMBER, which is an abstraction of a LOCATED_PART (see Figure 7.6). Each LOCATED_MEMBER_LOAD has any number of UNIFORM_MEMBER_LOADs and CONC(entrated)_MEMBER_LOADs. A uniform load has a starting point along the member's length, with three-dimensional moments and forces defining the loads at each end of the uniform load condition. A concentrated load is located at a distance along the member and six possible moments and forces defined at the point. A MEMBER may also have TEMPERATURE_LOADs that are uniformly distributed along their length. BASIC_LOAD_CASEs each have an associated ANALYSIS method, providing a small area to define the assumptions embedded in the load case.

Joint loads are based on LOCATED_JOINT_LOADs. They are either CONC(entrated)_ JOINT_LOADs or a PRESET_DISPLACEMENTs. The concentrated load at a joint is defined the same as a concentrated load for a member. The PRESET_DISPLACEMENT of a joint provides its fixed location to the ground or any other absolute condition and is defined as three translations and three rotations. These loading conditions can be packaged into LOADING_COMBINATIONs and reused within a design as LOAD_COMBINATION_OCCURRENCEs; each may be factored by a percentage of the BASIC_LOAD_CASE.

7.2.5 MEMBER AND JOINT STRESSES

After an analysis is executed, the results are loaded back into a model that carries the structural results. A structure in this part of the model continues to be defined as a set of MEMBERs. Each MEMBER has two MEMBER_ENDs. The MEMBER_ENDs have an associated JOINT that carries a JOINT_RESULT. A JOINT_RESULT is one of two types, a JOINT_DISPLACEMENT or a SUPPORT_REACTION, both of which are defined by six real values. A MEMBER_END also has MEMBER_END_RESULTS. The end results consist of a displacement or an END_FORCE, composed of moments and forces.

A MEMBER "is_submitted_to" various MEMBER_RESULTS. These consist of forces and displacements, each with six values. The information associated with members and joints is sufficient for associating member cross-sections and includes JOINT_DEP(endent)_FEATUREs that can be used to assign joints. Alternatively, custom joints may be defined.

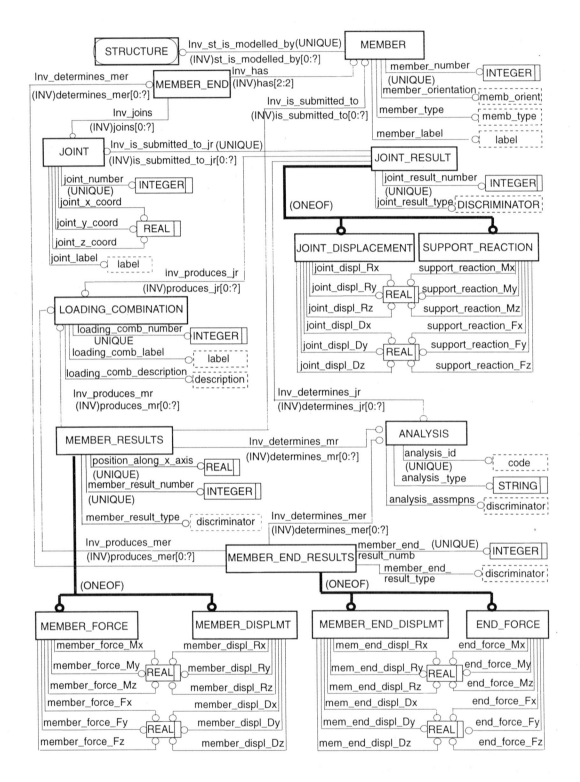

Figure 7.10: Resultant stresses and moments on members and joints.

7.2.6 DISCUSSION

Structural steel sections have become standardized by fabricators over the last fifty years. Functional behavior is managed by smelting and production standards that are followed reasonably well throughout the world. Thus the performance of steel within the elastic range is quite reliable. Given standard shapes, standards for performance and relatively regular types of use, standard procedures have been defined for steel structure design, for detailing of a particular project, and for erection. These procedures greatly facilitate the development of data exchange standards. There are strong indications of broad industry acceptance. Many vendors outside of the original group have joined and are building CIS translators. Beyond Europe, translators have been developed in India, Israel and the US.

In developing and launching CIS/1, CIMsteel made a strategic decision to take maximum advantage of the specifications and technologies emanating from STEP. This initially caused problems due to the poor availability of tools, particularly those running on the PC platform. This issue has now been rectified.

Currently, the Leeds team is about to release the Beta version of the CIS/2 specifications to software developers, which features a considerably expanded LPM with nearly 600 entities. CIS/2 represents a substantial advance on CIS/1 and will take some time to fully realize in software. The engineering scope is considerably expanded, while data management considerations are also addressed, giving CIS/2 the additional ability to support incremental data exchange and data sharing. In parallel, STEP Part 230 for Building Structural Frame: Steelwork is being pursued to establish a formal ISO version of the CIS.

The CIMsteel project is one of the first STEP-based projects to gain significant acceptance in the building industry, and as a pioneer it has addressed many practical issues. Key features of the project deserve credit in this regard:

- specification of the model based on wide collaboration with both end-users (steel fabricators, contractors) and software developers
- implementing a model using STEP tools that are publicly available
- working closely with the application vendors to develop translators to/from the interface
- providing training and working with end-users on the effective use of the model

The experience of Leeds University in carrying out these activities provides valuable guidelines on how other such building models may be effectively implemented in the future.

7.3 COMBINE

COMBINE was a major development project funded by the European Union (EU). It was part of the JOULE Programme, which focused on the rational use of energy. COMBINE had two phases: Phase 1, which ran from 1990 to 1992; and Phase 2, which ran from the end of 1992 to the middle of 1995. Together both projects involved seventy man-years of effort.

COMBINE was not focused on developing new IT capabilities through research, but rather on demonstrating the potential of existing capabilities. The stated goal was "to smoothly combine a number of state of the art energy and HVAC consultant's tools in an integrated environment, ... to demonstrate efficiency improvements in the design process and related energy efficiency improvements of the planned buildings that these new systems will enable."

The participants were

Technical University of Delft, Civil Engineering, The Netherlands - project coordinator, Godfried Augenbroe

University of Newcastle on Tyne, Architecture, United Kingdom
University College Galway, Mechanical Engineering, Republic of Ireland
TNO-BOUW, Delft (Netherlands Institute of Building and Construction Research)
CSTB - Centre of Science and Technology in Building, France
BRE - Building Research Establishment, United Kingdom
SBI - Danish Building Research Institute, Denmark
Fraunhofer Institute for Building Physics, Germany
University College Dublin, Energy Research Group, Republic of Ireland
Strathclyde University and Appleton Research Laboratory, United Kingdom
VTT - National Research Laboratory of Finland

Participants in the first stage only:
University of Liege, Laboratory for Study of Architectural Methods, Belgium
University of Edinburgh, Architecture, Scotland
COSTIC - a French consultancy
University of Ulster, Building and Environmental Engineering, United Kingdom

The project team was purposely composed to include a mixture of expertise from energy performance, building engineering and in information technology, especially STEP technology.

7.3.1 COMBINE-1

The first phase focused on data integration, supporting a set of separate "actors" and their applications around a central data depository. The deliverables included a large conceptual building model, called the Integrated Data Model (IDM), an IDM implementation, and a set of ISO-STEP tools supporting interface development for a set of Design Tool Prototypes (DTPs). The organization of the functions is diagrammed in Figure 7.11.

Six design tool prototypes were developed and demonstrated (in varying degrees of completeness) in Stuttgart, in November, 1992. The tools were

DT1. *construction design of external building elements*: This application involved the development of a neutral building model for energy analysis, from which a variety of specific analyses would later be developed. It was written in Actor, an expert system developed by the Whitewater Group.

DT2. *HVAC design:* This design system was an existing application that integrated two applications: DOE-2, a highly tested and respected dynamic building energy simulation model that derives internal energy changes, and MEDIA-LC, a locally written application. MEDIA-LC was used to simulate the mechanical system, given the loads generated by DOE-2.

DT3. *dimensioning and function organization of inner spaces:* This is a graphical application allowing a user to develop a space allocation and floorplan, based on an accessibility graph.

DT4. *thermal simulation in late design stage:* DT4 integrated two applications, *tsbi3*, developed by the Danish Building Research Institute and SUNCODE, developed by the Solar Energy Research Institute. Both applications calculate thermal, indoor climate parameters and energy consumption. They both used the same front-end shell and data model, written in C++.

DT5. *energy analysis in early design stage using the L-T method*: The L-T method models rough energy gain, based on building shell geometry, glazing areas, atria and lighting. The interface was written in MOLE, a logic programming language.

DT6. *radiator network design*: This application allows 3D layout of a radiator design system, using an object-based extension to AutoCad and dBaseIV to handle attributes.

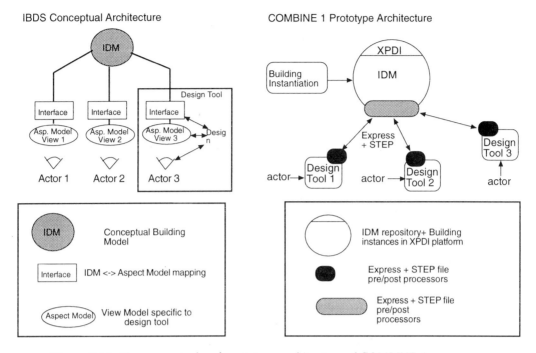

Figure 7.11: The conceptual and prototype architecture of COMBINE-1.

The IDM integrated all the separate data views needed by the applications, resulting in a single, large, comprehensive building model, specifically developed to support the six applications. The building description to be loaded was <u>not</u> part of the final project and was hand loaded into the model. At the demonstration, the team was able to download from the IDM repository the needed building data subsets to each application. Limited ability to accept updates from some applications was also apparently provided.

Given the task and final capabilities, it is clear that a major aspect of this project involved integrating each application view into an overall building model. The team used NIAM as the data modeling language, with a carefully defined data dictionary of attributes. The resulting building model was quite straightforward. It required many attributes, however, to support the different applications. The main building objects and their structure were

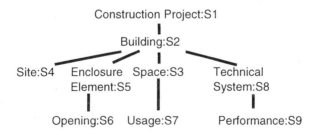

The final building model consisted of about 400 entities (NOLOTS in NIAM), 500 attributes (NIAM LOTS) and 600 relationships. Because the model was so large, tools were needed to construct it and to interrogate and manage its entities. The group at CSTB used the MIPS tool, derived from ILOG, to provide these functions, allowing navigation and querying of the data model and data dictionary.

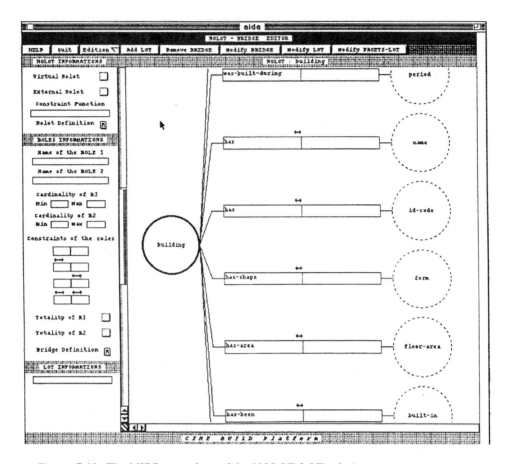

Figure 7.12: The MIPS control panel for NOLOT-LOT relations.

Two examples of menus embedded in the MIPS tool are shown in Figures 7.12 and 7.13. The first figure shows a window with a menu at the left side for defining object types in the widnow on the right. Figure 7.13 shows a window with a menu on the left providing a list of all NO-LOTs, which are selected, composed and related to each other in the screen on its right.

The input to the MIPS tool was generated by interactively defining a NIAM diagram. (The MIPS tool was later extended and renamed XPDI.) Upon request, it translated and output an EXPRESS representation of the NIAM-defined model. The EXPRESS model then relied on STEP-based tools to map the EXPRESS model into the STEP physical file format. This physical file format provided a simple syntax for EXPRESS descriptions, allowing easy parsing for reading or writing. Given the EXPRESS file format, the COMBINE developers generated an Interface Kit of C++ classes that facilitated writing translators from the STEP physical file to an application. The library was used to write custom application interfaces for each Design Tool.

The demonstration in 1992 showed off these capabilities. Applications ran on a heterogeneous set of machines and files were transferred by floppy disk. Much was learned from this large exercise. It was found that the interface tool-kit was hard to use and insufficiently conceived and developed for the intended goals. Much time was spent writing and debugging the application interfaces. In retrospect, Phase 1 showed that data exchange is workable; a common model and set of exchange facilities could be developed and used. However, these facilities did not support a real design development scenario, but just physical exchange of data. That is, the work as conceived and completed did not address tactical design issues: how to coordinate design actions among users or how to execute an application multiple times on different entity instances. The work also pointed to strategic-level design issues: how to coordinate multiple designers' actions, both against design goals and to coordinate their processes. Within the model, weaknesses were found in the geometry and shape descriptions, which were particularly limited in the initial representation.

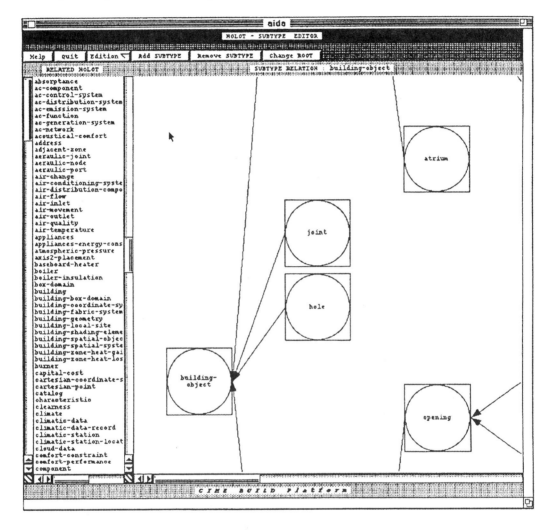

Figure 7.13: The Subtype MIPS control panel.

7.3.2 COMBINE-2

COMBINE-2 was initiated in November, 1992, to expand upon the efforts realized in the first phase. It was funded not in order to address information technology research as such, but rather to apply and assess the current state of the art with regard to generating practical results. Beyond this, the second phase had several objectives. They included integrating a broader range of coordinated design tools, so as to produce a more usable system. A key aspect was to integrate a regular CAD system and to use it to strengthen the representation of geometrical information available to all applications. The building model implementation was augmented by storage within an object-oriented database system, rather than a flat file format as used in Phase 1. Around the database were to be developed better tools for interfacing applications.

Also participating in the project were representatives of three architectural firms, four building services firms and one public works agency. They were to be potential test users of the Phase 2 system. More important, they provided useful advice and feedback with regard to criteria and capabilities for the system throughout its development.

The new set of applications to be integrated involved two CAD programs and a number of analysis tools. They were organized into two suites. One suite focused on the building shell. It involved

1. shape design and space layout using an architectural CAD system (specifically, the Microstation Architectural ArchiCAD package) - TU Delft
2. on-line document browsing - CSTB
3. component database access, for building fabric - BRE
4. building regulation compliance checking, including
      - BREAIR - ventilation and infiltration tool
      - WG6 summer overheating risk tool   BRE
      - Average daylight factor program
      - BRECOND - condensation risk checker
5. PrEN832 annual energy consumption tool - SBI
6. cost estimation tool - VTT

The second suite focused on the design and analysis of mechanical equipment:

1. HVAC design with CAD support (using an AutoCad HVAC application) - UC Galway
2. on-line document browsing - CSTB
3. component database access, for HVAC building components - VTT
4. SUPERLINK lighting design program - FIBP
5. tsb13 building energy simulation program - SBI
6. DOE-2 thermal/energy/comfort simulation - UC Galway
7. ESP-r building energy simulation tool - Strathclyde

The main element of the system is called Data Exchange System (DES). Within this system is the Data Exchange Kernel (DEK), which provides persistent storage of building models. The ObjectStore object-oriented database system was selected as the platform for implementing the DEK. EXPRESS was used to define the building model, and the ISO Standard Data Access Interface (SDAI) was used to provide access to/from the DEK.

An important issue was whether to "compile" the Integrated Data Model (IDM) into the DEK or, alternatively, to store the data as primitive elements of the EXPRESS language. (This distinction is similar to the early and late binding alternatives in SDAI.) In the first case, access to IDM is through high-level IDM classes, which are compiled and thus fixed. In the second, the IDM class structure is interpreted at run time, allowing changes to be made without recompilation (at some cost in performance). In the second approach, all knowledge about the IDM structure is embedded in the read and write modules to/from the DEK. COMBINE-2 opted for the second approach. A cost of this "late binding" approach is that each application needs to know the semantics relied on by all the other applications. Thus if the model changes, probably all the application interfaces need to be updated also. This cost was accepted because of its slightly better flexibility.

The general architecture of COMBINE-2 is shown in Figure 7.14. The main element is the Data Exchange System (DES). Within this system is the Data Exchange Kernel (DEK), which provides persistent storage of building models using the ObjectStore database system as the implementation platform.

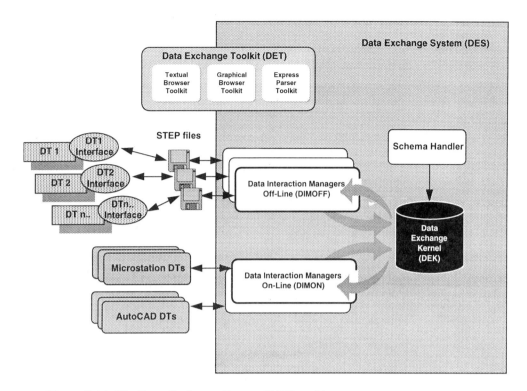

Figure 7.14: The Data Exchange System (DES) architecture.

Two forms of interaction with the DEK were provided. One was interaction through STEP physical files, which was referred to as "Data Interaction Managers - offline" (DIMOFF). The other was direct access to the data in the DEK, especially for the CAD design tools. These were called "Data Interaction Managers - online" (DIMON). For DIMOFF, a C++ library was used to write an interface to extract the data needed for an application, which was then stored in the STEP physical file format. This was then read by the application (translated) and executed. All the analysis applications were of this type.

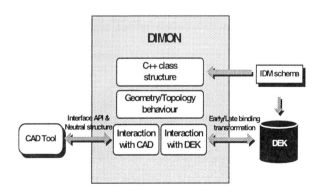

Figure 7.15: The DIMON configuration.

The online interaction was realized through ObjectStore's Meta_object Protocol, which allows C++ classes to be defined and stored in one format (the early binding format), then later mapped to another format (the late binding one). Thus this capability allows the CAD tools to directly write IDM objects and later to "uncompile" them into EXPRESS structures. This later format was used for all other application interaction. Two C++ libraries were used to do the direct mappings: one that supported interfacing to the CAD entities, the other for reading and writing to/from the DEK. These were compiled to provide the DIMON interface, as shown in Figure 7.15.

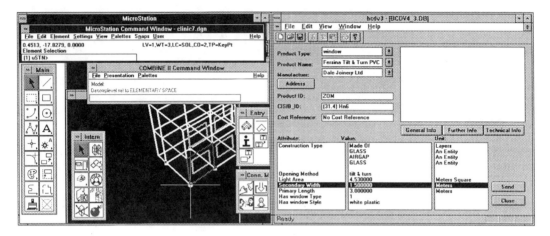

Figure 7.16: Performance display: the Microstation display is on the left; an application user interface on the right.

The CAD applications were used to enter product data. In the COMBINE-2 user scenario, a design may begin as a sketch or as a CAD drawing. In both cases, these must be restructured to allow interpretation. The COMBINE-defined semantics required all entities defining certain elements (i.e., walls, windows, door, ducts) to be on specific layers. This allowed internal applications to identify and process them and to define more detailed and tagged models on other layers. After a model was defined in the DEK, a limited set of operations were implemented in Microstation (for the Architectural application) and in AutoCad (for the HVAC application) to allow direct editing of the data within the DEK. The examples written up were to "Move Door/Window", "Resize Door/Window" and "Change Duct Layout". Each one of these operations involved direct operator mappings from the CAD system through the DEK to the database. Thus only a few such operations could be implemented. These operations appear more to give the impression of simple functionality, rather than as an effort to move toward a practical solution for making updates. An example of an application window is shown in Figure 7.16. It shows the representation editor displaying the rm-model geometry on the left window and the property assignment taking place in the right window.

A key component of COMBINE-2 was the concept of a "Project Window". The COMBINE Project Window provides two interrelated capabilities:
- incorporation of the basic schema for storing and extracting data to/from the DEK
- structuring of the sequential execution of applications

The first capability has already been described. The second capability was needed to allow management and coordination of the overall state of a project. The Project Window provides a graphical display of the sequence in which operations (that is, execution of tasks through an application) can be executed, defined by a process scenario. The logical structure used was a type of Petri net, a common representation for complex task structures. In COMBINE, it was called Combi-Net. Combi-Net represented an operation as a Design Tool Function (DTF), consisting of an input state and an output state, and designated the precedence order regarding the sequencing of tools. This was needed to identify when a particular application had its input data ready for execution, as prepared by previous applications. Executing the applications in arbitrary sequence would quickly lead to most data being nonsense, because it was generated using obsolete or inconsistent input data. Interactive CAD transactions and exchanges with off-line applications were both depicted in Combi-Net. All combinations of actions were thus diagrammed in the Project Window along with the status of each application. The Project Window allowed the user to review what operations were possible at different points in time. An example of a project Window is shown in Figure 7.17.

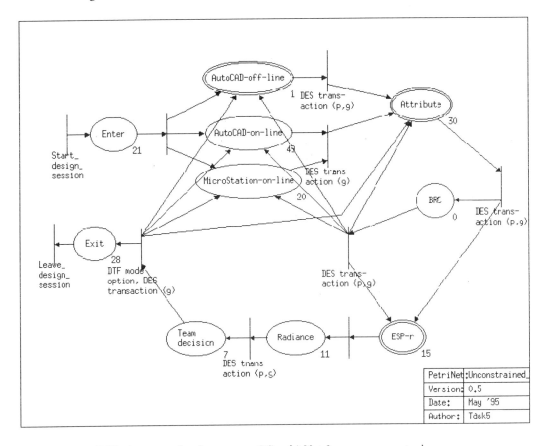

Figure 7.17: An example of a screen of Combi-Net for process control.

The Combi-Net structure defined the Actors and their relationship to DTs and DTFs, the IDM subschemas used for input and output, and the allowable precedence structure among DTFs. Three different operating modes were provided:

Mode 1: Precedence order displayed for user guidance, but with no restriction on operation invocation. In this mode, design tool operations are monitored for state, but no control is imposed.
Mode 2: Precedence order displayed, with tool invocation only allowed in the order presented. However, the functions allowed within a tool are unrestricted.
Mode 3: Precedence order displayed, with tool invocation only allowed in the ordered presented. The functions allowed within a tool are restricted to the order allowed within Combi-Net.

These controls were implemented within an Application Knowledge Handler (AKH), written in Prolog, called the Project Manager. The process state of the model received updates at the invocation and completion of each DT. The resulting design state was captured and displayed in the Combi-Net screens. Data exchange operations were also controlled by the Combi-Net diagrams, in a program called Exchange Executive (ExEx). ExEx used the schema information to detect possible clashes between DTFs, e.g., those that have to be rerun because their input has potentially changed since their last invocation. These tests were made using timestamps of the input and output datasets.

The design process models defined by AKH were editable, allowing new processes to be added and existing ones to be reconfigured. However, no means were developed to check the logical characteristics of a new process or of changes made to an existing one. Since it was easy to define a process that was nonsense, the process model in Combi-Net had to be carefully organized and tested before it could be made available to designers. One demonstration in the final work showed how these processes could be structured to guide users in making updates to evaluate and progress toward a predefined goal. The project assumed that a specific set of Project Windows would be defined for a project, and possibly revised as design proceeded. This work on project windows introduced a new set of issues concerning the coordination and management of data exchange among applications using a shared building model.

For the final review, a fairly elaborate scenario of use was shown, moving control from geometry layout to attribute value assignment, to simple evaluation tool, to further design refinement. A schematic specification for a building was defined, followed by HVAC layout and analysis. In this way, the second phase of COMBINE was able to show tactical design process management, and contained strong suggestions regarding strategic process management. Some aspects of the final work including project reports are available on the World Wide Web.

The Project Window and Combi-Net operates on task and class-level data, not instance-level data. Thus a task is identified as requiring iteration, operating on some class of objects. The Project Windows could not specify which instances of objects should be acted upon. Because of this, each data exchange transfer to a tool mapped the whole dataset and never just a part of the design. The user had to figure out what changed, just as the user must today. At the same time, this project recognized the importance of integrating process coordination with those of data exchange and made major strides in this area. It defined a set of central issues to be addressed in future work.

7.4 COMBINE-II BUILDING MODEL

CIMsteel and COMBINE are among the few building models up to this time that have attempted to develop a general model that also was able to support specific off-the-shelf applications. COMBINE attempted to be more than an exchange model for a fixed set of applications; it aspired to have rich building semantics, capable of extension to support additional future applications. Procedurally, the core model was developed first as an ideal *de novo* model, to which the various application view models could be appended later.

The COMBINE building model is quite extensive, defining many energy-related components and properties, including HVAC equipment. It took advantage of the already existing STEP resource entities, in particular using some definitions from Parts 041, 042 and 043. It consists of forty-four schemas that, when combined, represent an overall building model. The building model consists of over 400 Entities and as such is one of the largest models developed to date. A breakdown of the various schemas follows:

        idm_core ----------------- basic physical model of the building (10 schemas)
        idm_performance ------- interface views to the various applications (9 schemas)
        idm_site ------------------ representation of site and climate (3 schemas)
        idm_space_function ---- space functions and occupant activities (3 schemas)
        idm_tech_system ------- mechanical system and HVAC components (19 schemas)

It is not practical to review the full COMBINE model here. As with the other models, we review selective parts, attempting to make visible its more interesting and innovative aspects.

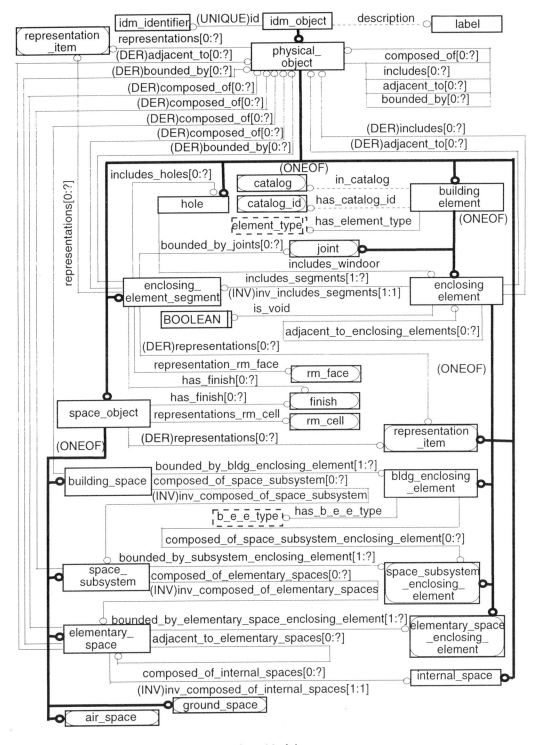

Figure 7.18: Overview of the IDM Core Model.

7.4.1 IDM CORE MODEL

The Integrated Data Model developed by the COMBINE team was based on several, specific design philosophies. One was that a functional definition of a building should be specified separately from its geometric representation. This allows object descriptions to be replaced with other descriptions, with little effect on the overall model structure. Another key concept was the development of the building spatial hierarchy. It was developed in detail, in order to identify building spaces and their multiple uses, and was articulated as needed to define how energy zones may overlap and possibly decompose the spaces. The various spatial elements in this hierarchy also needed to access the built-up construction that bounded the spaces, so that the lighting, energy and other behavior of the spaces and flows between them could be analyzed. The building model was carefully organized with its internal relations to capture this structure. In order to realize this structure, a new, space-based building topology was developed to define the relations between spaces and their boundaries. Each of these aspects is discussed in more detail as the building model is reviewed below.

A portion of the IDM Core Model is shown in Figure 7.18. All entities are subtypes of `idm_objects`, which assign unique identifiers and optional descriptions to entities. One main class of `idm_objects` consists of physical objects, which in COMBINE include spaces as well as walls and solid objects. The left-hand side of Figure 7.18 depicts the class hierarchy of `space_objects`, while the right-hand side depicts the class hierarchy of `enclosing_elements`. These two hierarchies are closely interlinked.

All `space_objects` have a set of finishes that serve as their bounding surfaces and a `rm_cell` that defines their geometry. The top-level `space_object` is a `building_space`, which is the total space formed by the building and is located on the building site. `Building_space` is composed of `space_subsystems`, which are groupings of `elementary_spaces` that may be used to define a section of the building space, such as a wing, a story, or an apartment. `Elementary_spaces` may be composed of lower-level `elementary_spaces`. Elementary_space is a singular space formed by physical boundaries, located on a single story that is composed of `internal_spaces`. `Internal_space` is a part of an `elementary_space` that must be distinguished for various reasons, for example, because it has a different function. Notice that `internal_space` is not a subtype of `space_object`, but rather a subtype of `physical_object`. It does not inherit `rm_cell` or `finish`. `Air_space` is the outside space where the building is or will be built and is used as a reference for outside walls that separate two spaces. The `ground_space` is the ground around the building. All of the space types are subtypes of `space_object`.

The three larger `space_object` subtypes carry general attributes that allow a variety of relations to be specified. These include `composed_of`, `bounded_by` and `adjacent_to` relations with other `physical_objects` (often constrained to a specific subtype). These provide general ways to define relations between `space_objects`.

The enclosing elements are shown on the right side of Figure 7.18. They are subtypes of `building_element`. `Building_element` can be of two types, `standard` or `custom`, as defined by the `element_type` attribute. If `standard`, the element is specified in a catalog with a `catalog_id`. Otherwise, it is a `custom` element whose definition is unique. `Enclosing_element` is a subtype of `building_element`. It can be either a void or made up of one or more `enclosing_element_segments`. It may be adjacent to other `physical_objects` and also to other `enclosing_elements`. `Enclosing_element_segments` are the basic part of `enclosing_elements`. They have a shape represented by `rm_face`, with orientations to spaces (as we shall see). They do not have any interior joints, may have zero or more holes and are made up of a uniform layer structure, with one or two finishes and are uniform on one or both sides. The composition of enclosing_element_ segments into the subtypes of enclosing_elements allows them to be aggregated into both space_subsystem_enclosing_element and also bldg_enclosing_element.

Closing_element has as subtypes bldg_enclosing_elements, space_subsystems_enclosing_elements and elementary_space_enclosing_elements. These correspond to elements that enclose building_space, space_subsystems and elementary_space, respectively. As shown, there is some redundancy; it is unclear whether enclosing_element_segments are carried at each of these levels of aggregation (they can because of subtyping) or whether the elementary_space_enclosing_elements carry the full wall structure and are aggregated upward, only referencing the lower-level aggregations in a non-redundant fashion. An application was written that explicitly loaded consistent model descriptions (redundantly) to all levels of aggregation. The structure of space_subsystem_enclosing_element and elementary_space_enclosing_element are partially shown in Figure 7.18 and fully in Figure 7.20.

Holes are defined at a high level, carrying a relation with all enclosing_elements. This structure was defined because the various cases where a hole may exist. Most often, a hole (a window or door) is within a segment making up an elementary_space_enclosing_element. At other times, a hole is more naturally represented between other segments. To allow different conditions, it was defined so as to be part of an enclosing_element at any aggregation level.

Both space_object and enclosing_element have their own domain-specific aggregation hierarchies, as already described. That is, they both have composed_of relations that show how the building space object is decomposed and how an enclosing_element includes other elements. In addition to these, however, there is also the general-purpose set of relations at the top right of Figure 7.18. These provide general aggregation-disaggregation relations and boundaries to ALL physical objects. The inclusion of two types of aggregation hierarchies was a result of different modeling philosophies of members of the COMBINE team. The problem with a general aggregation hierarchy, as defined using general composed_of or includes relations is that in building and managing a model, it becomes difficult to functionally distinguish the different levels of aggregation and to manage data at these different levels. For example, what level of enclosing element bounds an elementary space? Such a question can only be answered by distinguishing clearly the levels in the aggregation hierarchy.

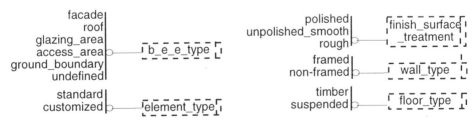

Figure 7.19: Four enumeration types defined in the idm_core models.

Some enumeration types were defined, shown in Figure 7.19, that allow distinguishing various types of construction. B_e_e_type has as values different types of building envelope elements. Surface treatments, wall-types and floor-types are also defined as enumerated types, allowing their values to be carried for easy run-time checking.

Figure 7.20 shows the detail structure of elementary_space_enclosing_element. It carries a length_measure to indicate its linear length along the floor, a direction to indicate siddeness of the enclosing element and a `construction_type`. The `construction_type` is defined as a series of `layers`, each with a `material` or designated as an `airgap`. Last, `zones` are defined as collections of `elementery_spaces`. `Story` is also made up of multiple `elementary_spaces` and carries a story number.

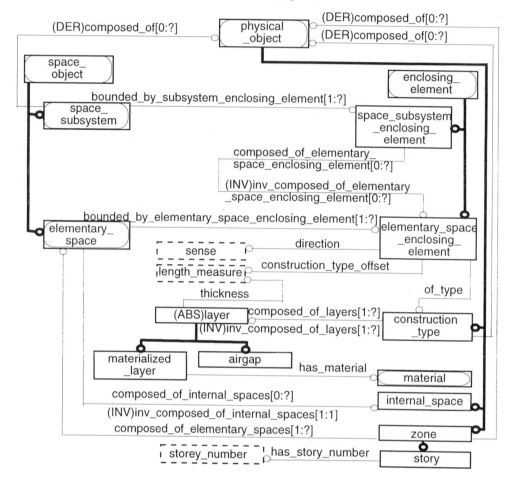

Figure 7.20: The detailed level of enclosing elements and construction.

7.4.2 IDM GEOMETRY AND TOPOLOGY

An important contribution of the COMBINE effort was the development of the Relational Model representation (rm-rep). The rm-rep development grew out of a review of STEP Part 042 and the recognition that its emphasis on boundary representation and constructive solid geometry was not adequate to the needs of energy analysis in building. In COMBINE, the essential quality of a building is that it provides a set of spaces that are enclosed by building elements. Rather than being construction element centric, the COMBINE model is space centric. The spaces need to be represented because they are the shapes through which energy, air and people flow.

The IDM provides an elaborate structure for carrying the boundary conditions of walls and a definition of the hierarchy of spaces bounded by walls. Spaces are aggregated upward, from `internal_spaces` and `elementary_spaces`. Elementary_spaces have enclosing_elements associated with them and (we assume) these are aggregated upward to define the enclosures for the aggregated space_objects. Notice that so far, we have not defined much geometry. Only a few lengths have been included. However, all physical_ objects have an associated representation. Enclosing_element_segments have an associated rm_face and all space_objects have associated rm_cells. These provide the linkages to geometric representations.

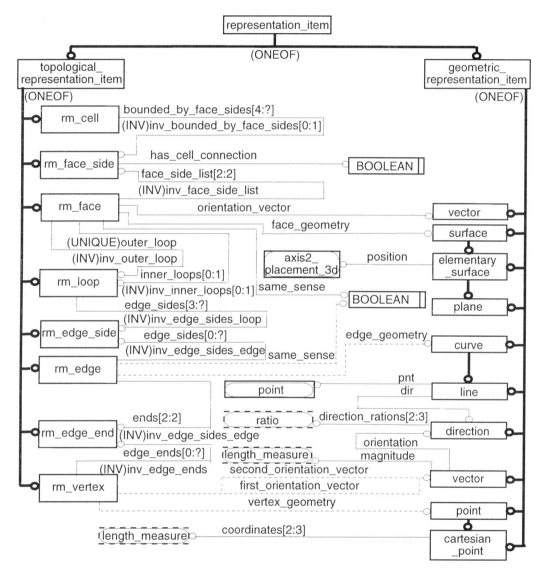

Figure 7.21: The EXPRESS-G diagram of the topology and geometry used in the COMBINE building model.

Figure 7.21 shows both the rm_model topology and the geometry. The geometry was adapted from Part 042. Rm-cells are the topological entities defining space shapes. They are bounded by rm_faces, each of which has exactly two rm_face_sides. Each rm_face_side is the partial boundary of one rm_cell. The rm_face is defined by one outer and multiple inner rm_loops, where each inner rm_loop designates a windoor—a door or window. Each rm_loop is defined as a sequence on rm_edge_sides, one or more of which corresponds to an `rm_edge`. Each `rm_edge` has two `rm_edge_ends` that are connected to an `rm_vertex`. An example of a relational model is shown in Figure 7.22. It shows two spaces, defined as `rm_cells`, that are bounded by sets of `rm_faces`, as outlined above.

The `rm_face`, `rm_edge` and `rm_vertex` each have corresponding geometrical descriptions that carry their geometric definition in 3D space. A small, planar subset of Part 042 was adopted for the geometrical descriptions.

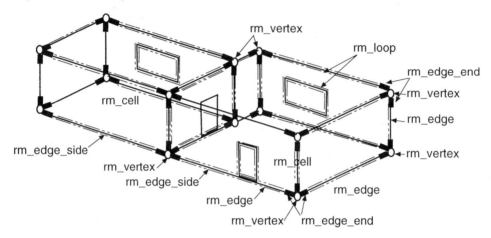

Figure 7.22: The material properties carried for use in energy studies in COMBINE.

In its pure form, the `rm_model` abstracts the construction elements that define the space boundary of a building as a set of planar faces. A wall is a set of one or more `rm_faces`, similar to the segments in the example of a building wall introduced in Chapter Three. That is, one `rm_face` is defined for each region of a wall (or ceiling or floor) that has its construction defined as a homogeneous sequence of layers. A `rm_face` and the corresponding `enclosing_element_segment` has an associated `elementary_space_enclosing_element` that carries a `construction_type`, made up of material (or airgap) layers, that defines the construction of the segment. These are aggregated into `space_subsystem_enclosing_element` to define

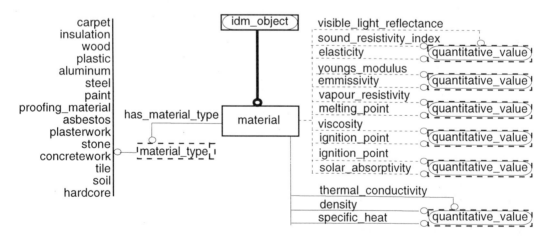

Figure 7.23: The material properties carried for use in energy studies in COMBINE.

the complete separation between two spaces. These are aggregated again into `bldg_enclosing_elements`, bounding the entire `building_space`. It is assumed that other physical objects can be associated with an `enclosing_element` or `space_object`, allowing the depiction of complex forms of construction. However, the focus here was to support energy-related design and analysis.

In order to support a range of energy studies, COMBINE incorporated a range of material properties. These are the properties used in the various energy analyses integrated with the model. Most were optional; `thermal_conductivity`, `density` and `specific_heat` were mandatory. Material also carried an enumerated type designating the material name. Only a subset of the enumerated material names that were included in COMBINE model are shown in Figure 7.23.

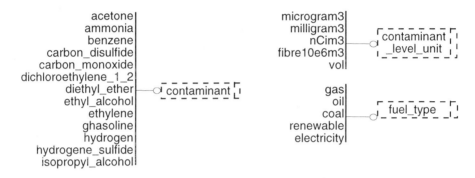

Figure 7.24: Enumerated types for contaminants and fuel types.

7.4.3 SPACE USE

An important consideration in the development of energy analyses are the uses of the building spaces. These uses include load-generating elements that are within the space, such as lighting, equipment that may generate heat and the activities of people. COMBINE also specifies the environmental conditions—lighting, air ventilation, air quality, temperature—that the environmental entities should realize.

Some enumerated types are defined first. These are shown in Figure 7.24. The space usage specification is shown in Figure 7.25. The space usage entity is characterized by six sets of requirements—lighting, air quality (in terms of contaminants), ventilation, allowed heating and cooling temperatures—and activity schedules for the two classes of heat generators in a space—people and equipment. The activity schedules and indoor air quality requirements are optional. Each of these requirements has a fairly detailed specification. For example, the `ventilation_requirement` has both a `total_required_space_ventilation` scheduled value and also a process ventilation scheduled value. Each scheduled value consists of a schedule and quantitative value. `Ventilation_requirement` also has a ventilation measure, consisting of measures for ventilation air per person, a required ventilation schedule, and an optional process ventilation. Occupants can be defined in terms of their latent heat gain, their sensible heat gain and their radiative heat gain. The `space_usage` subschema provides a first level, useful specification for specifying the quality of a human occupied space within a building.

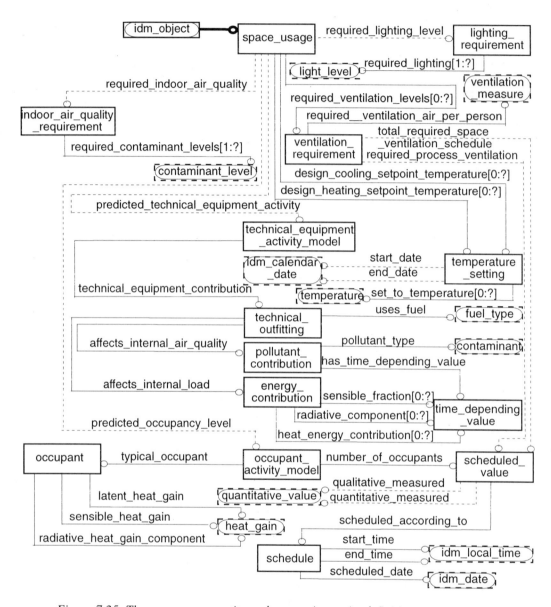

Figure 7.25: The space_usage entity and supporting entity definitions.

7.4.4 MECHANICAL EQUIPMENT

Into the building structure and spaces defined above is installed various mechanical equipment. The different classes of equipment are each small schemas; there are nineteen thus far. We will sample them here, focussing on heating, ventilation and air-conditioning equipment (HVAC).

The HVAC equipment is defined abstractly as a connected topological structure represented as a *graph*. Thus there are edge and vertex elements. The graph edges carry various kinds of flows. Some vertex elements have multiple connections, where the properties of the connection are defined by *ports*. Figure 7.26 presents some enumerated types that are used in COMBINE to define network properties, the level of decomposition of the system, the flow direction of ports and the kinds of flow within an edge.

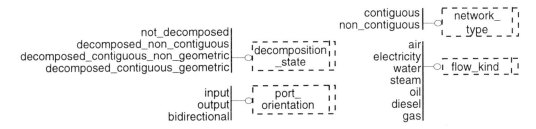

Figure 7.26: Some enumerated types defined for representing equipment.

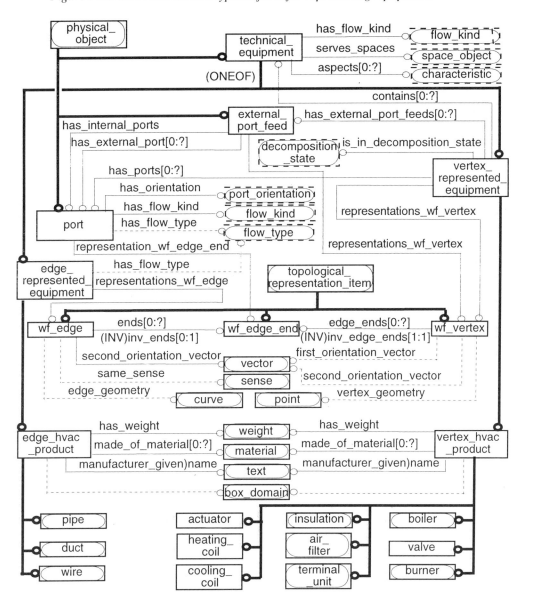

7.27: The overall structure of the HVAC equipment modeling in COMBINE.

Figure 7.27 presents an overall view of the organization of the mechanical equipment representation. The topological structure is shown in the middle of the figure, referenced from the idm_core geometry representation. (This portion of the geometry was not presented earlier.) It consists of three wire frame (wf) elements: `wf_edge`, `wf_vertex` and `wf_edge_end`. An edge references zero or more edge ends, which are also referenced by vertices. Edges and vertices have optional geometry. (The optional orientation vectors define direction and possibly the rotation about the direction axis.)

The mechanical equipment is defined as `technical_equipment` and its subtypes. `Technical_equipment` defines the kind of flow involved (only one), `aspects` which consist of text notes, and the spaces to which the `technical_equipment` applies. The `space_object` is the supertype space shown in Figure 7.20. `Technical_equipment` is specialized into either `edge_represented_equipment` or `vertex_represented_equipment`.

`Vertex_represented_equipment` is defined as having a set of `ports` and `external_port_feeds`. It has a `decomposition_state` (shown in Figure 7.26) and refers to a `wf_vertex` that represents it. Each `vertex_represented_equipment` may be decomposed into the lower-level `technical_equipment` that it contains. Each `port` has a connection location, for wiring, piping, ductwork, and so forth. Each has a flow-kind, -type and -orientation (direction). A port is represented by an `edge_end`. `External_port_feeds` have their own vertex location plus one or more external and internal ports. `Edge_represented_equipment` has an optional `flow_type` and references the `wf_edge` that represents it.

`Vertex_represented_equipment` is specialized into `vertex_hvac_product`. It maps the functional and topological entity into a physical description. `Vertex_hvac_product` has a `weight`, a set of `materials`, a box geometry and a description. This general product is specialized into different types of physical elements, such as `boiler`, `heating coil` or `terminal_unit`. Some of these entities are further elaborated in the IDM, but are not described here. The `edge_hvac_products` are given similar attributes, then are specialized into `pipes`, `ducts` or `wiring`. Each of these has more detailed definitions.

7.4.5 INTERFACES TO APPLICATIONS

The building, space and mechanical equipment models presented so far were defined to support a specified set of applications listed in Section 7.3.1. These were defined in the performance schemas of the overall model. The applications interfaced to the IDM through a set of application views. The application views were EXPRESS formatted files that were populated from data carried in the IDM and exchanged through maps. The maps were translations of the entities needed in the application view extracted from the IDM. Use of mapping languages, such as those reviewed in Chapter Eleven, were considered, but at that time were found inadequate for the job. The maps were written in C++ as an application interface library. General-purpose facilities were developed in the library and reused for many of the interfaces. These were then incorporated in an interface coded into an application, or into a separate program that prepared a file that the application could read.

All applications interfaced directly to the model for information dealing with the building envelope and layout. The application views dealt only with special-purpose performance and requirements data needed by the application.

One application view is presented in Figure 7.28, for the Superlink lighting analysis program. It supports the analysis of lighting in one or more zones. The zones are specializations of the building zones, defined in Figure 7.20. It allows assessment of the amount of artificial lighting required to satisfy light-level requirements within a space so that alternative natural lighting schemes may be tried and compared.

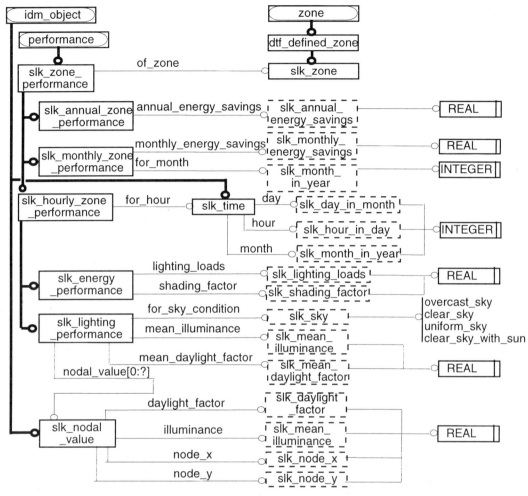

Figure 7.28: One of the Application Views developed in COMBINE. This one is for the Superlink lighting analysis program.

Lighting levels are defined at various locations, defined by `slk_nodal_values`. The application identifies the mean lighting requirements for each lighting zone on a hourly, monthly or annual basis. Similar application views were developed for each of the applications. Some were larger than the SUPERLINK example, such as for DOE-2 and the BRC energy code assessment program. Both CAD application programs—AutoCad and Microstation—were directly attached to the backend database.

A review of COMBINE is presented at the end of this chapter.

7.5 BUILDING ELEMENTS USING EXPLICIT SHAPE REPRESENTATION (PART 225)

The main aspect represented by current generation computer-aided design systems is the representation of geometry. Currently that geometry is mainly two-dimensional, depicted in drawings. Many experts expect that an early extension of current technology will be to support three-dimensional geometric representation of buildings, in order to support the modeling of different systems, such as structural, thermal and acoustic modeling, as well as to support construction planning. A fundamental component of data exchange is the shape and arrangement of the elements of which the building is composed. Part 10303-225 specifies a standard that allows the exchange of data between systems. Part 225 is currently the only fully defined and approved ISO-10303 STEP draft standard in the AEC area.

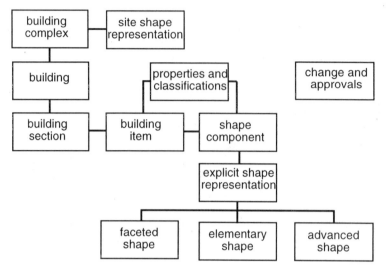

Figure 7.29: The Part 225 planning model.

7.5.1 PLANNING MODEL

The general planning model upon which Part 225 was planned and scoped is shown in Figure 7.29. It illustrates a building complex that may consist of one or more buildings on a site, where the site has a shape. A building consists of building sections, which are composed of building items. Each building item is composed of a shape component that may be explicitly represented by one of three types of representation: a faceted shape, an elementary shape or an advanced shape. A faceted shape consists of lines and planes. An elementary shape consists of lines, circles, conic curves, b-spline curves and planar, cylindrical, conical, toroidal and spherical surfaces. An advanced shape can consist of any kind of shape represented within the Part 042 Integrated Resources. These three shape classes are at increasing levels of complexity. Shape is just one of several properties and classifications. Part 225 also includes data for changes and approvals.

This simple structure was elaborated by defining classes of building entities. Generally, these are broken out into the classes of structure enclosure elements, building components, fixtures and equipment, service elements, and spaces. Several groupings are provided, including sections and levels, assemblies and groups.

The anticipated uses of Part 225 are for concurrent design processes, integration of building structure design with building systems to allow analysis, building design visualization and specifications for construction and maintenance. It includes the shape of the ground upon which a building is erected, specification of enclosing spaces and floor levels, specification of assemblies, such as roof and stairway, specification of material composition and specification of classification information. For these uses, it includes global positioning system (GPS) location data, building element components, component shape and location, positive and negative components, openings, and recesses. Part 225 does not address functional or performance characteristics of building systems or the information required for analysis of systems.

Part 225 relies heavily on Entities defined in Parts 041, 042 and 043. Many of the Entities defined there are specialized into more restricted definitions, allowing exchange translators to restrict the cases with which they must deal. Some issues are included regarding design administration, ownership, approvals, document and item references, lifecycle status and change requests.

This review presents an overview of the AIM short listing. That is, it covers the Part model after it has been integrated and resolved with Integrated Resources. However, it does not expand the definition to fully include all the relevant Integrated Resources, but instead refers to them. Thus it shows how a Part model utilizes the Integrated Resources.

7.5.2 ADAPTATION OF INTEGRATED RESOURCES

The top-level definitions of Part 225 are referenced from Part 041. They include entities from the application context schema, product definition schema, product property definition schema and product property representation schema, some of which are shown in Figure 6.1. From these general, high-level Entities, it adapts the Entity structure to the needs of building.

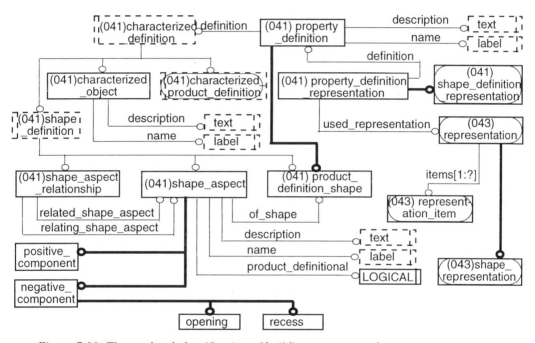

Figure 7.30: The top-level classification of building component data in Part 225.

Utilization and adaptation of Part 041 and Part 043 Entities are shown in Figure 7.30. The Part 041 Entity shape_aspect is specialized into positive_component and negative_ component. Negative components are further distinguished as opening and recess, where a recess does not pass through a structure_enclosure_element, but an opening does.

The characterized_product_definition shown in Figure 7.30 is articulated in Figure 7.31. Characterized_product_definition is a select type, referencing either a prod- uct_definition or a product_definition_relationship. Product_defini- tion is subtyped into a disjoint set of specific building product types. Notice that these subtypes of product_definition do not carry any additional attributes or relations. However, each Entity has a set of associated WHERE clauses that apply rules to each of its instances. These check that the instances define a legal schema. For example, a building_complex instance is checked that its related_product_definition is a building instance, a building instance is checked that its related_product_definition has one or more building_ sections and that its relating_product_definition is a building_complex instance. A building_section is checked that it participates in one product_ definition_relationship, where the relating_product_definition is a building, that it has a name and that its related_product_definition. Entity instances are of type building_level, building_element, fixture_equipment_element, service_element, space_element, or structure_enclosure_element. Each Entity instance is checked that it has a name.

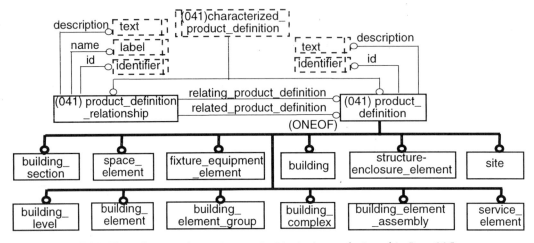

Figure 7.31: The subtypes of product_definition designed in Part 225.

The bottom-level elements are building_element, fixture_equipment_element, structure_enclosure_element and service_element. Each is checked that it has a product_definition_shape that is referenced by exactly one positive_component with a description 'main component'. This guarantees that there is one shape for each solid building element. Thus, Part 042, and, by adoption, Part 225 rely on the general hierarchical relations of relating_product_definition and related_product_definition, which are parent and child, respectively.

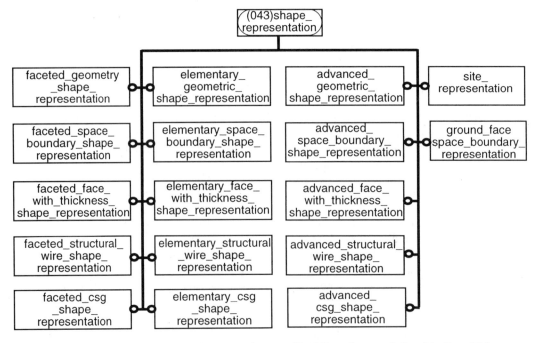

Figure 7.32: The seventeen different classes of building shapes, defined in Part 225.

7.5.3 SHAPE CLASSIFICATION

`Shape_representations` are specialized into seventeen different abstract shape classes, as shown in Figure 7.32. Fifteen of these classes are defined by a three by five matrix of classes, composed of shape classes—faceted, elementary and advanced—and five geometric modeling

	Faceted shapes	**Elementary shapes**	**Advanced shapes**
Wire frame shapes	Defined as one item of `composite_curve`, `polyline`, `trimmed_ curve`, `mapped_item`, possibly placed in space with `axis_placement_3d`.	Defined as one item of `composite curve`, `circle`, `ellipse`, `trimmed_ curve`, `mapped_item`, possibly placed in space with `axis_placement_3d`.	Defined as one item of an `advanced_curve`. An `advanced_curve` is defined as a `composite_ curve`, `circle`, `ellipse`, `offset_curve_3d`, `curve_replica`, `trimmed_curve` or `mapped_item`. All references are to valid `advanced_curves`
Face with thickness shapes	Defined as a `face`, bounded by a `poly_loop`, with a 'thickness' defined as a `measure_represen- tation_item`, possibly placed in space with `axis_placement_3d`.	Defined as `elementary_ surface`, bounded by `edge_curves` whose edges are `conic`, `line`, `poly- line` or `b_spline_ curve`, trimmed with `cartesian_point`, with a 'thickness' defined as a `measure_represen- tation_item`, possibly placed in space with `axis_placement_3d`.	Defined as an `advanced_face` (see above), with a 'thickness' defined as a `measure_ representation_item`, possibly placed in space with `axis_placement_3d`.
Space boundary shapes	Defined as one item of `closed_shell` or `mapped_item`, bounded by a `poly_loop`, possibly placed in space with `axis_placement_3d`.	Defined as one item that is an `elementary_space_ boundary_shape_re- presentation`, which is composed of `elementary _surface` whose edges are defined as `edge_curve`. Face bounds may be `conic`, `line`, `polyline` or b- `spline_curve`, trimmed with `cartesian_point`, possibly placed in space with `axis_placement_3d`.	Defined as `closed_shell` or `mapped_item`, defined by `advanced_faces`. An `advanced_face` is defined as `elementary_surface`, `swept_surface` or `b_spline_surface`, bounded by `lines`, `conics`, `polylines`, `surface_ curves` or `b_spline_ curves`, trimmed by `cartesian_point`, possibly placed in space with `axis_placement_3d`.
Geometric shapes	Defined as one item of `faceted_brep` or `mapped_item`, bounded by a `poly_loop`, with the `face` specified as a `plane`, possibly placed in space with `axis_placement_3d`.	Defined as one item of `manifold_solid_brep` or `mapped_item`. Each face is of type `elementary_ surface` bounded by `edge_curve` of type `conic`, `line`, `polyline` or `b_spline_curve`, trimmed by `cartesian_ point`, possibly placed in space with `axis_ placement_3d`.	Defined as one item of `manifold_solid_brep` or `mapped_item`. Each face is of type `advanced_face`, as defined above.
CSG shapes	Defined as a `csg_solid`, with each `mapped_item` being another `csf_shape`, possibly placed in space with `axis_placement_3d`.	Defined as a `csg_solid`, `extruded_area_solid`, `revolved_area_solid` or `mapped_item`, with each component shape being of these types, possibly placed in space with `axis_ placement_3d`.	Defined as a `csg_solid`, `extruded_area_solid`, `revolved_area_solid` or `mapped_item`, with each component shape being of these types, possibly placed in space with `axis_ placement_3d`.

Figure 7.33: The matrix of shape classifications used in Part 225.

classes—structural wire shape, face with thickness, geometric shape, boundary shape and csg shape. All classes are three dimensional. The shape classes identify different ranges of geometrical representation. Faceted shapes are those defined by lines and planes. Elementary shapes are those defined by regular planar and curved surfaces: sphere, torus, conics. Advanced shapes are all the rest allowed in Part 042, but are packaged into advance face and advanced curve representations.

The five geometric modeling classes address different functional capabilities of geometric modelers. These are shown in Figure 7.33.
- *Structural wire frame* representations are those defined by edges and vertices, without faces.
- *Face with thickness* representations are those that consisting of a face, defined in various ways, that is extruded some specified depth; the shape remains unevaluated.
- *Space boundary* representations are those for spaces within a building, represented by a variety of means.
- *Geometric shape* representations are the Breps and include those that are defined by evaluating the Boolean combinations of simple shapes such as planes, cylinders and cones, as well as more complex extruded and rotated shapes.
- *CSG representations* are those that are defined by various shapes combined with the Boolean union, intersection and difference operators, carried in unevaluated form.

The two extra representations in Figure 7.32 address terrain.
- *Ground face* representations that represent terrain as a set of poly_loops corresponding to contours, with an associated thickness of each.
- *Site* representations represent terrain as a connected face set or geometric curve set, defined by poly_loops or polylines, respectively.

By making the strict shape classes shown in Figure 7.33, the exchange of building part geometry between CAD systems of similar capability is facilitated. The geometry classes greatly reduce the range of geometry that would have to be addressed by an AEC CAD vendor developing an exchange of building model geometry, in contrast to a full Part 042 exchange. Part 225 also includes standard approval assignments for defining the approval status of building parts. It also includes portions of Part 041 dealing with people and organizations.

7.5.4 DISCUSSION

Part 225 is an attempt to provide sets of geometry that facilitate the exchange of geometric model data between CAD systems of similar capability. Those with similar geometry may target their geometry to a particular subset of the whole of Part 042. However, the specification of Part 225 does not define the scope of these subsets.

Part 225 deals with individual parts and their hierarchical assembly and group organization, but not their functional relations. Such a model seems most appropriate for layout checking and spatial conflict testing and, to some degree, for developing visualization models. Part 225 purposely omitted functionally specific connectivity relations, such as those defined in CIMsteel or COMBINE.

In contrast to CIMsteel and COMBINE, Part 225 relies completely on STEP technologies and was initially scoped and specified according to ISO-STEP procedures. An IDEF0 process model was developed to describe the architectural design processes being supported. This process model was very general, not addressing the details of actual practice discussed in Chapter One. An initial part model was then developed as an ARM, in EXPRESS, then interpreted so as to be consistent with the Integrated Resources. The ARM is much clearer, in terms of characterizing the intentions for the Part Model, than the later AIM. Mapping tables from ARM Entities to AIM Entities were defined.

Part 225 demonstrates how Integrated Resources may be adopted to achieve integration goals without defining a large model from scratch. The short form AIM defines forty-five new Entities needed for the Part. On the other hand, adaptation of this sort leads to less clarity of definition, because the Entities being used have to accommodate a variety of engineering domains and procedures. Funding for the development of Part 225 came primarily from European sources.

7.6 OVERALL ASSESSMENT

Both CIMsteel and COMBINE firmly addressed the issue of integrating a heterogeneous set of applications in a specific design domain. Both involved a consortium of specialists, who defined a model that had to accommodate diverse viewpoints regarding modeling strategies and techniques. Both relied partially on STEP technology, but found many places where they had to go beyond what was available. Neither fully followed STEP procedures. In contrast, Part 225 relies entirely on STEP technologies and followed STEP procedures.

The COMBINE model addressed squarely the issue of representing both spaces and their enclosures—a traditional issue in architecture that has proved vexing to many CAD developers and researchers. The issue is, should a building be represented explicitly as spaces that are manipulated directly from which are derived the enclosing walls, floors and ceilings, or should a building be represented explicitly as construction elements, which are edited directly and from which enclosed spaces are derived? Their space layout application designed both, and both were represented equally and consistently in the IDM. However, they implemented a special program that generated the consistent representation. Part 225 also represents both, but without defining the relations that must exist if the solids and the spaces representing a building are consistent. It is assumed that this would be done by a separate spatial conflict testing application.

While the CIMsteel developers defined their model around a fixed process plan, the COMBINE effort attempted to abstract from a single process plan and develop the information technology tools to allow various process plans to be implemented. They identified the issue that arises in an environment in which many design tools exist, possibly one for each task. Their tool integrated process coordination and data exchange within a single interface. The Combi-Net windows addressed the issue of "what information must be available in order to allow an application to execute and what previous application will have generated that information?" The result is a partial ordering on applications, but when we realize that the sequence of applications can be decomposed into individual sets of data, i.e., the data that is needed to run Superlink on one of several zones, then the management of application sequences that may be both very detailed and at other times aggregated to sets of operations can be seen to be a very complex issue, only partially addressed by COMBINE. CIMsteel restricted its process to that shown in Figure 7.1 and did not allow decomposition and partial iterations.

Part 225 defines a completely general aggregation hierarchy, using upward and downward relations. Both CIMsteel and COMBINE use domain-specific aggregation hierarchies, though they both also include in their schemas general-purpose aggregation relations for groupings not included in the fixed hierarchy. The incorporation of two different abstraction strategies points out the philosophical conflicts that can occur within any modeling effort. The issue is whether a model should use general or functionally specific compositional structures. For example, the general hierarchy in COMBINE was defined by `composed_of` and `includes`, referring to children and parents in a hierarchy, respectively. These relations can form an abstract hierarchy of any number of levels. They need to be flagged, however, by the associated function (structural, mechanical, electrical hierarchy), using an enumerated type. A specific hierarchy might consist of three fixed levels of aggregation, such as floor_level, zone and space. Certain functions are associated with each level. Initial work by the COMBINE and CIMsteel teams on these two approaches suggests that general-purpose aggregation hierarchies were more difficult to work with and to interpret. Both relied on the fixed-level hierarchies in their implementations.

One design objective in the development of the COMBINE IDM was the separation of geometrical entities from the functional entities and their relations. A review of the IDM shows that this objective was realized quite clearly. However, this separation has not been assessed. That is, did the model achieve greater flexibility or process functionality because of this separate structuring of form and process? The strong relationship of physical form to function has been an issue in architecture since the days of Louis Sullivan. Work in other fields of design has added a third information component to engineering models called behavior. Behavior is the performance of a product, according to some function. Form, function and behavior have been used to organize product models in a variety of engineering domains.

Considering these building models in relation to the early models defined in Chapter Two and the more recent models referenced at the end of that chapter, the current aspect models differ in several ways:

- Most early modelers placed a central emphasis on geometry and how to represent building shapes, while recent modelers consider geometry a "solved problem" and use existing STEP solutions. In the CIMsteel and COMBINE building model efforts, geometry is defined implicitly, as a set of variable dimensions and parameters for a building part, in place of a much lower-level definition based on pure geometry as defined by current CAD systems. Part 225 was defined to represent explicit geometry (although face with thickness and CSG are not truly explicit). It remains an open question regarding the degree that building geometry can be adequately defined using implicit, parameterized shapes.

- Most early building models were implicit and fixed within a larger application environment, while now the building model is explicit and accessible through a high-level language, allowing redefinition, and potentially, change.

- Recent work on building modeling places increased emphasis on the conceptual structure and organization of data, and on the underlying data model used to represent aspects and concepts.

7.7 NOTES AND ADDITIONAL READING

The CIMsteel web site is http://www.ac.uk/civil/research/cse/cis. Further reading on the CIMsteel structural steel building model is available in Watson and Crowley [1997], Crowley and Watson [1997], Watson [1995], Watson, Crowley, Boyle and Knowles [1993] and Crowley, Watson and Christodoulakis [1997]. The newly formed Part 230 Building Structural Frame: Steelwork model site is http://www.nist..gov/sc4/step/parts/part230/current/ .

The two major reports on COMBINE are [Augenbroe,1993] and [Augenbroe, 1995]. Much of the COMBINE technical documentation is available on a website http://erg.ucd.ie/combine.html. The COMBINE II Final Report is available at a TU Delft server: http://dutcu15.tudelft.nl/~combine/. It also has the COMBINE-2 building model. An important task report is [Lockley and Sun,1995]. More general reports on COMBINE include Augenbroe and Rombouts [1994], Augenbroe and de Wit [1997], Amor and Augenbroe [1997] and Clarke and MacRandel [1993].

The Part 225 reference is ISO/DIS 10303-225[1996].

There are several papers dealing with the structure between spaces and their enclosing construction. In addition to the work done by the COMBINE team, there is the work of [Björk, 1992b], Eastman and Siabiris [1995] and Eckholm and Fridquist [1996].

In parallel with the major efforts reviewed in this chapter, there are a number of other efforts, most smaller in nature that have attempted to develop aspect models for particular phases of the building lifecycle. Work developing models supporting *design* are reported in Eastman and Siabiris [1995], Rivard, Fenves and Gomez [1995] and Turner [1997]. Work addressing the *building construction* lifecycle phase are reported in Luiten [1994], Warszawski and Sacks [1995], Froese [1995], Fischer, Luiten and Aalami [1995] and Froese, Rankin and Yu [1997], Tarandi [1998] and Jägbeck [1998]. An aspect model for precast facades is reported in Karhu [1997]. A good overview paper addressing building models for *facility management* is Staub [1995]. Another is Ojwaka [1995]. A building model for this phase of the lifecycle is reported in Svensson[1998].

Work that addresses the articulation of design knowledge into a two- or three-way organization of form, function and behavior (often called *structure*, function and behavior) includes the work of Navathe and Cornelio [1990], Gero, Lee and Tham [1991], and Goel, Gomez et al. [1996].

7.8 STUDY QUESTIONS

1. Define a simple structure made up of eight to twelve structural steel members. Dimension, give preliminary size and assign an identifier to each of the structural elements. With recognition of the lifecycle phase in which you are working, represent this structural system in CIMsteel. Select the entities needed within the model and define multiple instances of them, either by copying the entity definitions for each instance and assigning values, or by using the STEP instantiation tools. What ambiguities do you encounter?

2. Define a simple, two- or three-room building, such as the one shown in Figure 7.23. Locate doors and windows. Carefully dimension the layout and identify the enclosure entities with IDs. Assume all walls are homogeneous. Using the approach proposed in Question #1, develop the subset of COMBINE ITM entities needed and instantiate the model.

3. Lay out a small run of ducts that feed air to two diffusers, limiting the number of entities to less than ten. Dimension the layout in terms of its centerlines. Outline the structure of the COMBINE model to represent this duct run, in terms of the wireframe centerlines. Associate entities with the vertices and edges.

4. Look up a diffuser in an equipment catalog and discuss how the port entity would be useful in describing it.

5. Examine closely one low-level physical entity in either COMBINE or CIMsteel. For example, identify a girder in CIMsteel or a wall in COMBINE. Flatten the specialization structure of entities that are inherited into the selected entity, so that all its attributes are made apparent. Discuss any redundancies you find and propose how the model could be simplified.

6. Examine the structure of a wall using Part 225, assuming faceted shapes. Define its component parts as well as its overall shape. What structure does this carry, in relation to the example wall defined in Chapter Five? In comparison to a wall in COMBINE?

7. Review the papers and data models referenced above that structure spaces and their enclosing construction. Assess the alternative models and discuss their differences. Keeping all the

viewpoints presented in mind, propose any one of the models with refinements, so as to produce the best of them all. In developing your refinement, consider the model from a particular perspective, such as that of an architect, energy specialist, or contractor.

CHAPTER EIGHT
Building Framework Models

My main task has been to show that there is a deep and important underlying structural correspondence between the pattern of a problem and the process of designing a physical form which answers that problem. I believe that the great architect has in the past always been aware of the patterned similarity of problem and process, and that it is only the sense of this similarity of structure that ever led him to the design of great forms.

The same pattern is implicit in the action of unselfconscious form-producing system, and responsible for its success. But before we can ourselves turn a problem into form, because we are self-conscious, we need to make explicit maps of the problem's structure, and therefore need first to invent a conceptual framework for such maps.

Christopher Alexander
Epilogue
<u>Notes on the Synthesis of Form</u>
1964

8.1 INTRODUCTION

Chapter Two presented early efforts to develop integrated building models. The efforts reviewed were based on older software development concepts; they incorporated new concepts, invented as needed, in such areas as geometric modeling, object representation, attribute handling and data modeling. Since then, design and engineering software has become a major industry. Software is a staple, in the sense that it is used daily and taken for granted as a necessity of work. In some product areas, rich sets of applications have been adopted by strong user communities, those, for example, associated with electronic chip and circuit board design and fabrication, and with piping and petrochemical plant design and construction. In these areas, product modeling efforts have been led by the user communities themselves, who are pushing for the integration of existing sets of advanced applications and demanding both easier use of the applications and the benefits of data exchange to support improved processes. At the same time, software application vendors have a strong, vested interest in seeing their products tie into the emerging integration framework.

In contrast, the development of an integrated building model has not had the support of a strong user community. No significant community of users exists that has adopted advanced applications; the applications widely used currently are largely generic and have low semantic content. The industry has not demanded data exchange in order to support process re-engineering. Organizational diffusion of the building industry and the reliance on drawings as a culturally defined representation inhibit many forms of representational innovation (layer management in electronic drawings being a notable exception). Instead, the product modeling effort in the building industry has had to address a much more complex agenda, developing both the rationale of building product models as well as the technical details, including

- developing new areas of application development, such as CAD/CAM or robotics or energy efficient design, that strongly add to the need for an integrated building model

- showing ways that building industry organizations may re-engineer themselves, using building modeling as a facilitating technology
- addressing unique technology issues involved in the product modeling of buildings that stem from the project-oriented structure of the industry

These issues require that building modeling be a research and proselytizing activity as well as a standards effort.

Possibly the strongest rationale for building models exists in Europe, as a result of the melding of the separate national building industries and markets into the European Union (EU). The integration of EU building industries under a common framework requires—to some degree—homogenization or rationalization of building codes, building products, contracting and legal processes, among many other things. As a result, some degree of process re-engineering is accepted as necessary by the members of the European building industry. Anticipating these needed changes, the EU has promoted several large projects to demonstrate the use of product modeling toward these goals. The COMBINE project, reviewed in Chapter Seven, is one such project. These efforts are bearing fruit, with more firms and organizations in the building industry in Europe using advanced products and data exchange than in the US.

Another issue thought to be inhibiting building product models is a technological one. The structure of the standard STEP architecture was presented in Chapter Five and diagrammed in Figure 5.2. It defines an application protocol on three levels—the reference level, the interpreted level, and the instantiation level—and augments the protocol with a library of integrated resources, which are shared and reused among multiple application models.

Focusing on application data, STEP addresses the modeling of the components of large systems, where most of the components have a single, primary function. This emphasis is made apparent by reviewing the application protocols developed or proposed so far, as listed in Figure 5.3. Shipbuilding, for example, is partitioned into *ship arrangements* - Part 215, *ship molded forms* - Part 216, *ship piping* - Part 217, and *ship structures* - Part 218. In this case, the STEP methodology involves choosing middle-level units to model, supporting piecemeal integration. The overall integration is left to be achieved later, when a set of application models has been fully realized and tested. The assumed mode of integration is currently through sharing of Integrated Resources.

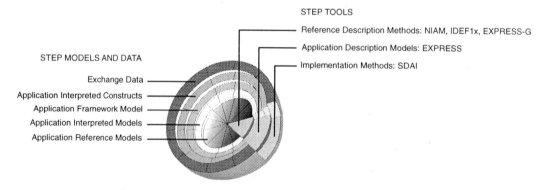

Figure 8.1: A proposed revision to the AEC STEP architecture. An Application Framework Model is added.

In contrast, the building industry's major concern is coordination across systems, in part because many of the building components—walls, roofs, floor/ceiling assembles—are multi-functional. These components define spaces and have structural, thermal and acoustic performances. Even at a small scale, multiple functions are affected by the location, size or materials of a wall or ceiling. As a result, many in the industry have thought that before developing models for individual systems, a broader framework dealing with common geometry and relationship issues should be defined first, into which the various aspect models could be placed. For example, while the CIMsteel and COMBINE models can be developed independently, they each derive their loads from each other. The building envelope and mechanical equipment specified in COMBINE define the structural loads in CIMsteel, and the building envelope of COMBINE must fit within the structure defined by CIMsteel. Since they cannot interact, each has to define the needed context in an *ad hoc* manner.

Many in the building industry think that the integration between application domains is the major information technology problem facing the industry. This new level of integration has been called the *building core* or *building framework model*. It is considered to fit as a new component into the ISO-STEP architecture as shown in Figure 8.1. Framework models are also being taken up for consideration in other industry sectors, such as in process plants and shipbuilding. Efforts are underway to introduce the concept of a product framework into the STEP vocabulary and product structure.

Since 1988, the STEP AEC Committee has undertaken various studies and trial efforts to develop a building product model structure capable of capturing the information needed to represent an overall building. Two important early efforts in this regard were (1) James Turner's report on the development of an AEC Building Systems Model, and (2) Wim Gielingh's study of a General AEC Reference Model that came to be called the GARM. Turner's work included the development of a NIAM model that partitioned a building functionally—that is, in terms of the systems that supported different building functions. The functional subsystems of the building were classified as:

> Active: lighting, communication, structural, spatial, conveyance, electrical
> Mechanical: plumbing, air-conditioning, ventilation, heating
> Automation: alarm, security, fire, energy
> Passive: acoustic, interior, enclosure

Systems were abstracted into their general components. Active systems have the following classes of components: sources, paths, controls, measurements, storage, terminals. Mechanical and Automation are subclasses of Active. Passive systems have no explicit controls. Turner's AEC Building Systems Model showed that a general-purpose building model could be defined—at least at a high level of abstraction—to be used as a framework for more detailed application models.

Gielingh's GARM addressed quite different concerns. It was not concerned with a building model *per se*, but rather a framework for addressing building model issues. It laid out the lifecycle issues, organized by the milestones that are the transitions between lifecycle phases. This classification was adapted for use in Chapter One. The GARM report also pointed out that we rely on different kinds of descriptions in the representation of building information. One way to look at these different descriptions is to distinguish form (geometry) from function and to define the different kinds of functions that are of interest. Each form and function description requires its own view of the building model.

In terms of product definition issues, the GARM report makes the practical distinction between (1) the specification of a parametrically varying product, (2) a class specification of a product of a particular model (and size) and (3) instances of that model installed at different locations within a building. We rely on all three levels of specifications and must manage the relationships between them. The GARM report pointed out how much information is hidden in the practices of different design disciplines—the attributes of interest, the performance dimensions and their measurement.

It also pointed out how the project team approach to building design practice results in decomposing the design and resolving it in parts that are merged back together to generate the solution, with the parts having diverse "ownership" by the various designers responsible.

One of the lasting concepts of the GARM report is the *GARM Hamburger model* of hierarchical design. It proposed that as design proceeds, usually in a top-down manner, a system is defined as being made up of some high-level components, each with a functional specification. At the next level, each functional specification (Gielingh called them *Functional Units*) was resolved by a *Technical Solution (TS)*. A TS is another design (or purchase) comprising both a form and one or more functions that fulfilled the higher specification and which together defined another (lower) level of Functional Units that were the parts needed by the Technical Solution. An example of such a hierarchical decomposition is shown in Figure 8.2. Each system has its own attributes. At one level, this structure has similarities with the system model proposed by Turner. However, it is unique in recognizing that at each level, there are two sets of values—one set defining the Functional Unit, and the other defining the Technical Solution.

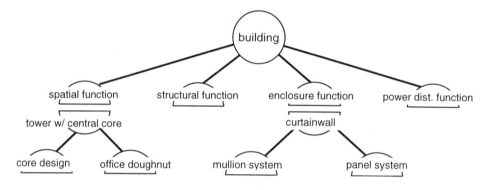

Figure 8.2: The Hamburger model of FU and TS in a hierarchical structure.

For assessment, both sets of values are needed. The Functional Unit (FU) attributes provide the goal specification at each level of abstraction, while the Technical Solution (TS) below it provides the design response. At each level, the TS may fulfill some functions, not fulfill others and over-achieve other functions. The Hamburger name comes from the appearance of the FU and TS pairs. In retrospect, we can see that the Hamburger model involves an aggregation—"building"—followed by a specialization, "enclosure function" specialized to "curtainwall"—followed by another aggregation, and so forth. The ideas put forward in Gielingh's GARM report are an excellent introduction to the many issues that must be addressed in a building product model.

In this chapter and the next, we review three recent, important standards and industrial efforts to define an overall building model. We start by reviewing the Finnish, RATAS national building model project, an early effort to resolve the basic problems of a national building core model. Second, we review Part 106, the Building Core Construction Model (BCCM), currently under development within the ISO-STEP community. It has been iteratively refined from the early works of Turner and Gielingh. In Chapter Nine, we review the current state of work by the International Alliance for Interoperability (IAI) to develop the Industry Foundation Classes (IFC) as a kernel building model. This is a strong, industry-led effort attempting to quickly make progress toward a unified building model.

These different efforts frame the current capabilities for defining an overall building model and reflect the special situation of building models in the construction industry.

8.2 RATAS

In the 1990s, some of the efforts in building modeling have been in support of Computer Integrated Construction (CIC). CIC addresses both the Construction Planning and the Construction Execution stages of the building lifecycle outlined in Chapter One. The need for CIC comes from the recognition that many aspects of a building's construction are essentially information driven, from the acquisition of materials to the details of certain fabrication activities (as versus labor driven or construction technology driven). CIC applies to all information technology applications in construction. In practice, its scope varies, depending upon the country and the nature of its building industry. In part, CIC draws upon computer-integrated manufacturing (CIM) for inspiration.

The fundamental information received by a builder and which forms the basis for planning a building project is the building description. The history of building descriptions is almost five centuries old, and for most of this time, drawings—particularly plan and elevation—were the main forms of representation. Later, sections, isometrics, and other projections were added. In the late nineteenth century, specifications were added to drawings, as a means to convey material properties and finishes. Each of these representations requires manual interpretation and merging with the conventions of construction practice to derive a construction plan. Errors and omissions inevitably arise in this process. If such a representation can be made in digital form and read electronically, then its interpretation and reorganization could be automated, with the potential elimination of copying and interpretation errors. Such a representation also supports further automation, for example, with regard to construction planning, material ordering, and potentially, construction robotics. Thus an important goal of CIC has been to define an appropriate representation or model with which to describe building projects, one that can support a range of computerized applications.

An influential effort to develop the concepts, structure and partial content of such a construction-oriented building model was undertaken at the Technical Research Center of Finland (VTT) by Bo-Christer Björk and his colleagues. As defined by Björk, the problem centered on one question:

> *How should we structure digital information describing a building in order to facilitate as much as possible the exchange of information between the computing applications in construction that produce or utilize this data?*

The VTT effort was preceded by a series of studies carried out by cooperating researchers and practitioners and executed by a committee appointed by the Building Information Foundation, known as the RATAS committee. (This committee continues its operations and deals with all standardization efforts related to IT use in construction. Its work includes such topics as EDIFACT and CAD layering.)

The building product model framework defined in 1987 is the original RATAS model. It should be noted that the framework presented below builds partly on the results of the earlier RATAS work but was carried further by Björk in subsequent work and does not represent officially approved RATAS guidelines. The framework defined the requirements for such a building model as follows:

1. *comprehensiveness:* The model should contain all the information needed to construct and maintain the associated building.

2. *coverage of all phases in construction process:* The same building product model should be capable of expansion and alteration throughout the briefing-design-construction-maintenance process by all the different participants. The model should address each of the phases of a

construction by offering "recipes" for producing different types of documentation. The phases, however, will not form part of the model itself.

3. avoidance of redundant information: Each item of information should be represented only once, eliminating replications. Current building documentation invariably includes inconsistent information.

4. output documents and formats: It should be possible to output the data contained in the model in very flexible ways, defined at any time during the model's lifetime.

5. independence from software and hardware systems: The model should be a logical structure that can be implemented using several different programming and database techniques. This requirement means that concessions should not be defined for a particular CAD or hardware system.

These requirements evolved somewhat as the VTT group proceeded.

The VTT group approached their work as having a layered framework architecture, similar to the ISO-STEP work which was being conducted in parallel. The group recognized the complexity of information and realized that it could only be effectively defined in conceptual layers. Their particular layered scheme was as follows:

(1)   At the bottom level is the *information modeling language.* This provides a low-level, domain independent declarative structure for defining a building model. Examples are the relational model, an object model, frames, the entity-relationship model and EXPRESS-G.

(2)   The next level is the *generic product description model* that defines high-level abstractions that are generally useful in describing a product. Such abstractions include shape information, location, aggregation-decomposition, types and attributes.

(3)   The third level is the *building kernel model,* defined within a given language structure and with some generic constructs. At this level, the model defines generic information structures common to most phases and disciplines in the construction of buildings. Some examples are high-level abstractions for site, building shell and systems, and spaces.

(4)   The next level defines multiple *aspect models,* which correspond to specific construction disciplines and/or phases in the construction process. Examples are the COMBINE IDM, some RATAS prototypes, and the CIMSTEEL model.

(5)   *Application models* make up the fifth level. They use the first four levels to define the conceptual schema for a particular application, defined generically, separate from a particular building.

Each level defines a generalized domain of possible definitions, or UoD, at some level of abstraction. Each of the higher-levels uses the lower levels as resources to define its UoD. At the top level is the UoD of the particular application domain, here a generalized building. This layered structure describes a product domain model starting with the low-level details and showed how details are built up to define high-level objects.

8.2.1 INFORMATION MODELING LANGUAGE

The RATAS team outlined several issues that affect the choice of an information modeling language:

•   capabilities for modeling the semantics of the universe of discourse without simplifications caused by the information modeling language
•   capability for modeling the designer's intentions
•   support for the evolutionary process of design and problem solving during construction
•   usefulness for the exchange of data between heterogeneous computer applications in construction

- technical feasibility for implementation using current commercial software
- realistic possibilities for achieving standardization, in terms of reaching consensus in standardization bodies and expenditure of time

The first three points emphasize the use of powerful and innovative modeling languages, while the latter three give priority for simpler, well-known ones. The RATAS team initially adopted the entity-relationship modeling language, extended to include *inheritance*. Later, they adopted EXPRESS as their modeling language, along with EXPRESS-G.

8.2.2 GENERIC PRODUCT DESCRIPTION MODEL

The RATAS model is based on a four-level conceptual definition structure. At the bottom level are *domains*. A domain defines the range of values allowed for an attribute. The base level domains are numeric values, text strings, bitmap or video pictures, and lists of such entities. Special purpose domains may be defined to address particular national building regulations and standards. For example, the State Building Board in Finland has defined a Quality Level, which, when associated with a building object, can take on one of four possible values: I, II, III, or IV, lowest to highest. In addition, RATAS extended these initial specifications to include composite properties, such as `space_use` and `material_type`, which are attributes of other objects.

The second level of information deals with an object's properties, which are carried as *attributes*. Attributes have a domain or class that defines their possible values. Examples of attributes for an office are shown within the heavy-lined box in Figure 8.4. The RATAS studies noted that attributes related to geometry include location, orientation and shape. A property of an object class may be defined using different combinations of attributes, and in many models, multiple definitions may be needed. Rather than develop a new way to define these attributes, RATAS chose to use existing, well-developed methods, including IGES and EXPRESS.

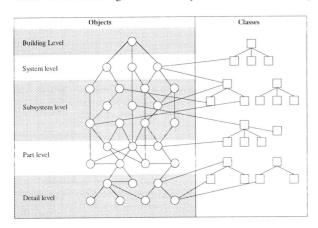

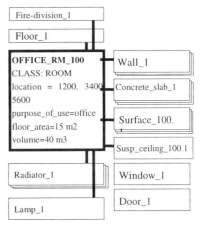

Figure 8.3: Characterization of the abstraction hierarchy of objects and classes.

Figure 8.4: An example of an object instance and its relations.

Specific aggregations of attributes in RATAS are defined as *classes*. Objects belonging to a class automatically receive all the attributes that make up the class. Each object instance belongs to at least one class. Examples of classes include `joints`, `partition_walls`, `surfaces` and `rooms`. A model of an actual building requires a large number of classes, and each class requires definition of all of its attributes. Many attributes are common to large groups of classes. However, if classes are defined hierarchically, with common attributes defined in higher-level classes and inherited by those beneath them, the high-level classes can be used to provide many of the basic definitions for lower-level classes. Thus all *physical objects* might be defined as a high-level class with location, rotation and shape attributes. Different types of *physical objects* can then be defined as children of the general *physical object* class and automatically receive these attributes.

If a building model is defined as a set of classes, and instances are created to define a particular building, then the building seems to be simply a collection of parts, without definition of how the parts are related. The fourth level of the RATAS building model includes relations between objects. Two types of relations are defined. The first relation, *part of,* is used to indicate that one object belongs to or is a constituent of another larger object. For example, a window is typically *part of* a wall, and all the individual pieces of construction in a building are *part of* the class building. The other type of relation that RATAS defines is *connected to.* This relation is typically between objects at the same level of abstraction, and defines how they are composed, possibly into a larger assembly. Examples of these two types of relations are shown in Figure 8.4. A particular object is shown; above it are the classes it is *part of,* and below it are the classes that are *part of* it. *Connected to* relations with the object are shown at the right.

The complete description of a building and all its parts results in a very large and complex model. It is seldom necessary to deal with this whole. Rather, only certain aspects of the building need to be considered at any one time. These subsets of the overall building model are called *views.* Views can be associated with particular subdisciplines or specific design stages.

The specifications for the RATAS product domain model are similar to the typical object-oriented definitions used in many other conceptual models of products. Given the need for a structure that could be quickly adopted by the Finnish construction industry, there was good reason to develop a simple and easily understood model.

8.2.3 BUILDING KERNEL MODEL

The RATAS building kernel model was defined primarily as a decomposition hierarchy, using *part of* relations. The model focused on the definition of certain critical aspects of a building kernel model. It was also outlined as a series of layers, structured as the abstraction hierarchy shown in Figure 8.3. Here, the layers in the hierarchy of the building kernel model are defined by aggregation. Each layer is defined in terms of the classes provided in the information modeling language and the product modeling domain model. The building kernel model is structured vertically as a directed acyclic graph (DAG), with the *building* as the top class, which is then decomposed into *systems.* Each system is made up of *subsystems.* The subsystems are made up of *parts,* which are further decomposed—if needed—into *details.* The elements defined at each level are specified as classes. When populated with the data describing a particular building, each class is instantiated with a varying number of instances.

An instance carries the relationships specified in the model. For example, Figure 8.4 shows an example of an office room. Vertical links indicate what *office_room_100* is spatially *part of (Fire-division_1, Floor_1)* and what is spatially *part of* it *(Radiator_1, Lamp_1).* Horizontal links indicate the room's *connected to* relation *(Wall_1, Concrete slab_1, Surface_100.1, Suspended ceiling_100.1, Window_1, Door_1).*

Later, VTT's research group recognized that a building model should include the site, which generally defines the context for the building project. The site, in this case, is meant to include weather, orientation and other such properties. This later kernel model is diagrammed in EXPRESS-G in Figure 8.5. The building-level model includes four levels, similar to those diagrammed in Figure 8.3 but more precisely defined. Detail is added at the building-part level to include spaces and joints. Joints were found to be particularly important for the specification of parts and their proper assembly.

8.2.4 ASPECT MODELS

The studies that made up the VTT project led to the development of subsets of an overall building model. The aspect models developed do not address a set of applications and functionality in the same way that the models in Chapter Seven do, rather they focus on identifying the relationships in particular ambiguous and complex areas. For example, a *model defining spaces and their enclosure* was developed. After reviewing other models of enclosure, Björk observed continued conflicts arising from the ambiguity of the words used to label classes. An important facet of the aspect models that were developed was a complete dictionary of the entities used in the model.

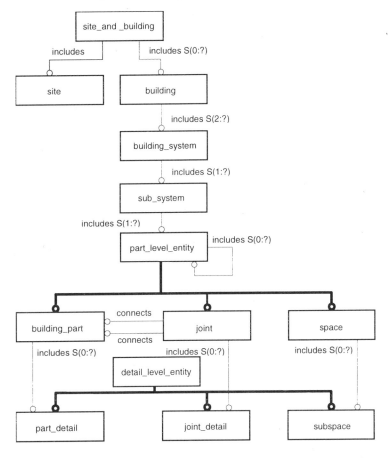

Figure 8.5: The RATAS building kernel model, defined as an abstraction hierarchy.

The space and enclosure model is presented incrementally. At its center is a space and subspace model, as shown in Figure 8.6. The model responds the need for both space assemblies, such as fire zones or departments, and the decomposition of a physically bounded space into subspaces, defined functionally. Thus a generalized space is specialized into space and subspace. A space may have any number of subspaces, and a space assembly may have any number of spaces within it.

The enclosure bounding a space consists of both physical boundaries and imaginary boundaries. The enclosure model is shown in Figure 8.7. From the perspective of enclosed spaces, interest is focused on the surfaces of the enclosing entities and surface finishes. Thus these are provided with a cross-link.

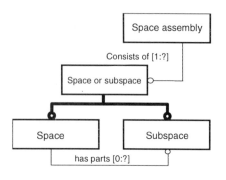

Figure 8.6: Abstraction hierarchy for spaces.

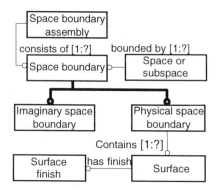

Figure 8.7: Abstraction hierarchy for space boundaries.

The enclosure bounding a space is a part of the `enclosing structure`, a main concern of construction companies (Figure 8.8). The model begins with individual enclosing entities and recognizes that there are several aggregations to define, for example, the enclosing shell of a building and the enclosing entities for each space. There are three specializations of an enclosing entity: `enclosing structure`, `enclosing structure section`, and `component`. All three have common attributes and may be referenced by an `enclosing entity assembly`. It is assumed that the structure and components are used to define entities such as `beams` and `columns`, while the section is used to define built-up entities.

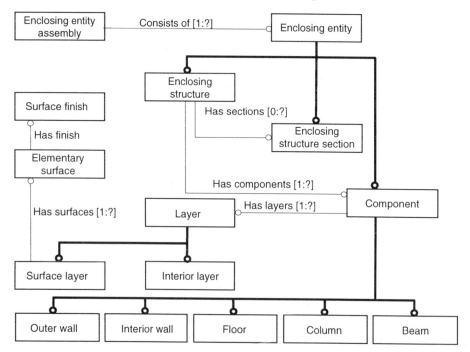

Figure 8.8: Abstraction hierarchy for enclosing structures.

In order to support both the construction element-oriented view and the space-centered view, all components are assumed to have a *layer*, which is referenced by the *space* as well as the component. Layers are specialized into *surface layers* and *internal layers*. *Components* are specialized into types of construction elements, some examples of which are shown in Figure 8.8.

Additional detail is provided regarding opening components in Figure 8.9. These provide both access via doors and light and ventilation via windows. They are considered part of the components making up the physical space boundary. Enclosing entities can have zero or more holes, which may be filled by an opening component. Spaces can then refer to the opening components.

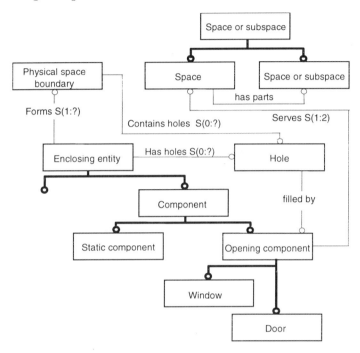

Figure 8.9: Schema for openings and their relationships to other entities.

Figure 8.10 presents the synthesis of each of these partial definitions. This model captures the general part of and connected to relations in a building enclosure model. It assumes that further refinement would support the definition of particular types of construction and construction elements. In contrast to other parts of the RATAS project, the model incorporates certain forms of redundancy, specifically dealing with derivations. Storing derived data allows certain applications to directly access and use that data. Otherwise, the application would have to access the base data from which the derivation was made, which may be beyond the scope of the application. (In the original work, elementary textual EXPRESS models were also provided, showing the derivations.)

In considering such a model, it is valuable to study the aspects that it includes and those that it does not include. In the RATAS model, spaces and subspaces are defined in terms of their boundaries. Electrical, water, air conditioning and other services are explicitly omitted. In the same way, enclosing entities are defined in terms of their section and components, which are assumed to be descriptions of high-level entities, such as wall, floor, column or beam elements. The model also omits services that are typically routed within or through enclosing entities. One of the shortcomings of the model, as Björk notes, is that at the abstract level of component definition presented, it is probably desirable to include joint definitions, as these are an important consideration, especially for prefabricated elements such as are commonly used in Scandinavia. A strength of the model is that by treating enclosing entities generically and defining holes in them, they can be used to define skylights, vents, chimneys and other elements (which

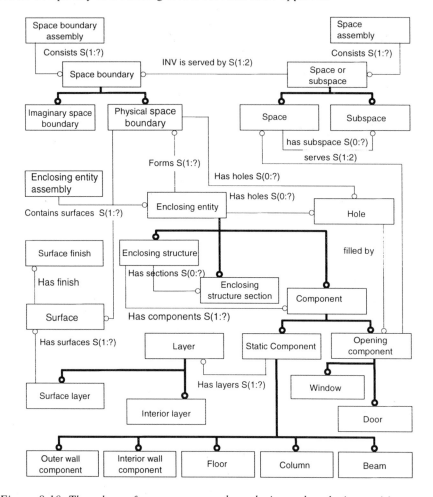

were also probably outside the scope of the model). When all these different considerations are combined, the complexity of a building model becomes more apparent.

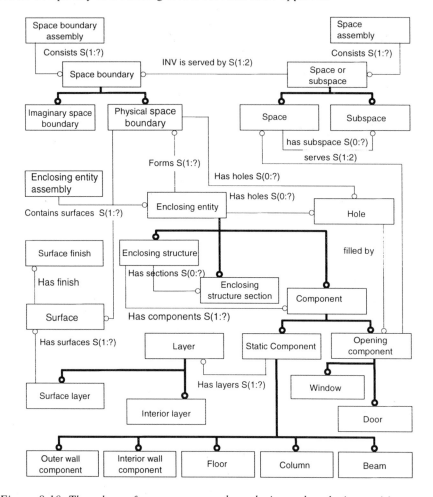

Figure 8.10: The schema for spaces, space boundaries and enclosing entities.

8.2.5 PROTOTYPE IMPLEMENTATION

The project undertook several small prototype implementations using different software tools. One such implementation, which is considered here, was developed in Oracle on top of MS/DOS. In order to make the prototype more realistic, a portion of a real, four-story project was modeled. The database was built up manually using SQL *Insert* operations. Twenty object classes were defined, with an average of eight attributes per object. In order to develop a homogeneous modeling strategy, all relations were defined explicitly as separate relational tables. (Normally this would only be required of relations with cardinality *1-to-m* or *n-to-m*.) Figure 8.11 shows an example of a table of definitions of both objects and the binary relations across objects.

The table defines sixteen of the object classes, involving sixteen relations. 1-to-m relations are defined in such relational tables using the same instance value on one side of the relation with different values on the other side. For example, all the rooms on *Floor-3* are identified as a relation between Floor-3 and the different *Room_Id*s. This allows any number of rooms to be on a floor. The relations defined in this table can be utilized in both directions. For example, it indicates *structural_sys - horiz_structure - horiz_structure_element* are three levels of structural definitions, using the *part_of* relation. *Walls - surfaces* and *walls/windows* are other *part_of* relations, while *walls - walls* and *beams - slabs* are *connected_to* relations.

OBJECT TABLES	RELATIONS	
	1st class	2nd class
FLOORS	FLOORS	ROOM-SPACES
ROOM-SPACES	ROOM-SPACES	ROOM-SPACES
HORIZONTAL_ STRUCTURE	ROOM-SPACES	SURFACES
HORIZ_STRUCTURE_ ELEMENT	ROOM-SPACES	WALLS
SLAB_FIELDS	ROOM-SPACES	HORIZ_STRUCTURE_ELEMENT
BEAMS	HORIZONTAL_STRUCTURES	HORIZ_STRUCTURE_ELEMENT
COLUMNS	HORIZ_STRUCTURE_ELEMENT	SLAB_FIELDS
WALLS	HORIZ_STRUCTURE_ELEMENT	SURFACES
SURFACES	SLAB_FIELDS	SLABS
WINDOWS	BEAMS	SLAB_FIELDS
APARTMENTS	BEAMS	SLABS
BUILDINGS	COLUMNS	COLUMNS
SPATIAL SYSTEM	COLUMNS	BEAMS
HEATING SYSTEM	WALLS	WALLS
TELECOMMUNIC_SYS	WALLS	SURFACES
STRUCTURAL_SYS	WALLS	WINDOWS

Figure 8.11: The Object Tables and the Relation Tables used in the relational database.

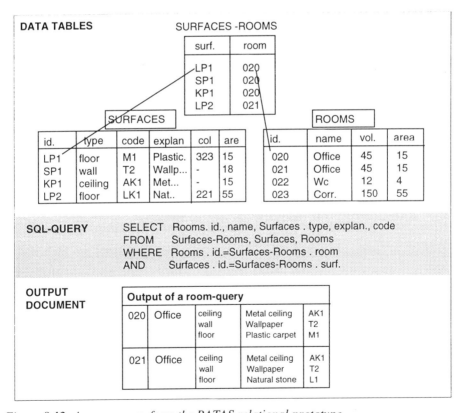

Figure 8.12: A room query from the RATAS relational prototype.

In order to test the capabilities of the relational database approach, twenty different queries were formulated and tested. An example of such a query, its structure and the output, is shown in Figure

8.12. The query takes data from two different tables, combining them according to the relation in the third table. The query finds all the surfaces in a building and sorts them by rooms. An architect collaborating on the RATAS project identified this as a common type of need: finding all the products of a certain vendor and identifying in which room they belong. Searching drawings is both error prone and time consuming. Such a query can be formulated in a few minutes and answered in seconds.

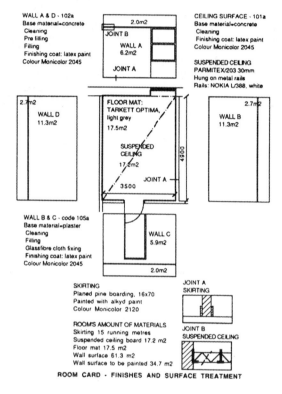

Figure 8.13: An example of a work task document that could be produced for each building trade, from the information contained in a building model.

The relational database prototype showed the usefulness of a database format for such information, especially for "bill of material" type of data.

A second prototype addressed the kinds of custom drawings that could become possible if a central building model came into everyday use. The example in Figure 8.13 deals with the information and drawings associated with a particular work task, showing the potential for improving the quality of information at the job site. A third prototype implementation involved a conventionally generated 3D model of a building combined with an object browser. Figure 8.14 shows a sample screen.

8.2.6 DISCUSSION

The intention of the original RATAS model was to define a building description to be provided to the builder, as a phase transition representation. This is in contrast to other models that support work within a phase which we will review shortly.

The RATAS and VTT effort made a significant contribution to building modeling research in providing a broad-based sampling of the issues and trade-offs that are involved in developing a building product model at an industry-wide level. It offered a kind of "roadmap" of how other nations or institutions might proceed. At a technical level, the work articulated a five-level architecture that clearly defined levels of abstraction and semantic functionality. These levels are similar to those used in STEP, but are defined more clearly and made more appropriate to the building industry. The RATAS work also pointed out the need for a building framework model and began to identify semantic relations needed to define construction-oriented building models. The approach emphasized simplicity and overlooked certain aspects of a completely realistic building model in an effort to gain general support for the building model concept. The focus was to define relationships rather than detail attributes. It was perhaps easier to develop such a model in a small but homogeneous nation like Finland, rather than in a larger, more heterogeneous country like the US.

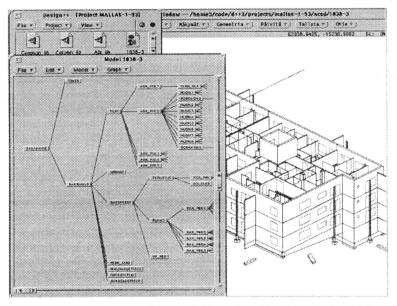

Figure 8.14: An example of a product model-based application offering both geometric modeling and object browser user interfaces.

The project addressed other issues not considered here, including representation of processes and the use of multimedia as a means of structuring and accessing building information. It avoided dealing with geometric models on the assumption that good definitions would evolve within the international standardization efforts, such as STEP. However, new types of problems would certainly emerge if such features were included. Within the model's five-level structure, it is unclear whether the different *aspect models* are all fixed and loaded at the time of model creation, or whether these are defined as libraries to be added when needed. Several of the papers refer to rules embedded in a building model, but they did not develop a mechanism for including them.

8.3 BUILDING CONSTRUCTION CORE MODEL (PART 106)

A STEP-based, core model for buildings was discussed by the STEP AEC committee since the early days of the standards effort. However, the role and placement of such a large-scale model was not anticipated within the STEP conceptual framework, even though similar needs were identified (less strongly) in other industries. Application protocols (APs) normally address a precise class of applications within some industrial context. The structural steel application model is a good example. A building core model, however, was not intended to support a specific class of applications, but rather to provide a common framework for data exchange across a set of more application protocols. Integrated resource models were given a similar role by the STEP organization. However, as presented in Chapter Five, integrated resources were conceived as parts of models developed in one application area that could be reused in other application areas. Moreover, integrated resources were assumed to be identified at the Application Interpretation level of development, not at the Application Reference Model level. The need for a framework model, however, exists in both the ARM and the AIM aspects of an AP.

A new STEP concept, an Application Resource, was defined to support framework models, both in building construction and other industries. Several requirements for an Application Resource were identified. It should be able to:

1.   provide a kernel equivalent to current application resources
2.   define common requirements equivalent to an AIC at a Reference level
3.   support pre-interpretation of an AP without restricting the ability to do post-interpretation
4.   develop integration at the ARM level in a standardized way as a precursor to a complete STEP interpretation
5.   be conformance tested at the ARM level
6.   enable fast implementation development and deliver a response to external challenges

The Building Construction Core Model was approved as a STEP activity in late 1994 and was identified as Part 106. It has gone through several iterations developed by a team of primarily European researchers and led by Jeffrey Wix. Presented below is a review of the most recent complete version of the Building Construction Core Model (BCCM). It must be emphasized that while the results have incorporated ideas and work of previous efforts, it is not complete. Data exchanges have not yet been made using the model.

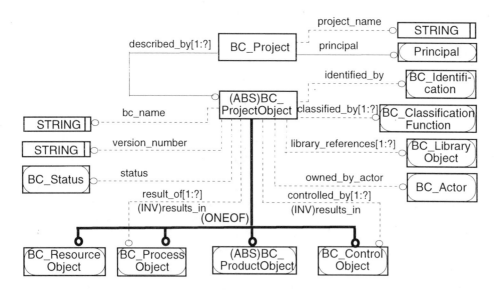

Figure 8.15: Project-level abstractions of the Building Construction Core Model.

8.3.1 BUILDING CONSTRUCTION CORE MODEL

The model focuses on information needed for construction planning and monitoring. Its units of functionality include project definition and location, space function, separation, distribution (piping, ductwork), furnishings, structure, construction costs and construction scheduling. Its scope statement proposes that it include

- *Products:* the tangible items that are the result of building construction processes
- *Processes:* the logistics and activities that are required in order to achieve the resulting products
- *Resources:* human, plant and constructed items employed and deployed to enable the processes to take place
- *Controls:* constraints which may be applied to the products, processes and resources during construction

In addition, building construction was defined to include the full building lifecycle: the Proposal, Design, Planning, Realization and Management of the product. (This is a very broad and inclusive scope.) To support this functionality, 245 core objects were defined. The objects and the relations between them have been the primary focus to date, with only a few of the attributes being defined.

The top level of the BCCM—shown in Figure 8.15—addresses project identification and its status. A `BC_Project` is made up of abstract `BC_ProjectObjects`, which carry an identification, are owned by some Actor and have a classification. `BC_ProjectObject` and its attributes are inherited into one of four subtypes: `BC_ResourceObject`, `BC_ProcessObject`, `BC_ProductObject` and `BC_ControlObject`.

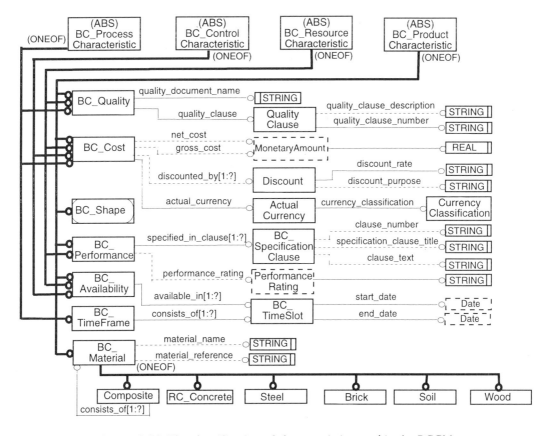

Figure 8.16: The classification of characteristics used in the BCCM.

Another top-level set of definitions is Building Construction Characteristics, shown in Figure 8.16. There are four general types: `BC_ControlCharacteristics`, `BC_ProcessChar-acteristics`, `BC_ResourceCharacteristics` and `BC_ProductCharacter-istics`. All of these types of characteristics share common attributes, grouped into sets. The sets are `BC_Quality`, `BC_Cost`, `BC_Performance`, `BC_Availability`, `BC_Timeframe`, `BC_Material` and `BC_Shape`.

The sets are defined down to a certain level of detail. Quality is defined by reference to a clause within a quality document, for example, ISO 9000. Costs are defined using several factors, including type of currency as well as discount. At this level, performance refers to the quality standard being applied, such as a building code and the relevant clause. Availability is defined by a begin and end time window. Material references a material document and a high-level classification (material is illustrative and not meant to be complete).

Shape is detailed in Figure 8.17. It is characterized in two ways: in terms of the geometry and in terms of the topology that define the shape and its presentation. Reference shapes are defined to a lower level of detail, to later refer to Part 042. At the bottom of Figure 8.17, the `BC_Class-ificationFunction` is defined separately. This property is used to classify an object according to common attributes. The classification function sets priorities based on which classification functions are primary, secondary and so forth.

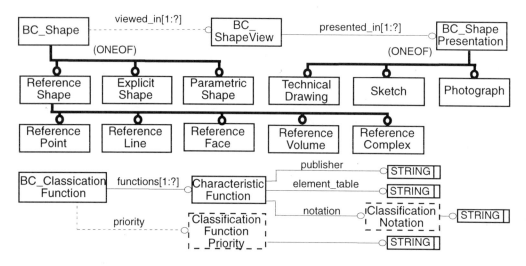

Figure 8.17: BC_Shape and its specializations.

The characteristics shown in Figures 8.16 and 8.17 are used in all the more detailed entity definitions that follow below. By centralizing these characteristics or attribute classes, the typical explosion in the number of attributes needed in different contexts is nicely controlled. At the same time, they may be combined in multiple ways, as needed in different contexts.

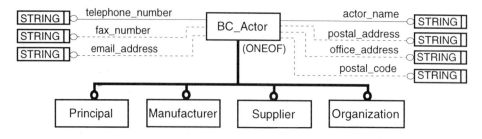

Figure 8.18: The actors in a building project are defined centrally.

`BC_Actor` defines in one place the various actors in a building project. It is shown in Figure 8.18. The attributes provide means to identify and contact the actor.

8.3.1.1 The Building Product Subschema

The `BC_ProductObject` is an abstract object that carries a number of product characteristics, including proposed, required, planned, realized, and managed attributes. These are used to define characteristics of each of the different lifecycle phases. Multiple Product Characteristics can be assigned to a `BC_ProductObject`, with the relations shown in Figure 8.19. `BC_Product-Object` carries a position and a flag telling it whether it is "made" or purchased. These attributes are further inherited into one of three subtypes: `BC_BuildingObject`, `BC_Site-Object` or `BC_FunctionObject`, each of which is described below.

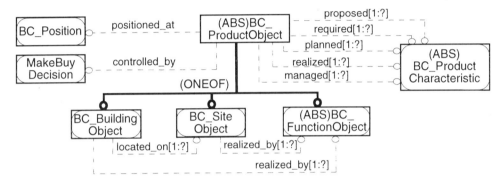

Figure 8.19: BC_ProductObject is an abstract object characterizing more detailed kinds of product objects.

The building object, shown in Figure 8.20, may be a whole building or set of buildings (a building complex). Alternatively, it may be a high-level part of a building, such as a building element or element assembly. `BuildingElementAssembly` may be hierarchically organized, shown by the loop in the upper right of the figure. `BuildingElementAssembly` is an aspect of the building, such as a `cellar`, `story`, `wing` or `block`. A `block` is defined as "an assembly of building elements where each element fulfills the same purpose." `Section` here is not a project or presentation but rather a "geographical part of a building." None of the bottom-level entities are further defined. However, they carry all the attributes inherited from their supertypes.

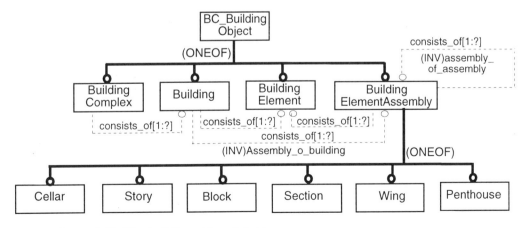

Figure 8.20: The building object definition.

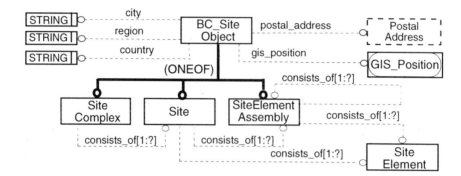

Figure 8.21: The BC_SiteObject defines the various types of site objects.

The BC_SiteObject in Figure 8.21 defines site objects at different levels of abstraction: SiteComplex, Site, SiteElement or SiteElementAssembly. They inherit their location and address from the site object. Currently, they do not carry information required for the planning, design or maintenance of the site.

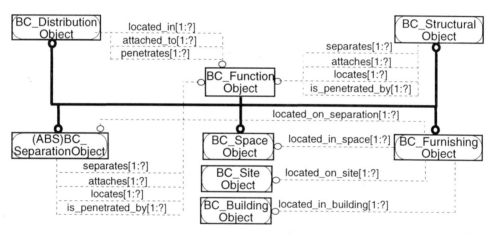

Figure 8:22: The BC_FunctionObject and its direct subtypes.

The BC_FunctionObject in Figure 8.22 defines objects that have a specific function—the system objects in Turner's model. They cover all the physical elements making up the building. The different subtypes carry relations back to the function objects that define separation, location, attachment and penetration relations. There are many other classes of functional objects in a building, but the scope of BCCM—for now at least—addresses structure and HVAC; thus separation must be defined. Notice that BC_SiteObject and BC_BuildingObject have been defined previously and are shown here to depict their relations. We review the Space, Separation, Furnishing, Structure and Distribution objects below.

The BC_SpaceObject classes are shown in Figure 8.23. All spaces have an area (net and gross for leasing), occupancy, an occupancy profile and connections with other spaces. Notice that the SpaceConnection is between two or more spaces. These properties are inherited into GeneralSpaceObject and SpecificSpaceObject. GeneralSpaceObjects are unassignable spaces, while SpecificSpaceObject defines those classes of space that are assignable, i.e., that are part of the net space within a lease. These space types could be articulated much further. The level of detail presented is meant to indicate the structure of space classification for the purpose of planning during the Proposal stage and the Management stage.

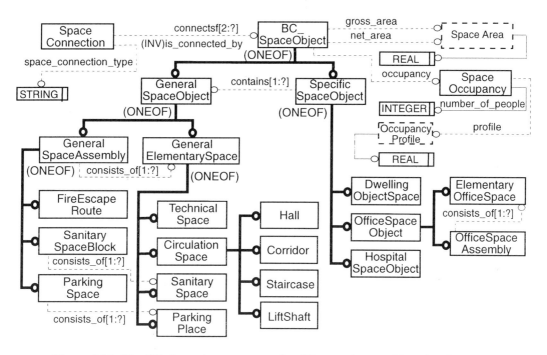

Figure 8.23: The BC_Space is a supertype for different classes of functional space.

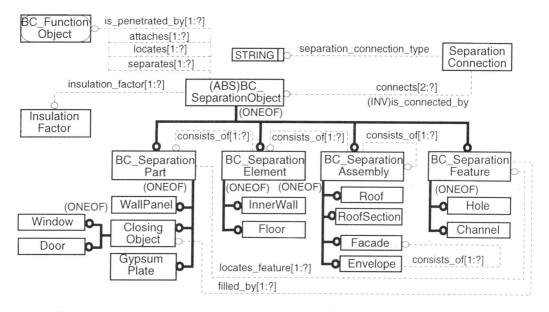

Figure 8.24: The BC_SeparationObject type and its subtypes.

The `BC_SeparationObject` is shown in Figure 8.24. Each separation object is optionally defined by two `SeparationConnection` objects and by relations with other `BC_Function Objects`, with regard to what the object separates, attaches to, penetrates and so forth. While these relations suggest that a separation object could separate a wide range of object classes, the subtypes show that the intention is to consider separations as boundaries between spaces. Separation objects are defined at three levels of aggregation: assemblies, or groups of separation objects; elements, or single objects; and parts, or significant aspects of a single object. Features deal with openings, which may be filled or not. Assemblies include roof, roof section, facade and the complete exterior envelope.

The BC_FurnishingObject class is defined in Figure 8.25. Furnishings are defined within the Units of Functionality as parts of a building that facilitate its use. They are not part of the separation assemblies and instead are added to the model after separations. Furnishing objects include indoor furnishings and also outdoor site-related furnishings.

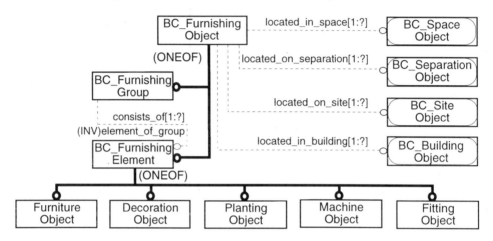

Figure 8.25: Building furnishing objects and their subtypes.

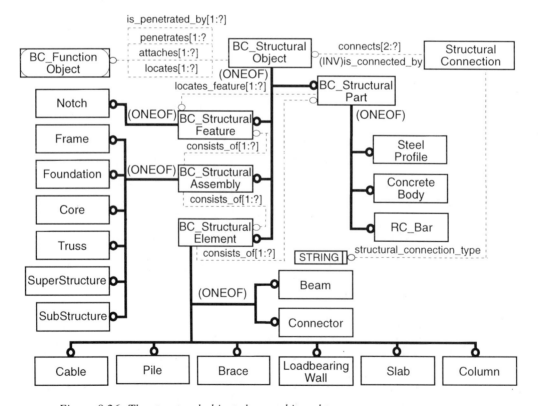

Figure 8.26: The structural object class and its subtypes.

The structural types and subtypes are shown next, in Figure 8.26. They evidently address at a high level the structural definition of both CIMsteel and reinforced concrete structures. These objects are not meant to subsume the CIMsteel definitions, but rather to allow a subset of them to be combined with other object classes, to address issues of integration and coordination. For example, in this figure, StructuralConnection is very simple in comparison to the joint definitions in CIMsteel. On the other hand, the definitions in Figure 8.26 introduce a variety of assembly definitions not found in CIMsteel, such as SuperStructure, SubStructure, Truss and so forth.

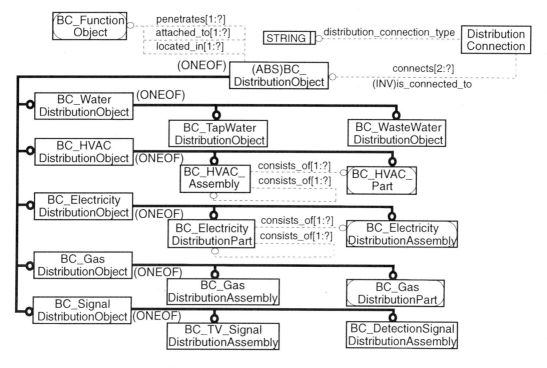

Figure 8.27: The distribution object classes.

The BC_DistributionObject class is defined in Figure 8.27. It defines five types of distribution objects: HVAC, electrical, water, signal and gas. Some of these have two subtypes, an assembly and a part. Some also have additional examples of parts, but have not been detailed. The critical relations with other function objects can be defined. The DistributionConnection requires further definition also, for specifying the wiring or piping used for the connection.

Putting the pieces of the building product subschema together, we can see that it has a regular structure that captures many of the relationships needed for the basic definition of a building. It separates the site and high-level aggregations of the building (BC_BuildingObject), which are understood to eventually reference various functional objects. The BC_FunctionObjects are the actual constructed elements of the building. The subschema carries the layout of spaces, separators (floors, walls, ceilings), furnishings, and structural and distribution systems.

8.3.1.2 The Building Resource Subschema

The BC_Resource_Object class and subclasses are shown in Figure 8.28. The figure shows the various classes of resources considered for inclusion in the BCCM. Resource characteristics include properties for quality, cost, performance and availability (as shown in Figure 8.16). A set of these properties is defined for each of the relevant lifecycle phases (except facility management). The resources shown have not been fully developed. For a framework model, it is necessary to define the partition line between those resources carried in it and those carried in specific application Part models.

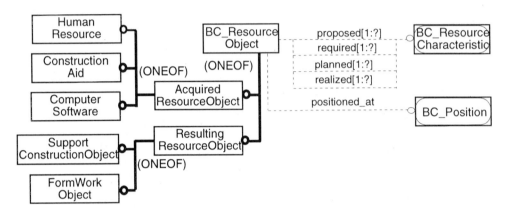

Figure 8.28: The BCCM resource object type and its subtypes.

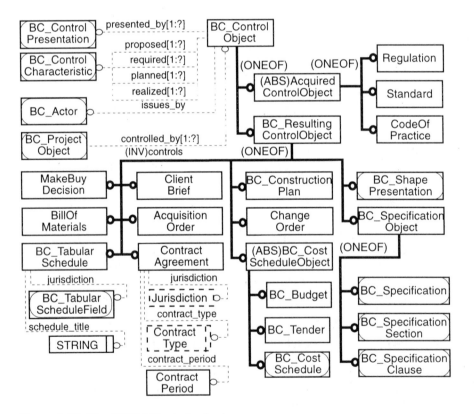

Figure 8:29: The BCCM Control subschema.

8.3.1.3 The Building Control Subschema

The control subschema addresses the various pieces of information used to control a building project through its various lifecycle phases. Two broad classes of controls are identified: Acquired and Resulting. Acquired includes all the general, externally defined and imposed controls, such as regulations and standards. Resulting control objects are those created during a project that are used to control the project's outcome.

Each control object has a common set of attributes, dealing with cost and availability (as expressed in Figure 8.16). They are defined separately for each phase in the lifecycle (except management). Each control object has an optional actor that implements the control and the project object that is being controlled. Resulting control objects include schedules, bills of material, client briefs, construction plans, specifications, purchase orders, change orders and contracts. One of the control objects—BC_ShapePresentation—was defined earlier in Figure 8.16.

8.3.1.4 The Building Process Subschema

The BC_ProcessObject is subtyped into three different processes: construction processes, design processes and logistic processes. (It is assumed that management processes will be added later.) Process characteristics that are inherited into all processes include quality, cost and timeframe. They are defined separately for each of the lifecycle phases.

In general, Processes are high-level entities, which are decomposed into multiple Activities. Activities, in turn, are decomposed into Tasks. This breakdown applies to both design and construction processes. Logistics processes have a different structure. A logistic process involves the acquisition and delivery of some object involved in design. It applies not only to the product objects, but also to the resource and control objects, as well as acquisitions and remittances. Two subclasses are defined, one for procurement and one for transportation. It is assumed that construction process objects are not articulated into usable task definitions because this is a framework model, not a more application-related building construction model.

In this review of the Building Construction Core Model, 194 of the 243 objects of the BCCM have been described. Omissions have been at the most detailed level, usually dealing with example entities for a class. This review is in terms of its definition as of mid-1998.

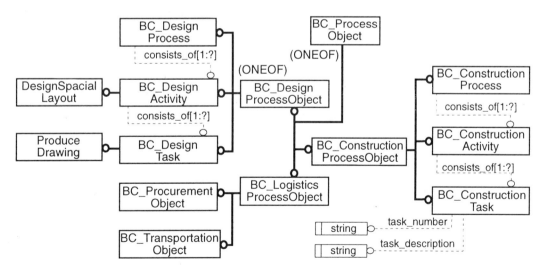

Figure 8.30: The Process Object subschema.

8.4 OBSERVATIONS ABOUT BCCM

It is important to recall that this is a framework model and not meant to be a complete and definitive model for any particular use, as would be supported by an AP. Rather it is meant to support exchange among different kinds of use, between different APs.

Because of its intended use, the model is oriented toward a taxonomic definition of objects and attributes. Grouping attributes at a high level, then inheriting combination sets of them into different classes of objects is one means to address the proliferation of attributes that need to be used in such a model. Almost all of these attributes are optional. In this regard, the model defines a framework for the various APs it is meant to support. Shape data may be defined geometrically, as a reference, explicitly or as a parametric shape. The documents that present shape data are paper-based, i.e., photograph, sketch, technical drawing. Detailed definitions of objects are not included, presumably because these are not needed for coordination among models.

The model is currently viewed as an Application Reference Model. It has not yet been integrated with the Integrated Resources reviewed in Chapter Seven. Rules or WHERE clauses are not yet defined for identifying what is a legal or meaningful set of objects. It is not yet complete.

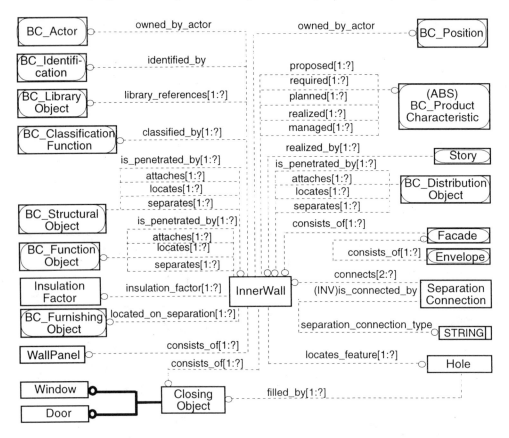

Figure 8.31: The expanded and flattened wall model, as defined in BCCM.

8.4.1 OBJECT EXPANSION

In order to fully comprehend the object definitions for this or any model, it is often insightful to expand an object type definition so that all the inherited attributes are shown with those defined within the object itself. We do that here for two different object classes. We expand the definition of a wall, since that has been an example used throughout this book, and we also expand the definition of a construction process object, so as to give an expanded view of how processes in general are handled.

The expanded BCCM wall model is shown in Figure 8.31. Some judgments are required to determine which of the many possible inheritances are meaningful for the wall model. The set included in Figure 8.31 is thought to be the most probable set. (Eventually, the required set may be better delimited by WHERE clauses that define the appropriate attributes for particular object classes.)

The Innerwall may be part of a Facade, Envelope or Story. It may be penetrated by zero or more DistributionObjects or StructureObjects. It may also be attached to, penetrated by or located by another object, or used to separate two or more function objects. The last are assumed to be two spaces. It may be made up of WallPanels. It may have zero or more Holes, which have Doors or Windows as ClosingObjects. The Wall may provide an Insulation Factor. It carries product characteristics dealing with shape, material (not defined in any detail), availability, performance, cost and quality and also a position (inherited from BC_ProductObject). It may be an object retrieved from a library; it may have a unique identification and various subclassifications (inherited from BC_ProjectObject). In summary, while it does not carry behavior properties—for example, for structure or energy—it does seem to provide the definition of a wall at a level for addressing general placement, adjacency, circulation, management and other uses.

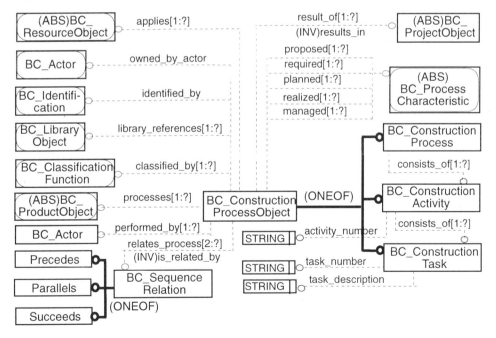

Figure 8.32: The expanded model of BC_ConstructionProcessObject.

Last, we turn to the expanded version of the BC_ConstructionProcessObject. It inherits a number of attributes from BC_Project_Object, including the classification function, identification, the owning actor, the library it comes from, and the relevant control objects. From `BC_ProcessObject`, it inherits more attributes, including the sequence relation, the applied resource object, the process characteristics of quality, cost and timeframe (from Figure 8.16), the actor that performs the process and the product the process produces. The construction process itself is inherited into one of three subtypes regarding process, activity, or task.

From the expansion, we can see that a construction process carries the essential attributes of a process, supporting precedence relations, resource and control requirements, as well as the product objects acted upon. It carries two actors: an process owner and the process performer. It carries a variety of identification and classification attributes.

Neither of these objects is fully expanded, in that the characteristic attributes are referred to indirectly. They have not been laid out because there may be a set of characteristic attributes combining the various types. This manner of assigning attributes is very general, though processing them requires much checking before they can be extracted within any particular translation task.

8.5 REVIEW OF BCCM

The BCCM specification focuses on definition of an exchange framework, not support for any particular application protocol. Rather, its focus is to define common objects, possibly at a high level of abstraction, that can be referenced, then inherited into particular APs. The aspect models to be supported are listed in the UoD of the report. They include the HVAC and CIMsteel APs. How they would be integrated through the BCCM is not stated. For example, it is not clear how the top-level CIMsteel model, defined in Figure 7.2, can easily fit into the BCCM structural subschema, as shown in Figure 8.26. Also, there is no indication how loads might be defined in the BCCM, such that they might be read, say, from the HVAC layout and then transferred to the structural steel model.

The centralization of properties, as shown in Figure 8.16, was an advanced idea that makes property proliferation much more manageable as a building model gets expanded. In general, these properties—all optional—are defined at a high level of abstraction, and the attributes can be used in any subtype model. This leads to very large proliferation of attributes in entity instances. While the number of distinct attributes has been greatly reduced, the number allowed in any object is left unnecessarily broad. It should be noted that in any implementation, all optional attribute fields are created, and if not used, they are just left empty. In a flat file implementation, the overhead of empty data fields is not great. In a database implementation, however, space is reserved for all optional attribute values. No rules have been defined yet (though they were planned) to specify the necessary subsets of data required in different object types and conditions.

The BCCM has purposely ignored geometry. However, geometry is one of the most important coordinating issues that applies across all building aspect models. Though not mentioned, it may be that Part 106 was assumed to be responsible for geometry coordination.

The implementation of a framework model introduces new processing needs that have not been fully addressed within ISO-STEP procedures. Currently, all STEP data exchange processes are assumed to take place between an external application and an application protocol. The STEP organization assumes that application vendors would implement the software supporting these exchange processes. However, the framework model makes the assumption that some data will move either directly from AP to AP or from an AP to the BCCM and then to another AP. It is unclear who will develop and maintain the computer applications supporting these exchanges. These new applications point out the role and growing importance of Product Data Management

(PDM) functions and the computer tools that have been introduced to the market to address PDM issues. Most of the PDM functionality has drawn on software engineering data management practices, such as library checkout with version control, read and write access control, and grouping of files into semantic units. The framework model adds a new function—mapping between application protocol formats.

8.6 NOTES AND FURTHER READING

The rationale for a building model framework model is presented forcefully in Björk [1995] and also by Junge and Leibich [1997].

James Turner's AEC System Building Model is presented in ISO TC184/SC/WG1[1989]. A research paper describing it is presented in Turner [1988]. Wim Gielingh's GARM is presented in Gielingh [1988a], as well as in a research conference, Gielingh [1988b].

The RATAS work has been represented in a variety of publications. These include Enkovaara, Salmi and Sarja [1988], Björk [1989], Björk and Pentilla [1991] and Björk [1992a]. An important summary of the RATAS model which also compiles all the related work is Björk [1995]. In Björk [1992b], the VTT notion of aspect model is presented. It deals with various parts of a building model that are ambiguous and complex, specifically spaces and their boundaries.

The rationale for a building core model, within the ISO STEP framework is argued in various documents available on the BRE Web site:
        http://www.bre.co.uk/bre/research/cagroup/consproc/itra/bccm/
The STEP Building Construction Core Model's current version is available on the NIST SOLIS Website:
        http://www/nist/gov/sc4/step/parts/part106/
An earlier version is available on
        http://maillist.civil.ubc.ca/~tfroese/rsch/models/directory/step_bc_core/str100.html
The current draft is ISO TC184/SC/WG1Part 106 Draft T100 [1996].

8.7 STUDY QUESTIONS

1.   The RATAS model includes a definition of wall that is similar in coverage to the example of a wall defined in Section 3.2. Compare these two wall models, in terms of (i) the internal structure of the wall and (ii) the relation of the wall to its boundaries and to the spaces it encloses. What seems to be missing from each of the models? Define a revised model of wall that takes the best features of both models.

2.   Compare the structure of relations for space enclosure developed in the RATAS model with that developed in the BCCM. How similar are they? How are they different? Propose changes that you think are important—if any—to the BCCM as a result of this comparison.

3.   All the models surveyed include aggregation hierarchies, often referred to as the part of relation. Compare the aggregation hierarchies defined in RATAS, COMBINE, CIMsteel and BCCM. Do they assume one aggregation hierarchy for all functions or do they allow for different functions—structures, mechanical systems— to have separate hierarchies? What is needed in a building model?

4.   Undertake a review of the interface conditions between the BCCM and the CIMsteel models. Identify the properties that could be generated in other applications that the CIMsteel model

might utilize. Similarly, identify the properties in the CIMsteel model that other applications might utilize. Propose changes to the two models that would allow these two sets of properties to be exchanged and that would harmonize their entity structures. Identify the proposed REFERENCE BY and USED BY entities that should be added to either the CIMsteel or BCCM models.

5.   Consider the BCCM model from a construction procurement viewpoint, with the intention of developing an application that will track material and equipment orders and relate them to when a construction task is scheduled. Define a subschema of the BCCM that could support this application domain. Are there any critical types of information that are missing—either data or relations?

CHAPTER NINE
Industry Foundation Classes

We must bear in mind, then, that there is nothing more difficult and dangerous, or more doubtful of success, than an attempt to introduce a new order of things in any state. For the innovator has for enemies all those who derived advantages from the old order of things while those who expect to be benefited by the new institutions will be but lukewarm defenders.

Niccolo Machiavelli
The Prince
1513

9.1 INTERNATIONAL ALLIANCE FOR INTEROPERABILITY (IAI)

We turn now to the newest and largest effort being undertaken to develop an integrated building model. In late 1994, Autodesk, Incorporated initiated an industry consortium to advise the company on the development of a set of C++ classes that could support integrated application development. Twelve US companies joined the consortium. Initially defined as the Industry Alliance for Interoperability, the Alliance opened membership to all interested parties in September, 1995 and changed its name in 1997 to the International Alliance for Interoperability. The new Alliance was reconstituted as a non-profit industry-led organization, with the goal of publishing the Industry Foundation Class (IFC) as a neutral AEC project model responding to the AEC building lifecycle.

Procedurally, the goal of the IAI has been different from that of ISO-STEP. The IAI has sought to define and release a new version of the IFC every year, encouraging quick, incremental development. In the fall of 1995, Version 0.9 was released, followed by Version 1.0 in early 1997. In November of 1997, Version 1.5 was released. In contrast to ISO-STEP, this process was designed to demonstrate quick progress and to maintain industry enthusiasm for the benefits of integration.

Each release of the IFCs was followed up with commitments and implementations by various vendors, whose demonstrations of operable data exchange showed the utility of the IFC. For example, in June 1997, several companies participated in a demonstration of Version 1.0 at the AEC Systems '97 show in Philadelphia. Autodesk, Muigg (Austria), RoCAD (Switzerland) and Softistik (Germany) all exchanged data from AutoCad Release 14. Bentley Systems' Microstation, Nemetschek Systems' Allplan, and Softech Spirit (the last three systems from Germany) all exchanged information between each other and their own AEC application systems. A number of other companies with CAD or AEC applications are IAI members, including ACADGraph, IEZ, IBM, Ketiv, Kozo Kleilaku, Lawrence Berkeley Labs, MC2, Battelle Pacific Northwest Labs, Primavera, R13, Timberline and Visio.

9.2 IAI ORGANIZATION

The IAI has had wide industry support. It currently has nine chapters in 18 countries, with over 625 member organizations. The general organization is shown in Figure 9.1. Each chapter has a board that plans its activities and coordinates them with the International Council. In the US, the chapter has two management groups, one addressing business and one technical aspects. All organization representatives participate in Domain Committees, each of which addresses one area of the Project Model. Currently, the Domains are

AR -   Architecture
BS -   Building Services
CM -   Construction: CM1 - Procurement Logistics, CM2 - Temporary Construction
CS -   Codes and Standards
ES -   Cost Estimating
PM -   Project Management
FM -   Facility Management
SI   -   Simulation
ST -   Structural Engineering
XM -   Cross Domain

By participating in a Domain Committee, all members have input to the product model that corresponds to their area of expertise. Different national Chapters are focussing on different Domains; the domains covered and chapter foci evolve over time. For example, there is discussion of adding a Libraries Domain to address product libraries.

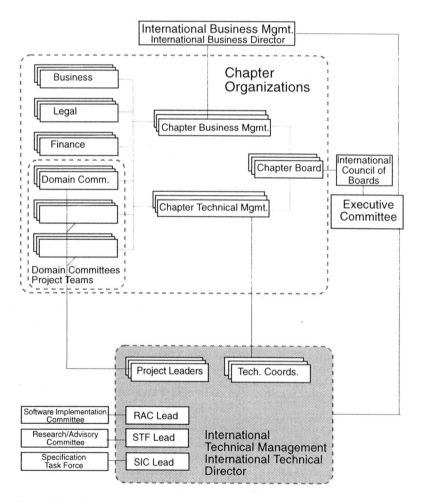

Figure 9.1: The organizational structure of the IAI.

The International Council includes the Chairs of all the Chapters. At the highest level, business activities are coordinated through the International Business Director, and the technical activities through the International Technical Director. The International Technical Director coordinates the development of the Project Model specifications, software implementation and certification testing with leaders of the Specification Task Force (STF) and Software Implementation Committee (SIC). A Research/Advisory Committee provides advice to this group. Currently, the International Business Director is Richard Groome and the International Technical Director is Richard See. As in STEP, most of the activities and services are supported by voluntary effort of the member firms. However, some model development is paid for by member dues and other forms of corporate sponsorship. There is a continuing low-level concern about whether or not the activities of the IAI can continue to be funded completely through such *ad hoc* arrangements.

9.3 DEVELOPMENT STRATEGY

The basic approach of the IAI is to develop quick model versions that can be tested and elaborated. Early release models are not supposed to be revised by later models, only added to, allowing vendors to begin using the early version IFC Models as a basis for products.

The development cycle involves overlapping efforts. While Version 1.5 was being released and tested, Version 2.0 was being specified and reviewed. At the same time, the Domain Committees were proposing the functionality to go into Version 3.0. Because of this overlapping development strategy, an assessment of the IAI work requires selection of a specific point in its development timeline. For our purposes, that point is Version 1.5, the most recently released version as of the time of the writing of this book.

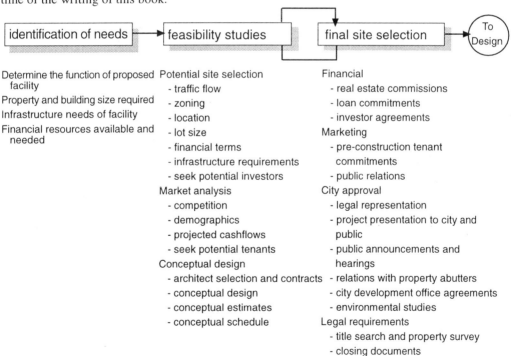

Figure 9.2: The IAI Feasibility Phase process model.

The Industry Foundation Class documentation for Version 1.5 consists of seven volumes:

  I.   IFC End User Guide
  II.   IFC Specification Development Guide
  III. IFC Object Model Architecture
  IV. AEC/FM Processes Supported by the IFC
  V.   IFC Object Model Guide
  VI. IFC Object Model Reference
  VII. IFC Implementation Certification Guide

The first three volumes are general and lay out the scope, organization and procedures of the IAI. The last four volumes are technical and deal with the IFC release. While this chapter draws from and interprets information from all volumes, it particularly reviews and presents information from Volumes III and V. The seven volumes are packaged onto two CD-ROMs.

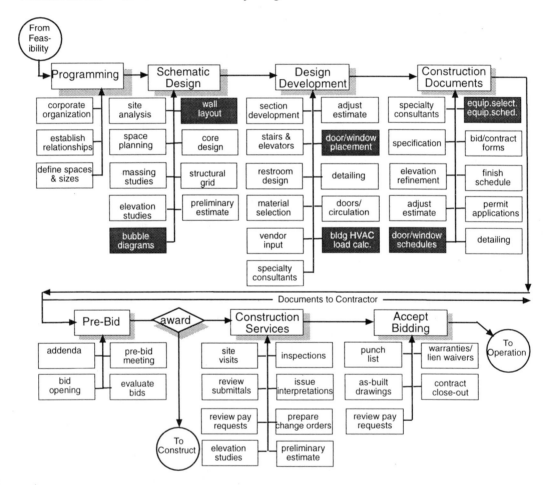

Figure 9.3: The IAI Design Phase activities.

The IAI recognized that a building project model must support the activities and processes carried out over all the different phases of the building lifecycle. It relied on process modeling as a basis for defining various domain-specific portions of the model, drawing inspiration and some techniques from the Total Quality Management (TQM) recommended procedure for systems development. A four-phase building lifecycle is employed, comprised of *Feasibility*, *Design*, *Construction* and *Operation*. Process models have been defined for each of these phases, with emphasis on identifying the activities that might be computer-augmented during each phase. These process models provide the semantic link and functional context for IAI model development activities and identify which aspects of the building model will be the focus of activity.

The Feasibility Phase process model is shown in Figure 9.2. It lists a broad set of activities that go into the planning of a building project. Conceptual rather than analytical activities are emphasized.

The Design Phase process model is shown in Figure 9.3. It breaks down architectural design into a set of detailed activities, grouped into seven development stages. In this diagram, each box is considered a design phase activity. At the top level are the major stage-level activities (shown with shadows) with the activities that fall within them shown below. The bottom activities may or may not be currently supported by computer applications. Rather, they are considered potentially viable areas for application development.

Within the Design Phase, *programming* focuses on space sizes and relationships. *Schematic design* supports applications for space planning, massing studies, elevation layout and initial wall layout. *Design development* addresses layout of special spaces—stairs and restrooms—and some detailing and material selection. *Construction documents* activities include specifications, finish schedule generation and detailing. The last three stages include cost estimation. Three further post-design stages include bid review, construction and activities leading to acceptance of the building.

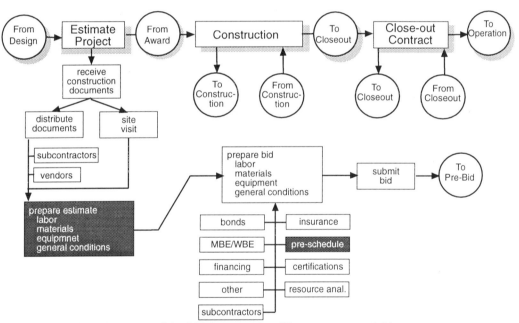

Figure 9.4: Overview of the IAI Construction Phase process activities.

The *Construction Phase* is shown in Figure 9.4. It is broken into three stages: *estimation*, *construction* and *contract close-out*. Estimation is shown in the same figure, while the other two stages are presented separately, in more detail. Estimation emphasizes document distribution, site visits for preparing estimates, preparing bids, submission of the bid and dealing with permits, construction financing and other administrative actions.

The *construction* stage activities are shown in Figure 9.5. Construction includes the preconstruction administrative activities, followed by generic issues of construction management—payroll, labor management, job costing, daily reports, and so forth. The last column addresses more administrative activities. In the middle are two (of many) interface activities. The emphasis of construction activities is on administrative activities rather than on-site activities.

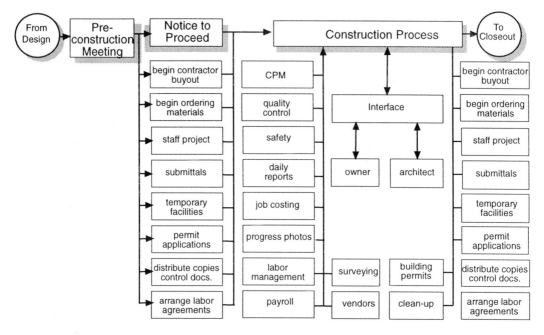

Figure 9.5: A more detailed presentation of activities in the IAI construction process.

The *close-out* stage of the Construction Phase is detailed in Figure 9.6. As can be seen, it identifies seven major activities, most of them administrative in nature.

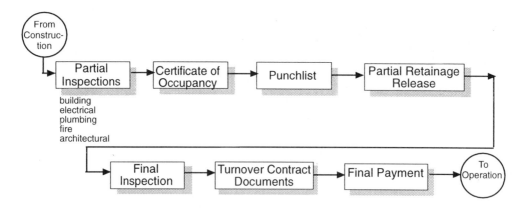

Figure 9.6: The construction close-out activities identified in the IAI process plan.

The last phase is *Operation*, shown in Figure 9.7. Two activities that take place both prior to and during occupancy are listed, followed by two more detailed lists of issues and activities, organized as "financial" and "physical facilities." The financial category emphasizes facility operation budgeting. The physical facilities activities additionally emphasize facility maintenance and repair records and planning—areas closely associated with the physical plant.

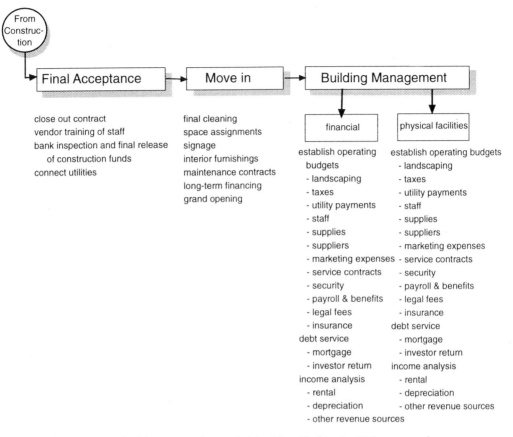

Figure 9.7: The building operation activities identified in the IAI process plan.

These six diagrams and the phases and activities they identify serve as a scaffold for incrementally building up the various details of the IAI project model. You will notice that several activities in Figure 9.3 are highlighted as black boxes with white lettering; for example, *bubble diagrams*, *wall layout* and *door/window placement*. The highlighted activities are those being developed and detailed. (Some development activities are not shown). Version 1.5 identified nine specific subdomains for detailing:

Feasibility Design
    1.   capture of client brief, bubble diagramming and space layout
    2.   wall layout
    3.   door and window placement
    4.   door and window schedules
HVAC
    5.   area take-offs for load calculations
    6.   manufactured equipment selection and HVAC equipment schedule sheets
Construction Management
    7.   cost estimating
       -   object identification for estimating
       -   estimating task and resource scheduling
       -   estimating quantity and cost modeling
    8.   construction scheduling
Operation
    9.   schedule generation for equipment, furniture and occupancy

More detailed process models have been developed for each of these targeted activities. Thus, the diagram in Figure 9.8 was developed for Capture of Client Brief, Bubble Diagramming and Space Layout activity. Out of such an analysis, the data requirements for particular activities were determined. (The process models for the other eight subactivities are not included here.)

These process description activities provide a framework for the object modeling activities of IAI. They bring to the attention of the Domain committees the variety of processes followed by different practitioners within a segment of the building industry and in different countries. Different processes followed to achieve the same high-level activity are made explicit. The exercise also encourages rethinking of those processes and encourages convergence toward more consistent processes.

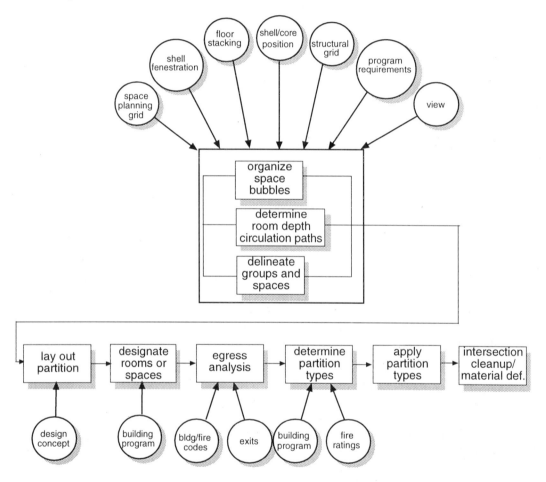

Figure 9.8: The process model for the Capture of Client Brief, Bubble Diagramming and Space Layout Activity detailed in Version 1.5 of the IAI Project Model.

The process models reflect a limited form of convergence. While some activities are ordered in a sequential form, most are listed in a manner to indicate that they are supporting or flowing into one major activity and can potentially be applied in parallel. The process descriptions indicate that the order of activities is varied in different organizations and/or that a particular order cannot be agreed upon. In addition, not all sequences of listed parallel activities are possible. In the design development stage, for example, door/circulation layout interacts with restroom design by constraining the layout of the restroom. There are many such relations that affect both activities and the contents of applications developed to support them. Such interactions between tasks need to be articulated before applications can be developed. The contextual assumptions regarding what information needs to be available for executing an application, considered in depth in the COMBINE project, are not considered here. Thus, further elaboration of the process models is called for if they are to serve as templates for application development.

Another obvious issue is the selection of the particular activities to be addressed in a product model versus other activities that have not been listed. Some obvious activities that have been omitted include the development of design phase applications to support the layout of specific activity spaces, such as laboratory spaces, office spaces, schoolrooms and kitchens. There are many high quality kitchen cabinet design and layout packages which have no role in the IFC process model. Similarly, several significant activities are omitted in construction—scheduling deliveries, planning on-site storage, and materials procurement planning and tracking, for example. In facilities management, the traditional activities of space, equipment and services allocation are not included, nor is equipment maintenance. Thus, the process frameworks for several stages are incomplete. There are many practical reasons for not getting to the details of such process models. Systems development based on voluntary donations of time is likely to have these sorts of problems. It will be important, however, to revise and elaborate the process models as the work of the IFC proceeds.

The associated documentation states that these activities are meant to serve as a framework for application development, as well as to provide a context helpful in gaining user involvement in further specification of the model. Thus, the detailed models, such as those for the Capture of Client Brief, Bubble Diagramming and Space Layout Activity, are meant as high-level specifications for application development.

Overall, this planning model is strikingly ambitious—apparently more ambitious than those efforts being attempted in other industries that have analyzed and provided automation support for more limited segments of the enterprise process. It recognizes the limited number of computer applications that exist today and attempts to define a roadmap for the development of future applications. It roughly shows how the different application areas would relate to one another. It appears to be more than an effort at information integration supporting existing applications; rather it is an effort to identify a long-range structure for automation within the building industry.

9.4 INDUSTRY FOUNDATION CLASSES (IFC)

Given this introduction of the process model developed by the IAI, we turn now to an examination of the building project model, as of Version 1.5. The IFC documentation provides three different views of the building model. Data model views for defining file or database definitions of the IFC are presented in EXPRESS and EXPRESS-G. Implementations take advantage of the EXPRESS tools reviewed in Chapter Five. Runtime Object definitions are also defined in Interface Definition Language (IDL). IDL has been defined by the Object Management Group (OMG) and provides implementation within Common Object Request Broker Architecture (CORBA), an object model interface definition language used as a standard within the software industry. The fullest definition is the textual one, covering a few issues not easily presented in EXPRESS. (These three views are not one hundred percent consistent.) The presentation and review here focus on the EXPRESS view, relying on EXPRESS-G diagrams for most of the presentation. Elaboration is offered to cover issues not dealt with in EXPRESS-G. For more detailed information, readers should refer to the original documentation.

9.4.1 IFC Architecture

The IFC is organized into sections that address different core areas and domain areas. These sections are organized into four layers, as shown in Figure 9.9.

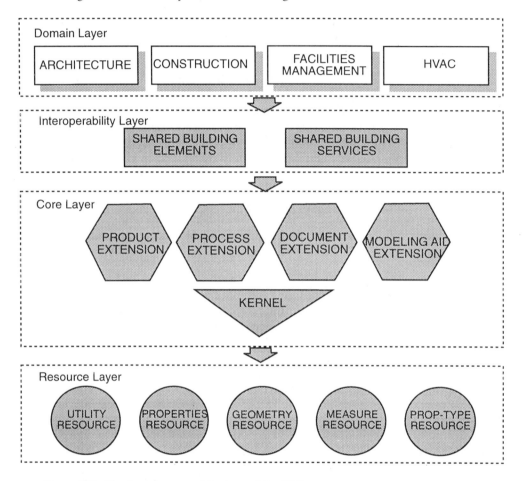

Figure 9.9: The four-layer architecture of the IFC.

The bottom Resource Layer provides common resources used in defining the properties used by the upper layers. It includes Utility, Properties, Geometry, Measure and Property Type Resources. These provide geometric properties, units of measurement, cost units, time units and so forth, corresponding roughly to the Integrated Resources of ISO-STEP. They are basic properties that stand alone, not requiring access to other data or definitions.

The second layer is the Core Layer. It includes a Product Extension, Process Extension, Document Extension, Modeling Aid Extension and Kernel. These provide abstract generic concepts that are used in higher-level definitions. They also access the Resource objects below them. The Kernel objects are required to be included in all higher-level models. The Product Extension supplies the majority of object classes making up the physical description of a building, defined at an abstract level. It includes generalizations for walls, floors and spaces, for example. The Process Extension provides definitions of classes needed to represent the processes used to design and construct a building. The Modeling Aid Extension provides those abstract elements used in developing a building design, such as grids, modules and centerlines. The Document Extension provides means to present project data in a particular format, useful for different needs in the building lifecycle. Future Core Extensions are planned for Controls and Resources.

The Kernel schema defines the most abstract part of the IFC architecture. It defines general constructs that are basic to object orientation, such as object and relationship. These are then specialized into constructs like product and process, which form the entry points of the next level, the Core Extension layer. The Kernel also handles some basic functionality, such as relative location of products in space, sequences of processes in time, or general-purpose grouping mechanisms. It also lays the foundation of extensibility of the IFC model by providing type-driven property definitions and property definition extensions.

The third layer is the Interoperability Layer. It defines objects that are shared by more than one application. These objects specialize the Core Layer objects and elaborate them for use by applications. Currently the Interoperability Layer objects are primarily building elements and building service elements. Later, they are expected to include Distribution Elements (ducts and piping), Furniture, Electrical Appliance, and Building Codes.

The fourth layer is the domain-specific application layer. It supports the applications used by architects, engineers or contractors, for example. Because of the careful specification of the lower layers, this layer should remain small and simple to follow in later releases.

This layered architecture identifies the different resources and incremental abstractions needed for the definition of objects that carry data in a building product model. It builds upon the experience of the ISO-STEP organization, not only at the AEC level of the organization, but also for dealing with the cross-industry abstractions found in the Integrated Resources. It also reveals the size and complexity of the building model effort.

The IAI Technical Committees developed a working set of guidelines for the development of models, to address base problems and to promote consistency of approach.

-   Recognizing that most objects in buildings have multiple functions, the function of an object and its form and material are to be treated separately. An object instance may have specialized performance properties added to it for uses within an application, but these properties are not intrinsic.

-   In the IFC, relations are recognized as first-class objects. That is, a relation has an object representing it. This supports complex many-to-many relations. It also allows relations to have behavior or performance requirements—for example, a weld connection of a particular strength.

-   The aggregation hierarchy also has been defined with care (aggregation is called *containment* in the IAI specification). A primary IFC model element hierarchy has been defined using "Has" as the attribute relation. Other relations are considered groups or systems and these are defined using "Part-of" attributes.

The primary IFC element hierarchy is based on the accessing structure,

Project   >   Sites   >   Buildings   >   Stories   >   Spaces   >   Elements

That is, a project is the top-level container made up of one or more sites. A site is a container of one or more buildings (and all of their parts). A building contains one or more stories (and its parts) and a story is made up of one or more spaces (and its parts) and spaces are defined of one or more elements.

The IFCs also include the building project program as part of the model representation, allowing applications to work with this information while generating floorplan and other layout data. The modeling aids also support grids and other guidelines and reference lines as layout aids for design—another aspect developed especially for the building industry. All EXPRESS Entities are considered objects in the IFC; we will follow that convention and refer to them as Objects. Because every object name begins with Ifc, we give the full name in all figures. However, in the text associated with the figures, we will drop the repeated use of Ifc. It is implied in all Type and Object names. We will review the IFC model from the bottom up.

9.4.2 IAI Resources

The four-layer architecture drew significantly from the parallel work of ISO-STEP. Its Resource Layer provides facilities similar to those provided by the Integrated Resources of Parts 041, 042 and 043. However, being targeted to the building industry, only a subset of these resources are needed and used. Also, some additional capabilities needed for building models have been added. Here, we review the resource layer, emphasizing areas that have been revised in response to building needs. As shown in Figure 9.9, there are five modules in the Resource layer, each defined as a separate schema. We will consider each separately.

The Utility Resource: This resource provides some basic bookkeeping facilities associated with work on a building project. Specifically, it defines and provides facilities for Identification, Ownership, History, Registration and the basic structure of Tables. A small part of the Utility Resource is shown in Figure 9.10.

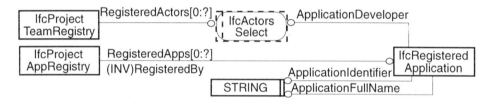

Figure 9.10: Actors and Applications are carried in two Project Registries.

To provide efficient access to model data, all applications and actors (individuals and organizations) are entered in a Project Register. The registered applications are assumed to be IFC compliant. The Integer index of the entry of the registry is used to identify them in later uses. Actors are defined in the Property Resource, to be reviewed next. All applications are defined with both a short name (16 characters), and a longer, full name and the ApplicationDeveloper Actor (developer organization).

The data models for Ownership, History and Identification are shown in Figure 9.11. Ownership, in IFC terminology, corresponds to the one or more Actors that are responsible for maintaining an object. The Ownership of an Object may change over time. History, in IFC terminology, captures any changes to the Owner/Application combination. OwnerHistory changes will be captured in the AuditTrail. In Version 1.5, the AuditTrail has only one instance (rather than a backward path in history). A version history will be implemented in a later release. Then, each update or stored change will be recorded as a Transaction, providing a full history of the recorded evolution of the design.

At the bottom of Figure 9.11, two identifiers are shown. One is a ProjectUniqueId and the second is a globally unique one. The globally unique one was implemented in Version 1.5.1, using an ID generator from Microsoft. The OMG ID generator was listed as another candidate.

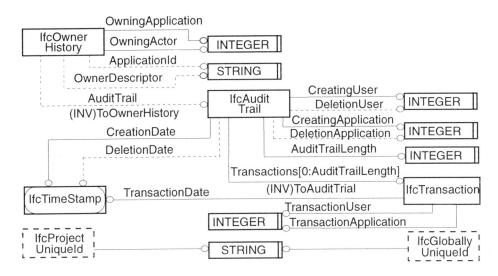

Figure 9.11: The history, ownership and identification data models in the IFC.

The Properties Resource: This resource defines the low-level nontechnical types and object classes dealing with Actors, Materials, Use Classifications, Cost and Time. The Cost objects define almost all national currency types and provide places to carry the conversion rates between them. The Time objects allow the coordination of time units across all time zones. Classifications allow ordered references to Materials or other properties for selection and use. New classifications may be added at any time. Actors include both persons and organizations. An Actor has a defined role and a standard address structure.

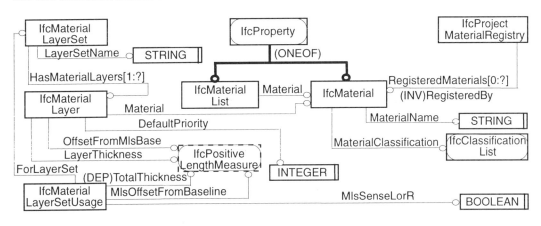

Figure 9.12: The Properties Resource structures for organizing materials.

Of special interest in the Property Resource is the structuring of materials, as shown in Figure 9.12. These are entered in a Material Registry, facilitating reuse. Because materials are often layered (as characterized in the core wall example used throughout this book), the IFC implementers structured materials in a layered section in a way that may be easily reused. Materials are referenced by a `MaterialLayer` and grouped into `MaterialLayerSets`. The layer sets are referenced by different usages. This provides a way to structure wall sections and floor/ceiling sections for possible reuse. Consideration should be given in a later release to linking these materials to property sets, which would allow them to have performance properties.

Properties also include the units of measurement. Those defined in the IFC are a large subset of those defined in Part 041 and reviewed in Chapter Six. As in the STEP standard, only SI units are specified.

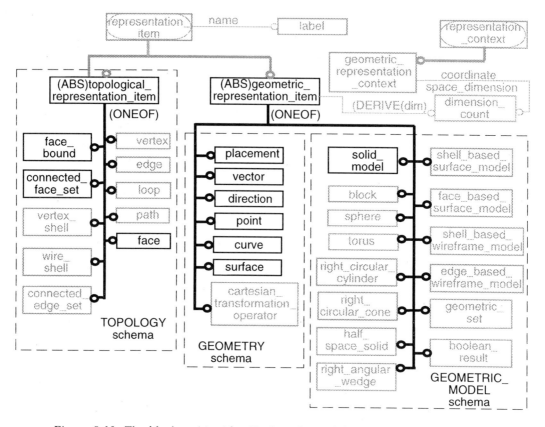

Figure 9.13: The black entities identify the subset of the Part 042 schema used in the IFC. The rest of Part 042 is shown in gray.

Geometry Resource: The geometry resource includes a subset of the entities defined in Part 042 of the STEP Integrated Resources. The top level of the subset is defined in Figure 9.13. The Geometry Resource adopts the base geometry entities of placement, vector, direction, point, curve and surface, restricting the surface types to only the planar specialization. It uses the solid_model, but moves it from the geometric model to the geometry schema. It adopts the face_bound for bounded planar faces, faces and connected_face_set topological entities, to depict, for example, typical triangulated meshes used in computer graphics and in site planning. The geometric model entities are the solid model entities, but with a somewhat different set of primitives. This subset excludes all b-spline and offset curves, hyperbolas and parabolas. It also excludes all the predefined CSG primitives, like cylinder, wedge and torus, and eliminates the distinction between geometric entities and topological entities carried in Part 042. Throughout, the entities have been cleaned up, with Abstract types defined where appropriate and new WHERE clauses added to refine the checking for illegal datasets.

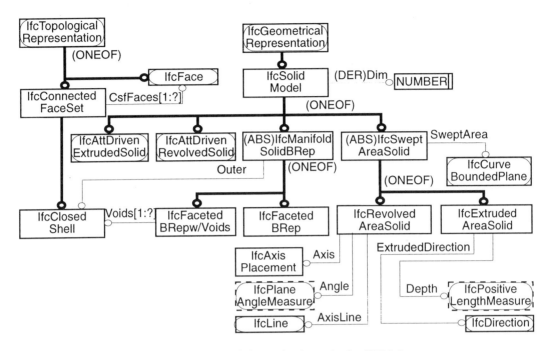

Figure 9.14: The revised solid modeling subschema in the IFC 1.5.

The IFC recognizes the importance of solid models in design and in representing building geometry and has modified them to better serve the needs of building. The revised definitions are shown in Figure 9.14. The `SolidModel` uses two of the Part 042 solid model entities: `SweptAreaSolid` and `ManifoldSolidBrep`. These use the `RevolvedAreaSolid`, `ExtrudedAreaSolid` and `FacetedBRep` of Part 042. The subset of topological entities from Part 042 is retained to provide the essential faces for the various `ManifoldBrReps` and `SweptAreaSolids`. These include `ConnectFaceSet`, `ClosedShell` and `Face`. The `ManifoldSolidRep` is an explicit boundary representation solid. It defines a shape as bounded by one or more `ClosedShells`. The default is a single closed shell; if there are several shells, it is a `FacetedBRepw/Voids` and the other shells are `ClosedShell` that depict interior voids. `RevolvedAreaSolid` and `ExtrudedAreaSolid` are adopted from Part 042 but are restricted so that the `CurveBoundedPlane` may only be a planar surface (its boundaries may be curved). `SweptAreaSolids` are examples of unevaluated or implicit solids, in that their faces are not explicitly represented. In general, these entities are the same as those in Part 042 with a cleaned up specification and additional `WHERE` clauses.

In addition to the above objects, the IFC Geometry Resource adds a set of entities to support parametric geometry. They are shown in Figure 9.15 and are worth reviewing in detail, because of their attempt to capture some unique aspects of building object geometry.

There are four new solid modeling types: `AttDrivenExtrudedSolid`, `AttDriven-RevolvedSolid`, `AttDrivenRevolvedSegment` and `AttDrivenExtrudedSegment`. These four objects reflect the two general classes of generative solids: those that are extruded along a direction and those that are revolved about an axis. The two types take as one parameter an `AttDrivenProfileDef`, which can be of various subtypes. This is the cross-section to be swept. The extrusion parameters are length and direction. The revolved shape parameters are an axis and plane angle measure, giving the portion of a circle used by the revolved shape. The sweep parameters and profile are grouped into a segment—`AttDrivenExrudedSegment` or `AttDrivenRevolvedSegment`. Multiple segments can be defined. If the end of one segment does not match the beginning of the next segment, a stepped shape results. Both types of sweep may result in hollows. The extrusion will have a hole if the profile being extruded has a hole. The revolved shape may have a hole if the revolution is 360 degrees.

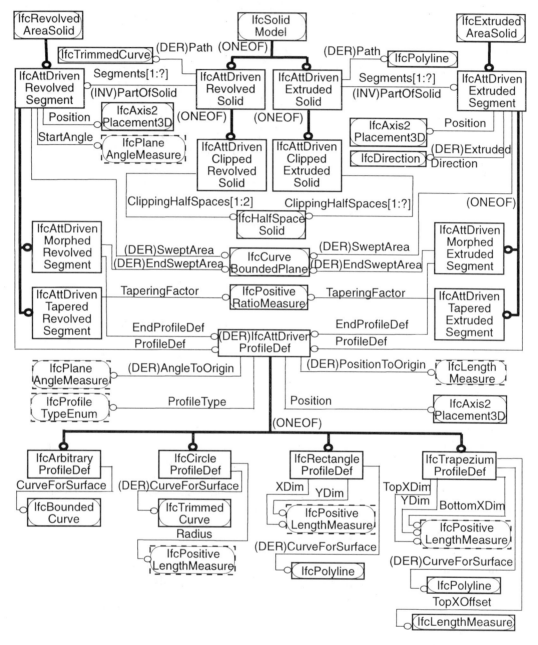

Figure 9.15: The new parametric modeling extensions in the Geometry Resource.

The different profile definitions include an arbitrary curve, a circle (with a radius and the derived circle), a rectangle and/or a trapezoid. Both the extrusion and revolved shapes may be tapered. If a different profile is defined at the ends of the sweep, then the shape is morphed. These shapes also can be trimmed with a cutting plane (technically called a halfspace). In each case, the parameters are stored and the surfaces and shape are unevaluated or implicit. At a higher level, they also may be combined using the Boolean operations described earlier. The parameters for the different pro-files are shown at the bottom of the page. Those objects not defined—such as Axis2-Placement3D, Direction, TrimmedCurve, Polyline—are as they are in Part 042.

The different parametric shapes provided are tailored to respond to particular types of 3D shape generation that may be quite useful for some 3D CAD packages. They seem to respond to the style of a single type of CAD package, rather than a response to a set of packages, as was explicitly the case in Part 042.

Measure Resource: The IFC Measure Resource has adopted the STEP Part 041 measurement schema without significant change. A quick tour of the schema was provided in Section 6.2. It provides the same unit measures (SI units only) and carries out the same or more rigorous consistency checks.

Property Type Resource: Property Types provide a means of grouping material and object properties into sets that correspond to those needed to describe the specific behavior of an object. For example, the properties needed to represent the aggregated thermal conductivity of a wall and the thermal conductivity of a roof are the same. By aggregating individual properties into property sets, these become effective for assigning sets of properties to objects. Property sets may be assigned to types that share the property set. They also may be assigned to object instances that have unique properties because of their special functions. Property sets (Psets) also guarantee that all the properties needed for some type of performance are loaded into the objects together. Psets may include material properties and also shape properties. The singular Material properties are those presented in Figure 9.12, reviewed earlier.

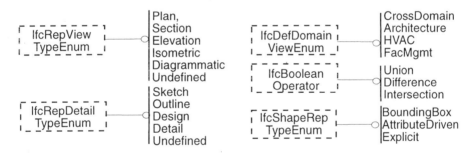

Figure 9.16: Four enumerated types defined in Property Types for use in shape property modeling.

Some enumerated types facilitate the assignment of Psets and provide some useful general definitions. These are shown in Figure 9.16. In theory, Psets may be defined at each of the three top layers of the IFC architecture. That is, some may be assigned at the Core layer, others added at the Interoperability layer, while others, particularly those associated with a particular design domain, can be assigned at the Domain layer. Psets may also be assigned to particular instances. This provides great potential flexibility in the assignment of Psets. To support such flexibility, objects can be accessed according to different view types; four domain types are defined in DefDomainViewEnum, shown in the top right of Figure 9.16. Presentation types are shown at top left. Another aspect of the Property Type schema is shape representation. Shape properties can be classified as to the type of projection or view that is defined, whether it is a crude outline shape (typically a bounding box), a parameterized shape or a fixed, explicit shape. Shapes are also classified as to the level of detail contained, whether it is a sketch, outline, design (assumed to correspond with design development stage) or detailed. Last, because one shape can be defined as a Boolean composition of others, using solid modeling and the union, subtraction and intersection operators—the Boolean operators are defined at this level to allow high-level implicit definition of these relationships. This makes it easy to define a shape as composed of a set of simpler elements—a window frame—or subtract a shape that is embedded in another—a wall with a window in it. That is, this relation can be used to define a window and to generate the opening in a wall, without the window actually being geometrically subtracted from the wall.

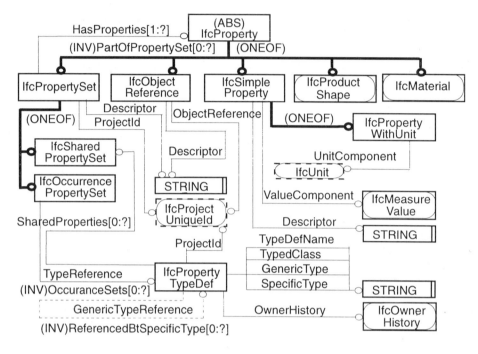

Figure 9.17: The Property Set Resource schema used to define and aggregate properties.

The schema for representing PropertySets Sets is shown in Figure 9.17. The subtype tree of the five property object classes is shown at the top. `SimpleProperty` is that with a label name, a `MeasureValue` and optionally a Unit for the value. `ObjectReference` refers to an object defined elsewhere. `PropertySet` is a list of any subtype of `Property`, including `Material` and `ProductShape` properties. `PropertySet` (Pset) is an important definition. It groups properties for assignment to objects. `SharedPropertySet` group properties are those assigned to an object type, where all instances have the same property values. `Occurrence-PropertySet` is the set of properties assigned only to instances and having different property values for each instance. According to the model definition, both `SharedPropertySet` and `OccurrencePropertySet` may be assigned at various levels in the inheritance hierarchy; in practice, they are assigned only to leaf object classes. Each `PropertySet` has a descriptor, a Unique ID, and a reference to a list of `Properties`. Each of its subtypes also has an associated `PropertyTypeDef`. This is the object type with which the `PropertySet` is associated. Since each `PropertyTypeDef` is defined as data, not as a class definition in computer code, they can be added or modified during execution. This means that properties may easily evolve over a project's lifetime. Thus any set of properties may be defined, reused in multiple ways and added or removed. Some of the uses of Psets will be presented later.

The Property shape schema is shown in Figure 9.18. A `ProductShape` is a more concrete and unique representation than pure geometry and topology. It has an ID and a Descriptor and is either a `ShapeBody` or `ShapeResult`, as defined by the select type, `ProductComponent-`

ShapeSelect. A ShapeResult is the combining of two or more Operands using the BooleanOperator. Because the Operands may themselves be ShapeResults, a nesting of Boolean combined ShapeBodys may be defined. The Boolean operators have been moved up in the abstraction hierarchy so that they may be applied to shapes of different objects, not just used to compose a single shape, as is the assumed practice embedded in Part 042. A ShapeBody is represented as a ShapeRepresentation, having an optional Descriptor and an enumerated type value identifying its use in the model. This allows different ShapeRepresentations for different uses. It references one or more GeometricRepresentationItems that define the actual geometric entities used to represent the shape. It also carries contextual information regarding its ProjectId and Precision, along with level of detail and view type enumerators.

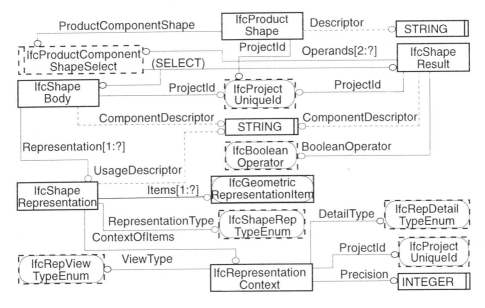

Figure 9.18: The Product Shape schema in the Property Type Resources.

In summary, we see that the IFC Resources have greatly reduced and simplified the Integrated Resources provided in Parts 041, 042 and 043, providing those that are most likely to be commonly needed in building type geometry. They are extended to deal with certain kinds of construction materials and geometry. They have also adopted a form of Property Sets different from those in the Building Construction Core Model that provide great flexibility in the assignment of functional properties to objects.

9.4.3 IAI Core Layer

The IFC Core Layer provides the first-level abstract objects built on top of the Resource Layer. The Core Layer includes five schemas: Kernel, Product Extension, Process Extension, Document Extension, and Modeling Aid Extension. We will quickly review each of them.

Kernel Schema: The Kernel schema defines abstract structures within the IFC information architecture. These address all user level objects (defined abstractly), different classes of relations and the most abstract level of modeling aid.

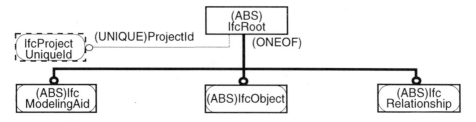

Figure 9.19: The Root structure of the IFC Kernel Schema.

The Root Object definitions are shown in Figure 9.19. The Root Object provides a ProjectID that is inherited into all other objects. It defines three classes of object: Modeling Aid, Object and Relation.

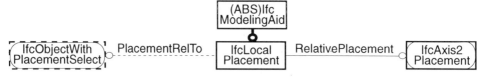

Figure 9.20: The Modeling Aid schema of the IFC Kernel.

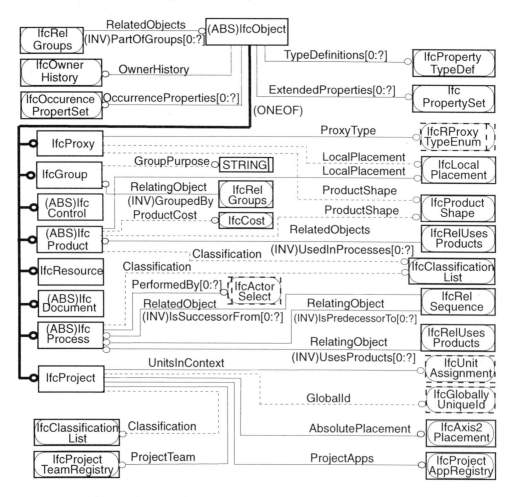

Figure 9.21: The abstract Object classes within the IFC Kernel.

The Modeling Aid Schema is shown in Figure 9.20. It is an empty abstract object with one subtype object, LocalPlacement. If there is no object, then the placement is global to the project origin, else it is a relative placement to the object referenced.

The Object definitions are shown in Figure 9.21. Object is the supertype of all abstract and physical objects in the IFC. This is the largest set of kernel definitions. Each Object has an OwnerHistory (giving who and what application it belongs to), a set of PropertyType-Definitions that identify type-specific properties, a set of OccurrenceProperties that overwrite at the instance-level properties that are defined for the property type, and a set of Extended Properties that are the instance unique properties that may be assigned by an application. These properties and relations are inherited by all more detailed Objects.

There are eight kinds of subobjects. The Project object is the top-level definition of a project. It carries a unique ID, a location defining the global coordinate system, a unit of measurement, a classification list and the Team and Application registries. (The ClassificationList carries the different means by which to classify objects; it is defined in the Property Resource.) The Process object identifies an Actor who executes the process and an optional ClassificationList; it is referred to by two durations (carried in RelSequence)—one for predecessors and the other for successors of the process. It also is referred to by the objectified relation with other objects. Group objects have an optional purpose and a relation defining the purpose of the group. Other object types, including Control, Resource, and Document, are not elaborated here. These definitions are inherited by the various subtype objects.

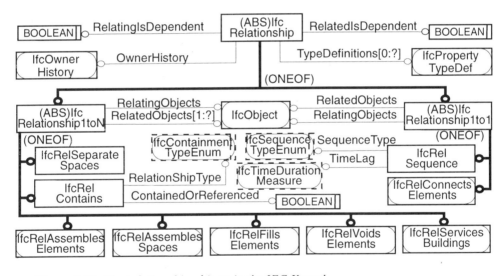

Figure 9.22: The relationship objects in the IFC Kernel.

Relationship object definitions within the IFC Kernel are shown in Figure 9.22. Relations are directional—from the relating object to the related object. All relations identify whether or not the relating and related objects are dependent upon the relation. That is, if the relation is deleted, are the objects that it refers to deleted also? Relations also have an owner history. There are two relation subtypes: a one-to-one relation and a one-to-many relation. Two one-to-many relations or multiple one-to-one relations can be used to define a many-to-many relation. A one-to-one relation may be specialized into RelSequence, with time lag duration and sequence type. Sequence type determines if the relation between the two processes is start-to-finish, finish-to-start, start-to-start (they start at the same time) or finish-to-finish (they end at the same time). As shown in Figure 9.21, each RelSequence also has a reference to the Related successors and the Relating predecessors. Product objects have a cost, a local placement, shape and optional classification.

A one-to-many relation may be specialized into a group (not detailed in Version 1.5) or specialized as a relation that uses other objects in realizing the relationship. The `RelContains` is characterized by a `ContainmentType` (see Figure 9.23) and whether the relation is Contain or Reference. All relations refer to a `RelatingObject` and to one or more `Related Objects` that the relation is between. The five Relations across the bottom of Figure 9.22 are defined in other schemas, mostly in the Product Extension schema. These definitions provide a rich and well-defined set of structures for objectified relations. So far, their use has been restricted to a small subset of those that are possible, because of the complexity that can easily arise.

The abstract object classes of the Kernel schema carefully structure the object classes that can be utilized later in the definition of a building. In particular, the relationship objects realize the objective of making relations as important as objects in the modeling of a building.

Product Extension: The IFC ProductExtension schema defines basic object concepts used within the IFC to capture the meaning of these concepts as used in the AEC/FM industry. The object concepts include Elements, Spaces, and the base aggregation hierarchy used in the IFC. This hierarchy was introduced in Section 9.4.1 and consists of `Site`, `Building`, and `Building-Storey`, `Spaces` and `Elements` within a Story. ProductExtension also handles some basic element connectivity and space boundaries.

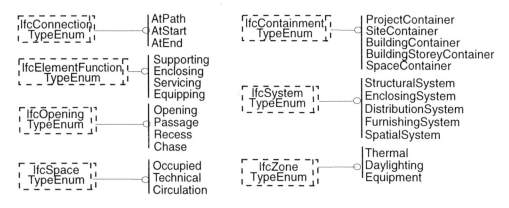

Figure 9.23: The Product Extension defines some types in classifying products.

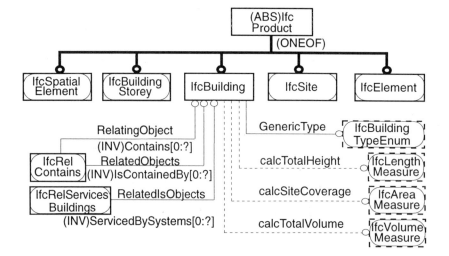

Figure 9.24: The top-level product definitions in the Core Layer of the IFC.

The enumerated types in Figure 9.23 define abstract relations used in defining a building composition. The ConnectionTypeEnum identifies three ways that one object may be connected to another. The ElementFunctionTypeEnum identifies four functional relationships, again to support particular views. OpeningTypeEnum identifies four types of passages within an object. SpaceTypeEnum introduces three initial abstract classifications for space. The ContainmentTypeEnum defines five different types of containment used in the IFC hierarchical access scheme. SystemTypeEnum presents five initial systems (several more are required). Last, the ZoneTypeEnum identifies three initial types of zones within buildings. (Some of these have not yet been used in the implementations.)

The Product schema provides a first-level set of definitions for Product objects within the IFC. Four subtypes of products are defined: SpatialElement, Site, Building, BuildingStorey and Element. These are shown in Figure 9.24. The figure also presents the Building definition. It includes a BuildingTypeEnum—not described above—that is declared but whose enumerated values are not defined. It includes some simple aggregate properties of the building—its total height (from the site elevation), a total building footprint area, and a TotalVolume. The Building, as part of the aggregation hierarchy, also is referenced by what the building contains and by what the building is contained in (at Site level). It also is referenced by a set of building services.

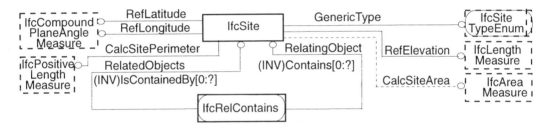

Figure 9.25: The Site Object class in the IFC.

The Site, shown in Figure 9.25, is the level between the Project above and the Building level below in the aggregation hierarchy. It includes a location in longitude and latitude and an elevation. It also includes a descriptor, a SiteTypeEnum (also not defined in terms of possible values) and a site area. Within the aggregation hierarchy, it has two relations to it: one to objects that it contains (site objects and zero or more buildings) and one to those in the Project that contains it.

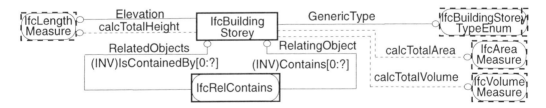

Figure 9.26: The Building Storey object type definition in the IFC.

The BuildingStorey is the next, more detailed level in the IFC aggregation hierarchy, shown in Figure 9.26. The Storey is defined according to two heights (height of story from site datum and floor-to-floor height), an area and volume measure, a BuildingStoreyTypeEnum (also without the value of the type being defined) and its two containment relations. The next lower-level objects are Spaces and Elements.

The IFC `SpatialElement` has two subtypes. `Space` is the basic spatial entity for all building spaces. It is shown in Figure 9.27. It carries as attributes a space type, a perimeter length, an area, a volume and an average height. Each Space is part of two hierarchical relations. One is the `Contains` relation. The other is the `AssemblesSpaces` relation, for grouping subspaces into higher level spaces.

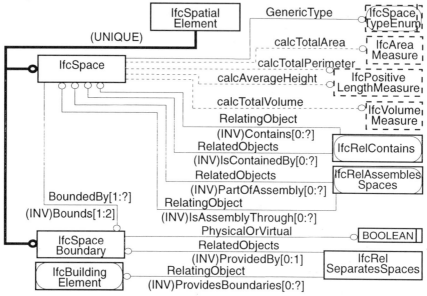

Figure 9.27: The abstract definition of a space and SpaceBoundary in the IFC.

`SpaceBoundary` is the other subtype of `SpatialElement`. It carries a flag telling whether it is a real boundary or a virtual one, i.e., with no physical embodiment, used for completing the enclosure. It also carries a relation that identifies the `BuildingElement` that realizes the separation.

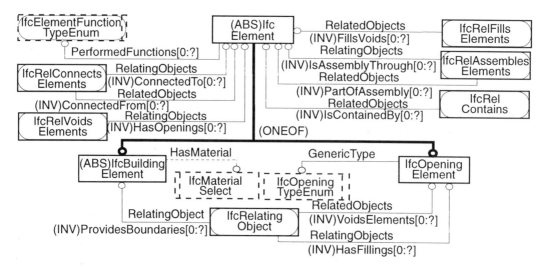

Figure 9.28: The IfcElement and its direct subtypes and some of its relations.

All components that are part of the building and are not spaces are considered Elements. The IFC `Element` description is presented in Figure 9.28. For instance, all furniture, ductwork and walls are Elements or subtypes of Element. The `ElementFunctionTypeEnum` in Figure 9.23 broadly classifies these uses. Elements are subtyped into two subclasses: `BuildingElement` and `OpeningElement`. A `BuildingElement` has a material and is subtyped into more detailed objects. An `OpeningElement` is a void of a kind defined by `OpeningTypeEnum`. It may have fillings or other void elements, which are `BuildingElements` related to it.

Element may be related in a broad number of ways. Connection, void, fill, assembly and container relations all reference the Element object and its subtypes.

Process Extension: The process extension provides the base-level object for defining various processes. Construction processes have been emphasized so far.

As do other Core Layers, it begins by defining needed SELECT and ENUMERATED types. One that is of obvious use is `TaskStatusTypeEnum`, shown in the upper left of Figure 9.29.

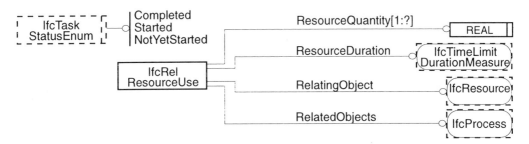

Figure 9.29: The main definitions in the IFC Process Extension.

The Relation Resource Use is a subtype of `Relationship1To1`. It defines a resource quantity, the resource used and the process to which it is applied. It also carries duration of the resource usage.

The Process Extension also includes definition of a WorkSchedule Entity. It includes all the standard time durations and dates needed to represent a schedule at some level of detail. It is shown in EXPRESS below.

```
ENTITY IfcWorkSchedule
  SUBTYPE OF (IfcControl);
     WorkSchedule : IfcWorkTaskOrGroupSelect;
     ActualStart : OPTIONAL IfcDateTimeSelect;
     EarliestStart : OPTIONAL IfcDateTimeSelect;
     LatestStart : OPTIONAL IfcDateTimeSelect;
     ActualFinish : OPTIONAL IfcDateTimeSelect;
     EarliestFinish : OPTIONAL IfcDateTimeSelect;
     LatestFinish : OPTIONAL IfcDateTimeSelect;
     StatusTime : OPTIONAL IfcDateTimeSelect;
     ScheduledStart : OPTIONAL IfcDateTimeSelect;
     ScheduledFinish : OPTIONAL IfcDateTimeSelect;
     ScheduledDuration : OPTIONAL IfcTimeDurationMeasure;
     RemainingTime : OPTIONAL IfcTimeDurationMeasure;
     FreeFloat : OPTIONAL IfcTimeDurationMeasure;
     TotalFloat : OPTIONAL IfcTimeDurationMeasure;
     TaskStatus : OPTIONAL IfcTaskStatusEnum;
     IsCritical : OPTIONAL BOOLEAN;
  END_ENTITY;
```

Document Extension and Modeling Aid Extension: The `DocumentExtension` outlines the needed cost schedules and groups for costing of materials and resources. It includes a `CostSchedule`, `CostElementGroup`, `CostElement` and a `RelCostSchedule Element`.

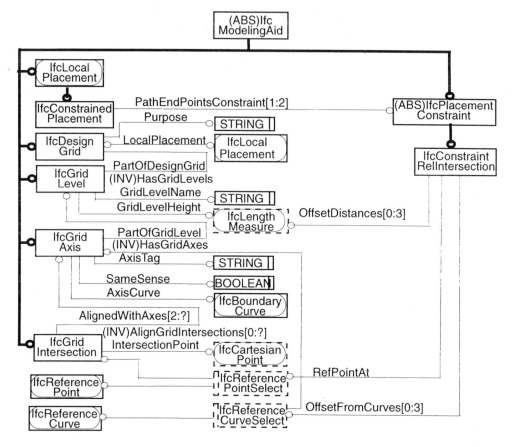

Figure 9.30: The Modeling Aid Extension Schema. It allows definition of grids and constraints to locate objects relative to the grid.

The Modeling Aid Extension provides definition capabilities for grids of any sort that may be used as a modeling aid. Their schema is shown in Figure 9.30. All objects are subtypes of `ModelingAid`. A `DesignGrid` is defined as a set of `GridAxes`. `GridIntersections` then reference the intersection of two or more `GridAxes`. Different `DesignGrids` may be defined for various `GridLevels`. Eventually a variety of placement constraints are planned. Initially, only one, `ConstraintRelIntersection` is implemented. It defines a placement constraint as an X-Y-Z offset from either a `ReferencePoint` or a `GridIntersection`. The grids appear quite general and potentially powerful. An example of a grid is shown in Figure 9.31.

9.4.4 Interoperability Layer

The next layer of the IFC architecture is the Interoperability Layer. It provides definitions for objects shared across applications. It consists of two schemas: `IfcSharedBldgElements` and `IfcSharedBldgServiceElements`. The entities defined at this layer are subtypes of lower-level objects and receive most of their definitions from these lower-level definitions.

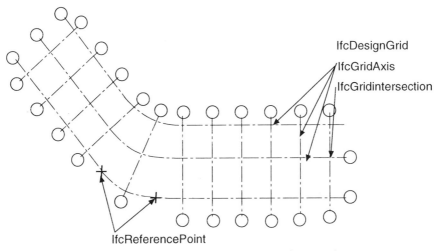

Figure 9.31: An example of a grid defined using ModelingAid.

Shared Building Elements: This section involves the definition of a set of high-level objects that inherit all their properties from `IfcBuildingElement` and are classified according to special type, using an enumerated type. We will sample of few of these here.

The enumerated types are shown in Figure 9.32. There are types for `Beam`, `Built-in` elements, `Column`, `Covering`, `Door`, `Floor`, `RoofSlab`, `Wall` and `Window`. They provide a preliminary definition of types for these entities, which can be added to and elaborated over time. In most cases, these types have associated Property Sets.

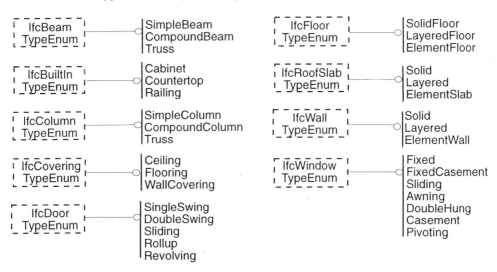

Figure 9.32: The enumerated types used to classify different objects in the Shared Building Elements schema.

Two examples of the shared building elements—`Floor` and `Wall`—are shown in Figure 9.33. They are classified according to the wall and floor types shown in Figure 9.31. They reference a Material Layer Set usage from the Product Extension layer. They also derive a new attribute from the parent object to `MaterialLayerSetUsage`. These move the material properties up the element level (evidently, for ease of access).

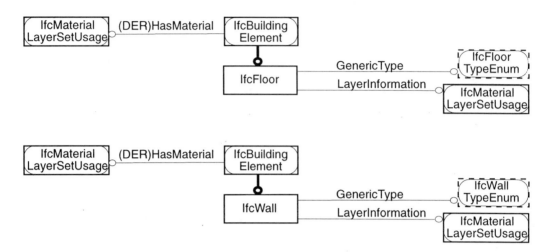

Figure 9.33: Two examples of objects from Shared Building Elements.

Each of the objects receives a predefined PropertySet (Pset), inherited from the `IfcObject` layer, based on the value of the associated enumerated type. There may be one Pset for the general type and an additional set for the specialized subtype. These are listed below, with the property name, followed by its type, followed by a description. For example, a `Floor` has a Pset made up of the four properties that also apply to `LayeredFloor` and `ElementFloor`:

TypeDescription	IfcString
	(note name is captured in the TypeDef object that references this PropertySet) describes the type of floor in words
Pset_StructuralAreaLoad	IfcObjectReference
	describes the structural loads assumed for this floor
FinishFloorMaterial	IfcInteger
	Index for a material in the MaterialSet used as the finish floor material
SubFloorMaterial	IfcInteger
	Index for a material in the MaterialSet used as the subfloor material

The Pset for walls has eight predefined properties:

TypeDescription	IfcString
	describes this `WallType`
ActualDimension	IfcPositiveLengthMeasure
	actual dimension of wall
PlanDimension	IfcPositiveLengthMeasure
	dimension of wall
ExtendToStructure	IfcBoolean
	does the wall extend to the structure above
FireRating	IfcReal
	rating of wall assembly. `0.0´ indicates value not set
ThermalRating	IfcReal
	rating for thermal transmissivity (`U´ value). `0.0´ indicates value not set

AcousticRating	IfcReal
	rating for thermal transmissivity (Sound Transference Factor =STF) for wall assembly. `0.0´ indicates value not set
PartitionDetailRef	IfcString
	detail reference for assembly of this wall

These also apply to the enumerated wall types. The general Window has the following sixteen properties in its Pset:

ProjectTypeReference	IfcString
	reference ID for this window type in this project (e.g., type `W-3´)
SpecificDescription	IfcString
	description for this type of Window
ManufactureInformation	IfcObjectReference (Pset_ManufactureInformation)
	reference to a SharedPropertySet -Pset_ ManufactureInformation, which defines information about the manufacture of this window hardware
ShapeProfile	IfcObjectReference (IfcShapeResult)
	object reference to the geometry object which represents this window
RoughOpeningHeight	IfcPositiveLengthMeasure
	height of rough opening to be provided for this window type
NominalHeight	IfcPositiveLengthMeasure
	nominal window height (rounded actual height)
RoughOpeningWidth	IfcPositiveLengthMeasure
	width of rough opening to be provided for this window type
NominalWidth	IfcPositiveLengthMeasure
	nominal window width (rounded actual width)
Frame	IfcObjectReference (Pset_DoorWinFrameType)
	reference to nested property set Pset_DoorWinFrameType, which defines information about the window Frame
Screening	IfcObjectReference (Pset_ScreenType)
	reference to a nested property set -> Pset_ScreenType, which defines information about screening for operable openings
FireRating	IfcReal
	fire rating of window assembly
ThermalRating	IfcReal
	rating for thermal transmissivity (`U´ value)
AcousticRating	IfcReal
	rating for thermal transmissivity (Sound Transference Factor = STF) for window assembly
HeadDetailRef	IfcString
	detail drawing reference for Head of the window
FrameDetailRef	IfcString
	detail drawing reference for the Frame of the window
SillDetailRef	IfcString
	Detail drawing reference for Sill of the door.

In addition, each of the window enumerated types has additional Psets. For example, the pivoting enumerated window type has three more:

```
WindowPanel                 IfcObjectReference
                            (Pset_WindowPanel)
                            Reference to a single, pivoting window panel, as
                            viewed from the finished (exterior) face
WindowPivotDirectionIndex IfcInteger
                            Integer index into the enumeration referenced by
                            WindowPivotDirectionEnum = index
                            specifying the pivot direction for this window type
WindowPivotDirectionEnum IfcObjectReference
                            (Pset_WindowPivotDirectionEnum
                            Reference to an enumeration of all possible window
                            pivot directions
```

The window example suggests the elaborateness and relative completeness of the properties that can be put together for an object. These are the only shared Psets assigned to the class level of object. To date, very few occurrence-level Psets have been defined at the instance level of object. Pset definition and management is an area of active refinement within the IFC.

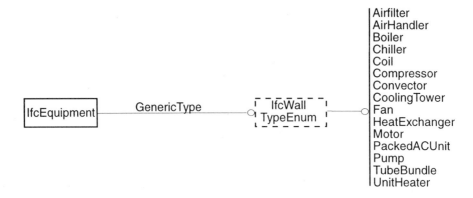

Figure 9.34: The Equipment definition distinguishes between different types of equipment.

Shared Building Service Elements: The Shared Building Service Elements are only outlined and not well developed. The main object that is detailed is for mechanical equipment. It is shown in Figure 9.34. It essentially types the piece of equipment. Associated with each enumerated type is a Pset for the type.

Fifteen types of equipment are identified. As an indication of the Psets for service elements, the Pset for `AirHandler` incorporates eight properties:

```
AirHandlerConstructionEnum ENUMERATION OF (ManufacturedItem,
                            ConstructedOnSite, Other, NotKnown,
                            Unset)
                            this enumeration defines how the air handler might
                            be fabricated
```

```
AirHandlerConstructionIndex    IfcInteger
```
index into the enumeration property set
Pset_AirHandlerConstructionEnum

```
AirHandlerFanCoilArrangementEnum ENUMERATION OF
                    BlowThrough, DrawThrough, Other,
                    NotKnown, Unset)
```
this enumeration defines the arrangement of the
supply air fan and the cooling coil
```
AirHandlerFanCoilArrangementIndex    IfcInteger
```
index into the enumeration property set
Pset_AirHandlerFanCoilArrangementEnum
```
DualDeck                IfcBoolean
```
does the AirHandler have a dual deck
```
Fans                    BAG [1:?] OF IfcObjectReference
                        (IfcEquipment)
```
bag of one or more references to an
`IfcEquipment` object of type Fan that defines the
supply, return or exhaust air fan(s) that are used by
the AirHandler
```
Coils                   BAG [1:?] OF IfcObjectReference
                        (IfcEquipment)
```
bag of one or more references to an
`IfcEquipment` object of type Coil that defines the
coil(s) that are used by the AirHandler
```
AirFilters              BAG [0:?] OF IfcObjectReference
                        (IfcEquipment)
```
bag of one or more references to an
`IfcEquipment` object of type AirFilter that defines
the air filter(s) that are used by the AirHandler

These Psets provide great detail for the definition of different building service elements.

9.4.5 Domain Layer

The Domain Layer carries definitions of objects that are needed by a particular application domain. At this time, this layer is only defined schematically. The Architecture Domain Layer has a bit of definition, dealing with the building program. In addition to a listing of individual spaces, a program sometimes has information about adjacencies. These are carried in the Architectural Domain, as shown in Figure 9.35.

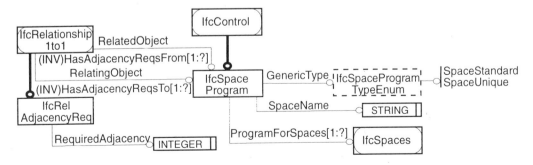

Figure 9.35: The Adjacency information carried in the Architectural Domain subschema.

A `SpaceProgram` consists of a list of spaces, each with their requirements. The `SpaceProgram` has a type and a name. Relationships are defined between pairs of spaces and are given a weight indicating how important the adjacency is. Each adjacency is directional, so there may be two adjacency relations between any pair of spaces.

Like the Shared Building Elements, the Architecture Domain objects also have associated Psets. The Pset for `SpaceStandard` applies to all instance of a `SpaceStandard` type. The Pset is

SpaceStandardDescriptor	IfcString
	description for this space standard
NumberOccupants	IfcInteger
	programmed number of occupants for this space standard. Zero means the value has not been set
CodeOccupancyType	IfcString
	occupancy type according to the building code for this project; see type empty
StandardArea	IfcAreaMeasure
	the area programmed for this space standard
StandardLength	IfcPositiveLengthMeasure
	standard length for spaces of this type
StandardWidth	IfcPositiveLengthMeasure
	standard width for spaces of this type

The Pset for `SpaceUnique` has no shared property values, rather they are all at the Occurrence level. They are similar to the standard, but with fewer properties.

There is also a small Facilities Management Domain, providing definitions of some furniture. These are the only two Domain schemas defined in Version 1.5.

9.4.6 Functionally Defined Psets

An important capability of the IFC is adding Psets to individual object instances, in response to functional performances. Thermal, acoustic, structural and lighting properties can be added to walls, roofs or other object instances that are relevant to the performance analysis being undertaken. Up to Version 1.5, only some Psets for thermal performance have been defined. We sample of few of them below.

Pset_24HourSchedule	schedule of usage for a 24-hour period, typically attached to an IfcZone, IfcSpace or IfcGroup object
Name	IfcString
	name of schedule
UsageList	LIST [24:24] OF IfcReal
	list of decimal fractions between 0 and 1 reflecting hourly usage intensity
Duration Schedule	IfcTimeDuration
	start and end dates and times
Pset_BoundaryThermalProperties	attributes related to thermal properties of building boundary elements; this property set is typically attached to objects of type IfcWall or IfcRoofSlab
BoundaryDescription	IfcString
	a boundary description used by the HVAC engineer (e.g., ASHRAE component type); this may or may not be the same material description provided by the architect

```
BoundaryThermalTransmittanceCoefficient IfcMeasureWithUnit
```
(IfcThermalTransmittanceMeasure)
overall thermal transmittance coefficient (U-Value)
of the composite materials used by the boundary
element

`BoundaryColor`   `IfcString`
color of the boundary (i.e., light, medium, or dark
for roofs)

Pset_AggregateLoadInformation   The aggregate thermal loads experienced by one or
many spaces or zones in a building or space. This
property set is typically attached to an IfcZone or
IfcSpace object.

`TotalCoolingLoad`   `IfcMeasureWithUnit`
`(IfcPowerMeasure)`
the peak total cooling load for the building

`TotalHeatingLoad`   `IfcMeasureWithUnit`
`(IfcPowerMeasure)`
the peak total heating load for the building

`LightingDiversity`   `IfcReal`
lighting diversity

`InfiltrationDiversitySummer`   `IfcReal`
diversity factor for Summer infiltration

`InfiltrationDiversityWinter IfcReal`
diversity factor for Winter infiltration

`ApplianceDiversity`   `IfcReal`
diversity of applicance load

`LoadSafetyFactor`   `IfcReal`
load safety factor

This is a small sample of the thermal property Psets. The other thermal Psets in Version 1.5 are
`Pset_AirSideSystemInformation`,   `Pset_ApplianceThermalProperties`,
`Pset_FillerThermalProperties`, `Pset_GlazingThermalProperties`, `Pset_`
`LightingThermalProperties`, `Pset_Load-DesignCriteria`, `Pset_Material`,
`Pset_OutsideDesignCriteria`, `Pset_SiteWeatherData`, and `Pset_SpaceEl-`
`ementInformation`.

9.4.7 Discussion

The IFC is the largest and most elaborate building model developed to date. Careful thought has
gone into defining structures that allow the high-level objects to be defined easily. At the same
time, it incorporates large and flexible property sets (Psets) that can carry the attributes needed for
specifying or selecting some types of objects. The IFC responds to many of the issues set out in
the earlier chapters; it objectifies relations, it provides well-defined accessing schemes, and it
allows extension of properties at the instance level in response to special functions. We will
examine some of the capabilities in the next section.

9.5 INTERPRETATION OF THE IFC BUILDING MODEL

A base accessing structure within the IFC building model is the container structure among objects,
illustrated in Figure 9.36. The Physical entities in the "contains" relationship are shown at the left
of the figure. The structure offers a well-defined accessing scheme within a Project. The Project
"contains" one or more sites and all objects that are relevant to a project but are above the site in
aggregation. This might involve schedules, budgets and costs. At the next level is the site, and it
"contains" one or buildings and other objects at the site level. These might consist of site planning

processes and site work, for example. The building level is next. The building level "contains" one or more building stories and other objects at the building level. These may include the service core and building-level mechanical system equipment, the roofing system and foundation, and other objects that apply to the building level. Each `BuildingStorey` "contains" `Elements` and `Spaces`, which make up the bulk of the objects, but might include other objects at the story level. Both `Elements` and `Spaces` may be nested into element assemblies and space assemblies (only the space assemblies are shown in Figure 9.36). This structure represents an effort to define containers that can hold any part of a building—whether it is a large aggregated part (exterior shell) or some detail piece (door hardware). It relies on general relations—`RelatedObjects` and `RelatingObjects`—but fixed levels of aggregation.

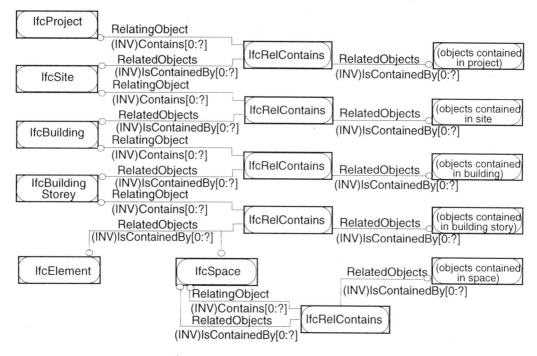

Figure 9.36: The container-based accessing structure in the IFC.

Because of its size—and especially because of its deep inheritance structure—it is difficult to see what a single object's definition really is. To see this, we must expand and "flatten" the object definitions. (This is an excellent exercise for readers wishing to understand the definition of some high-level object.) The process one follows to expand any object is to start with the object, list its attributes, identify the objects that are the supertype(s) of the current one, list their attributes, identify the next level of supertype for each of the existing ones—and so forth until there are no more supertype objects to identify.

If this process is applied to IfcWall, we see that Wall is a `BuildingElement`. `Building Element` is a subtype of `Element`, which is a subtype of `Product`, which is a subtype of `Object`, which is a subtype of `Root`. Based on this expansion, we can see that the full possible definition of a wall is as shown in Figure 9.37. It shows the full inheritance structure of objects used to define the `Wall` object.

The wall has clear conventions for referencing the opening elements and for accommodating any fillers for the opening. These are carried at the Element level so they can also apply to floors and ceilings, for example. The `BuildingElement` also is the supertype that passes space separation relations to the wall (in the `RelSeparatesSpaces` object). It references `SpaceBoundary` that then references the Spaces it separates.

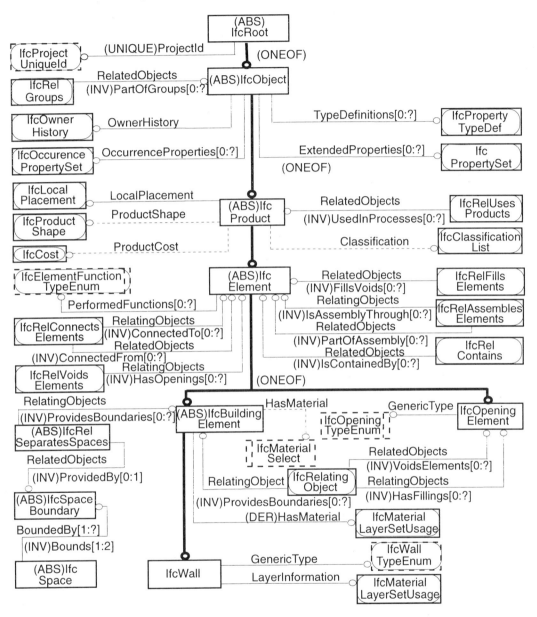

Figure 9.37: The expanded wall structure defined within IFC Version 1.5.

The `Element` level definition inherited into `Wall` provides reference to voids, connections with other elements and assemblies. The `Product` level adds shape, cost and placement. It should be noted that multiple shapes might be defined for different uses. The `Object` level provides general holders for Psets—for both shared and occurrence sets— and object `Groups`. The `Root` adds a unique identifier within the Project. In reality, this rich set of properties is associated directly with the Wall, being inherited by it. In addition, it has the Pset for wall, listed earlier. If the Wall is an external wall, requiring a thermal Pset, then Pset_BoundaryThermalProperties would be added.

Later, when they are defined, other Psets—for acoustical properties, for a bearing wall, or other functions—may be added.

In practice, while this set of properties may be needed by the universe of all applications, a specific application or set of applications needs only a subset. Even for the predefined properties and relations, not all are used in any one application, or even a set of applications. For the demonstration of the IFC, the implemented subsets were agreed upon ahead of time to allow time-effective implementation. A more formal management scheme for defining what Psets to model and exchange between sets of applications is being planned.

9.6 REVIEW OF THE IFC

Technically the IFC is not a standards effort, but rather an industry-led undertaking to develop practical user capabilities for data exchange. The IAI organization, funded through corporate donations, has not yet addressed how the results of their work will be maintained, published or updated over time. One possible outcome is that it will eventually become an ISO-STEP effort, with the ISO organization taking on these responsibilities.

Some of the aspects criticized here are being addressed in current work on new versions. The Reference Process Model, for example, discussed in Section 9.3, is being completely revised for the Version 2.0 release.

The IFC is an ambitious model, even at this early stage. It currently consists of a little over 280 objects and types. Of these, just about half have been adopted from the Integrated Resources of STEP. We have reviewed the majority of the new ones here. Of the details not covered here, most are types and objects in the `MeasurementResource` and `PropertyResource` schemas.

The IFC has adopted most of the good practices that have been identified in earlier efforts. Object definitions are extensible. Multiple geometric descriptions can be assigned to an object. Relations are objectified and can be quite complex; they can be qualified using properties. Properties can be dynamically added to an object instance, in order to support a particular analysis function. Wherever possible, the IFC includes an enumerated type for determining different cases, an improvement over using a `STRING` that must be matched at runtime and is prone to error. Both BCCM and CIMsteel rely heavily on `STRINGS` for defining cases.

The semantic capabilities of the IFC are richer than those expressed in current STEP models. While one or more Psets may be initially defined as a component of an object in EXPRESS, there is no way to define conditions for *adding* a Pset to an object class or occurrence. While people can understand these concepts and make the extensions manually, there is no straightforward way to enforce them inside an EXPRESS model. Also, some of the implementation assumptions in the IFC are different from those in STEP. In STEP, all attributes can be defined simply by looking them up in the object descriptions. Loading of properties in IFC requires a different type of lookup; the enumerated types that identify the Psets used must first be identified, then the Psets are assigned. Thus the SDAI-type tools require some extension to be fully usable in the IFC.

The IFC has made some strong assumptions about the type of geometry needed to model buildings that are different from the assumptions made at the end of Chapter Six. It relies heavily on a new set of attribute-driven types, but omits many more general types embedded in current generation CAD systems. It will be interesting to see how the geometry assumptions of the IAI work out. For example, can the proposed geometry deal with the space relations modeled in quite a different way than were implemented in COMBINE? Also, The IFC provides high-level Boolean operations, allowing one object to be subtracted from another rather than a single object shape being the result of subtractions.

Of course, the model is currently lacking needed functionality. It lacks stairs, many types of construction, and support for detailing special types of spaces, such as laboratories, auditoriums, restaurants, and so forth. Versions 2.0 and 3.0, currently under development, will partially fill in these gaps. The IFC has responded to most of the issues that have been put forward regarding the careful semantic definition of a building product model. With continuing careful development and extension, the generality of the IFC suggests that it could be adapted to support the needs of almost any set of building related applications. One unresolved question is its relationship to other parallel aspect models developed within the STEP organization.

With the IAI's IFC, we see the culmination of twenty-five years of effort in the development of building models. This volume has tracked the trajectory of building model development, starting with early efforts to define a set of applications around a common building data structure. We have seen elaboration of that structure, a growing sophistication with regard to separating the specification of the model from its implementation, and the handling of various issues arising from multi-function objects and complex geometries. IFC seems a solid response to the various specifications put forward regarding the needs of a building model and, with elaboration, able to support exchange of any well-specified data between a set of applications. In many ways, the IFC is the penultimate example of a model defined to support a suite of applications. It remains to be seen if it "solves" the data exchange issues in the building industry.

9.7 CURRENT CAPABILITIES OF BUILDING DATA EXCHANGE

The overview of building modeling systems presented in the last three chapters outlines the current technologies, their implementations and capabilities, defining the current state of the field. It is useful to summarize where we are, what we can and cannot do. We use the IFC as the most advanced example to date of a general-purpose building model.

The IFC and other models can be used for many forms of data exchange. It is a rich model that addresses the needs of different applications and provides a variety of ways to define the same part of a building. There seem to be at least two different ways the IFC model may be applied. One way has been demonstrated in the implementations developed thus far. The implementers reviewed the model and agreed upon the subset of objects, attributes and relations that the source and target would use for exchanges. These were decided by identifying the intersection of information used by both applications. These agreements were necessary so that the information generated by one application would be written to the Entities that were read by the second application, and that the attributes were defined in the units needed by the second application. If the information was not directly available, they had to sort out which application interface would include the necessary translation to put the data in the agreed to form. In addition, the sequence for running applications had to be determined, so that the application generating, for example, material properties and quantities was run before the cost estimation application. These assumptions about the order of application invocation usually were not codified. Only the COMBINE project developed explicit workflow structures that defined the order of application execution.

Given a subset of the IFC model, each implementer of an application interface developed one or both interfaces from the model to the application and/or from the application to the model. In this case, the IFC is used to define a relevant subset of a building model, useful for a set of applications of interest. The building model is custom defined, as are the interfaces for this application subset. Each interface usually takes several weeks to implement by an experienced programmer. If another application is to be added or substituted, then a revision of the model and the interfaces will be required.

Also, all exchanges made between applications currently must involve the whole project. One application can write the project data and another can read it, as a pairwise exchange. When the second one is done, it can write it back, providing access to another application. The current IFC model supports iterative use of the model data and can update changes, as long as object instances carry matching IDs. All exchanges must be between the application pairs originally defined in the process scenario. If the applications attempt a different exchange sequence, the model subset written and then read will not match. They also may use data at different levels of aggregation, with no way to convert between the levels. This is because a change at one level of aggregation cannot be automatically propagated within the model to other levels of aggregation. These limitations also exist in STEP models and in the current CIMsteel model. Both CIMsteel and the IFC plan to support incremental updates in later releases.

Propagating or noting changes that affect aggregated or disaggregated objects is only one example of the general issue of consistency management within a design model. More generally, after an existing model is updated, the changes made will invalidate other parts of the design, because other data was *derived* from the changed data. A range of additional updates may be needed, for example, involving functional properties and derived shapes. Without careful management, the changes that traditionally have been identified and monitored on paper using redline markups cannot be so easily tracked when there are multiple views and electronic communication.

Based on the current capabilities of product modeling, if an organization or set of organizations agrees to integrate a set of applications that are going to be used repeatedly in exchanges over a period of time (allowing the cost of developing the model to be amortized), a working interface can be developed. This use of a building model supports exchange between pairwise or a larger set of applications, based on:

1. *a priori* agreement on the subset of Entities to be used, the associated attributes and the relations between Entities
2. *a priori* agreement on the sequence in which the applications can be called, defining which application writes out the data for an exchange and what application then reads the data
3. delimiting exchanges to whole file exchanges, without incremental updates

The second potential way to use the building model is probably the way that people assume it should work. The whole model would be implemented, as seems to have been intended by the system developers. In this second case, each application interface developer would implement a general interface to the model, not knowing with which other applications it will work. The implementer of the application interface would identify the subset of objects, attributes and relations that best fit the data used in the application and write interfaces between them to and/or from the model.

In use, no predefined exchange protocol exists for this case. Because no conventions were set, each application interface reading data from the model would have to scan multiple Entity classes that may hold the data needed to see which of them provides a meaningful dataset on which to operate. An application interface may have to add data to the model or modify existing data so that it will work in the application. The interface may have to deal with alternative geometric descriptions, selecting the appropriate one to use. After searching through various Entity classes, the interface would have to identify what data are missing from what are needed to execute the target application, to determine if the data exchange is worthwhile. Alternatively, the user of the second application could repeatedly generate exchanges until one was generated that had sufficient data to execute the receiving application. To date, no application interfaces with this level of generality have been implemented, and it is unclear if they can be. This use of a building model encounters problems concerning the three issues addressed in the earlier exchange approach:

- It lacks a way to properly coordinate or match the data output of previous applications to serve as input to a specific application. We call this the *data subset problem.*
- It lacks a way to track or coordinate processes (applications) that have been applied to the building model to allow identification of which applications can be applied to the current building model. We call this the *process and data dependency problem.*
- It lacks a way to deal with incremental updates or to support use of the model as a repository over time. This is the *incremental update problem.*

These three problems apply both to integrating a set of existing—i.e., legacy—applications or to a new set of applications integrated around a building model. To date, there has not been a working building model that has not closed the set of applications and coordinated and tuned their interfaces so they will operate in a compatible manner.

These limitations posed by the three building model integration problems—the data subset, the process and data dependency, and the incremental update problems—suggest that the first approach is the only approach that is currently viable. Development of a building model using a subset of a full model and fixing the process flow between applications is acceptable for several types of building model data exchange. This type of data exchange can be served well by passing data from one central model used in one phase of building to a central model used in the succeeding phase—for example, from design to construction, or from construction to facilities management. It can also support adequately the needs of a company working with a fixed set of applications, possibly directed at a single building type, or within the construction phase. As long as file transfers are of the whole project—say from a component procurement package to a scheduling package—and the ordering of use can be agreed upon, current data exchange technology, such as the IAI, can support the exchanges.

The data subset problem, where source and target applications must agree on the subset of Entities used in an exchange is made worse when a model is defined to support data flows throughout all lifecycle stages. The model subset problem is greatly reduced in the aspect models surveyed in Chapter Seven, because they only define the objects needed in a limited range of exchanges.

Current data exchange technology does not easily support—as currently defined—coordinated work within a stage, such as is required by a team of different specialists working on a building's design, or by a set of contractors coordinating—possibly at a distance—on the planning of fabrication and construction of a building. In these areas, highly coordinated information flows are required, allowing different applications using different models to pass changes back and forth. Notice, for example, that if one object in the IFC—say a wall— has multiple geometric models, there are no facilities within the IFC to keep the two models consistent. The two applications may be working at different levels of detail, one with a highly aggregated wall covering multiple floors and spaces, i.e., a shear wall, while the other with a wall bounding a single room.

9.8 FURTHER READING

International Alliance for Interoperability [1997] *Industry Foundation Classes, Release 1.5* IAI, 2980 Chain Bridge Road, Suite 143, Oakton, Virginia 22124-3018 USA. The Web site for the International Alliance for Interoperability is http://iaiweb.lbl.gov/. There is a small but growing number of reports on efforts to use the IFCs. Among these are Aalami and Fischer [1998], Alexander, Coble et al. [1998] and Froese, Grobler and Yu [1998].

9.9 STUDY QUESTIONS

1. The process models developed by the IFC are very different from those developed in Chapter One. Select one of the lifecycle phases and review the different process descriptions. From these, outline a revised one that takes the best features of both. In doing so, think carefully about the criteria used in the definition of tasks or activities.

2. The Wall model defined in Figure 9.35 is quite different from the wall model defined in EXPRESS in Chapter Five (Figure 5.25). Compare the two wall models and discuss their differences. What changes, if any, might be made to the IFC Wall model to address any shortcoming you identify.

3. Define a building as a single space enclosed by four walls, a floor and ceiling. One wall has a door and two others have windows. Assume that no materials have yet been decided. Choose the IFC objects needed to represent these objects and lay out the minimal structure to define this as a complete but schematic model.

4. After several students answer question #1, compare the results and see where the differences have arisen. Based on the results of the study, discuss what the IFC accomplishes and does not accomplish.

5. The IAI and the BCCM have placed emphasis on representing different types of relations to be carried in a building model. (a) Compare the relations carried in each and suggest what you think are their different emphases. (b) Compare the relationships carried in the IFC model with those proposed in RATAS and discuss their differences. In particular, consider how relations are depicted in the RATAS Aspect models and those, for example, of walls.

PART THREE:

Research Issues

INTRODUCTION

Building modeling has certain capabilities today, even though it has not yet been widely deployed. These were summarized at the end of Chapter Nine. For some uses, these capabilities will be adequate and we will certainly see production uses of building models in the near future. However, for many uses in building, these capabilities are not adequate. And the potential for integration has not been fully realized. This last Part surveys some of the unresolved research issues in building product modeling arising from Part Two, and summarized at the end of Chapter Nine.

Chapter Ten deals with information exchange architectures. This chapter addresses how information flows between the different participants in a building process. Is the flow one-way or bi-directional? Is the information carried in a repository and updated, or passed back and forth as a batch file? Is there a single consistent model of the building that is being updated over time, or is there a need for several different models to be used in parallel, with incremental consistency checks? How can updates to a building model be coordinated over time? Within a distributed design environment, how can design work and the data exchanges needed to support that work be coordinated? Different processes need different information flow architectures and a range of such architectures are reviewed. Technical issues arising from the different architectures are surveyed.

Chapter Eleven deals with language issues and several limitations of EXPRESS and its use within the current ISO-STEP system architecture. Four different issues are reviewed. (1) What extensions are needed to allow EXPRESS to become a dynamic language, allowing schema evolution, rather than its current static form? (2) There has been strong interest in mapping languages—allowing transfer of data from one model or schema to another. What are the issues in the development of such languages? (3) Another limitation of EXPRESS is its lack of support for parametric modeling. A parametric model allows one property of an object to be derived from another or other properties. The Attribute Driven Geometry introduced in the IFC in Chapter Nine is an example of parametric modeling. How can such capabilities be generalized and built into the base modeling language? (4) EXPRESS provides a means to define the structure and content of a model and STEP provides semiautomatic means for implementing it as a file or database. However, no means have been developed for interrogating or reviewing an instantiated EXPRESS model, especially when it is being used as a repository. This chapter reviews these four issues and current research efforts for dealing with them.

The book ends with an Appendix listing current vendors supplying STEP tools.

CHAPTER TEN

Information Exchange Architectures

entia non sunt multiplicanda praeter necessitatem

entities are not to be multiplied beyond necessity

Willliam of Ockham
Occam's Razor
1285-1347

10.1 INTRODUCTION

Previous chapters focussed on the structure and semantics of a building model and on the concepts and technologies needed to define an appropriate model for production use. However, at the end of Chapter Nine, several collateral issues were raised. One set of issues involved the expected uses of data in a building model and the types of information flows and data exchanges needed to support those uses. That is, we are not just interested in the contents of that data, but in supporting the processes in which data exchange is embedded. These processes involve possibly complex information exchanges, beyond the pairwise exchange between two models. We call these information flows and the environment that goes with them different *information exchange architectures*. In the previous chapters, the work reported was relatively complete. Here and in the next chapter, the work and ideas are not complete but are open research issues.

In Chapter One, in the survey of lifecycle phases, some simple flow charts were diagrammed of information exchange. Most of the diagrams are either linear or network flows. Several identify complex information exchanges, involving multiple inputs and/or multiple outputs. But the diagrams in Chapter One were still at a high level of abstraction, not identifying the detailed-level flows involved in a process. These detailed flows include iteration involving multiple alternatives of a design or plan. They include multiple exchanges between groups of actors, in order to solve a shared problem. They might include small-scale exchanges of a portion of the building. That is, not shown in the diagrams of Chapter One are the detailed levels of information flow needed to support effective teamwork. This detailed level of data exchange is addressed in this chapter. We examine some different types of information flow used in building, identify new requirements and sometimes outline technological responses that respond to different information exchange scenarios.

Data exchange comes in all scales, from the exchange of a complete project to just a few variables. Flows may be linear or iterative. There are several dimensions to a data exchange architecture that are important to different building modeling contexts. We can illustrate some of these dimensions by considering different scenarios involving information exchange in current building processes. Four scenarios are presented below that detail some of the information flows outlined in Chapter One. The scenarios depict different issues arising in building model information exchange architectures. When we compare the requirements of these scenarios to the capabilities supported by the current technology, we can begin to identify the gaps in current data exchange technologies needed to support different building contexts.

10.2 INFORMATION EXCHANGE SCENARIOS

Scenario One: Design coordination. A principal of an architectural firm has laid out the schematic design for a high rise office building, defined in a set of release documents (or files). The firm's associates are working on the service core of the building and consulting firms are working on the structural and mechanical, electrical and plumbing (MEP) systems. A layout of the service core was defined in the initial release, which is now being refined by the design team members. One of the associates is detailing the restrooms, another the fire stairs, another the elevators. In parallel, other designers are working on the facade, lobby and other aspects of the building. As the designers modify and detail parts of the existing design, they coordinate with the other people being affected. This may involve a pairwise exchange ("can you give up three inches on the south wall of the restroom?"), or a group may form to determine the best solution to an issue based on their mutual perspectives (where to route plumbing lines). They collaborate on schedules ("I'll wait to detail the stairs since you are reconsidering the floor-to-floor height"). As changes are made, they are distributed to the other designers. If a change is proposed or made that cannot be accommodated by the others, then these are reviewed with the principal who adjudicates and resolves conflicts.

In this scenario, the building data is treated as a repository that is incrementally updated by the design team. Team members query the contents of the current design and select the information relevant for their tasks. Modifications are written back to the repository as incremental updates, changing or adding details. Updates made to shared data must be communicated to all whose work is affected. In general, participants need to know who has what data, so that coordination and the distribution of updates are facilitated. There is coordination of schedules as well as of the design.

Scenario Two: Energy consultant advises on design. The architectural firm has employed an energy consultant, in part because the client is concerned about ecological issues. The consultant is to advise on minimizing energy costs and to make sure that environmental conditions throughout the building are to a high standard. She has been asked to analyze potential issues of glare, both from lighting and from the expanse of glass on the south and west facades. The consultant examines the building orientation and typical sky lighting intensity (using data gathered by herself for this city). Asking the architect about expected surface finishes, ceiling finish and assuming typical internal lighting (these have not yet been defined by the architect), she derives the expected ambient lighting. Running the figure through a lighting analysis application, the exterior light coming in windows is found to be quite high and of high contrast with the internal lighting. After discussion with the architect, two alternatives are explored—using high levels of internal lighting to reduce contrast or reducing the external lighting level and relying more heavily on local light sources. The architect opts for the latter. The consultant examines louvers, overhangs and external sunshades by assessing their behavior in lighting simulations and identifies alternatives from which the architect should select. He chooses to use exterior louvers integrated into the building facade. The architect's layout is passed back to the consultant, who reruns the analyses, trying variations to enhance its benefits for certain days of the year, and passes these recommendations back to the architect.

In this scenario, the consultant relies on data supplied by the architect, but also retrieves data from a variety of other sources—weather information, her own data on sky conditions, material and finish properties and product information. The analyses identify unsatisfactory conditions and the consultant generates alternative design details. The architect selects and then further revises one of the alternatives and he and consultant pass it back and forth to incrementally improve it from multiple viewpoints.

Scenario Three: Construction scheduling. The general contractor has a person developing the construction schedule for a large-scale building project. The selected subcontractor for structural fabrication and erection is also developing that company's schedule. In order to speed construction, the agreed upon strategy is to finish the lower floors while the upper-level structure is still being erected. To do this, both organizations must closely coordinate. Week by week site and floor layouts have been printed to coordinate material stockpiling and the location of work assignments over time. Both groups proceed to develop detailed schedules independently, based upon a general schedule agreed to earlier. As they proceed, they identify all critical tasks that they assume to be completed by the other organization. They meet every other day to resolve issues and differences.

In this scenario, two separate schedules are being planned, based on a shared high-level schedule. Spatial conflicts and dependencies are checked, with conflicts identified and resolved during reviews.

Scenario Four: Pass off of the construction model to the facility owner/manager. Here, the as-built design is passed from the contractors to the building owners/managers. The contractors have been paid to revise the architect's drawing files, to reflect changes that were made during construction. The revised drawings are passed to the client organization, which uses a commercial facility management database to manage its facilities. The revised drawings are abstracted and simplified to obtain the general layout data required by the facility management package. Additional data are added, such as the enclosed spaces, organizational boundaries, HVAC control zones and other regions of relevance to facility management but not represented in the original model. So that proper maintenance schedules can be planned for mechanical equipment, specifications from the mechanical equipment suppliers are added to the model. Equipment specifications are currently defined in paper manuals, but will eventually move to an electronic format. Materials used in public spaces are identified so that maintenance schedules for these spaces can be developed.

In this scenario, there is a single pass-off of data. The scenario involves revising the available construction data both by subtraction and addition—the original design did not have organizational allocations. Others involve integration of data from heterogeneous sources—the equipment operating and maintenance manuals. The pass-off is done for the total project, in one direction.

These four information exchange scenarios are only a tiny sample of the many different exchanges regularly encountered around a building project. The scenarios presented are quite common and the type we might expect to be readily supported by current methods of data change.

Current Data Exchange Capabilities: In contrast to the scenarios above, the scenario supported by current neutral model data exchange capabilities can be characterized as follows:

> *A user extracts from a building application the data in the application that are to be exchanged with another application. An interface to the application writes out the data, using, for example, SDAI interfaces and an EXPRESS schema. That format is then interpreted by an interface to a second application that has been paired with the sending one. It extracts the data in the neutral format and loads it into second application, which can now be executed.*

The four scenarios have different information flows than those provided by current data exchange architectures. Some of these differences can be enumerated:

1. The data flow provided by current exchange capabilities is that one application sends a dataset and the other one loads the result. The data is not in a repository that can be added to, modified or deleted by different applications, as depicted in Scenarios One, Two and Three. There is no mechanism for two or more datasets to be merged for input to a third application. This issue is an elaboration of the incremental update problem, identified at the end of the last chapter.

2. Because in current exchange capabilities the dataset to be exchanged is prepared by the sender, another effect is that the sender determines what the receiver will acquire. The receiver cannot directly inspect and select the data needed, as the receiver can with paper-based exchanges.

3. The data from a building model is only one part of the information used in the development of another model. Particularly in the scenario of the energy analyst, the data regarding buildings, building products, users and local weather conditions are combined to define another model. From this perspective, building model data is only one of multiple types needed for most analysis applications. Input data extracted and reformatted from multiple sources is commonly required, resulting in a custom-configured dataset.

4. Existing exchange capabilities emphasize a single, neutral model for holding building information. However, Scenarios Two and Three point out the regular use of multiple, parallel but disjoint models. In these cases, the different actors, for both technical and professional reasons, do not fully integrate their results. Given current exchange technology, all the data used by the second application must be provided by the first. No provision has been made for multiple, domain-specific models or for consistency to be maintained between multiple models.

5. Data exchange is embedded in the more general flow of work. The energy consultant needed specific information to undertake her assessment of lighting. Final layout of the service core in Scenario One could be realized only after the structural beam location and sizing were fixed. Tasks are dependent upon each other, requiring identification of which tasks and applications are needed to exchange with which others. While the scenarios above reflect a workflow, only the COMBINE project addressed how multiple data exchanges may be coordinated. This is a new and large issue, characterized in the last chapter as the data and process dependency problem, only beginning to be addressed.

In addition to these differences, because of the effort required in programming and debugging, the development of an exchange capability is only worthwhile when the interface is going to be used repeatedly. For instance, let us imagine that an exchange format was devised for Scenario Four above that received a dataset from an architectural CAD application, such as Microstation® AutoCad® or Graphisoft's ArchiCad®, and output it into one or more facility management applications. After extensive debugging and testing, the interface between the two programs will become reliable enough to be used for production input of geometry. Such a scenario is appropriate for Scenario Four; it is probably appropriate for Scenario Three, where the exchange is also between fixed applications. Development of an expensive fixed interface is not likely to be helpful in Scenario One or Two, which involve a different mix of applications for each project.

The ideal for Scenarios One and Two is that the different applications can be integrated in a 'plug-and-play' manner, similar to hardware plug-and-play interfaces now available for personal computers. It should be possible to plug in the application; the environment would recognize it, build the necessary links to and from it, and allow it to operate as easily as, say, a new graphics tablet attached to a computer, or a new modem. The integration of a new application ideally would take minutes, not days. The field is a long way from realizing a plug-and-play application integration capability.

10.3 INFORMATION EXCHANGE FEATURES

In the following sections, we survey in more detail the five varied data exchange features identified above. Some of the theoretical, technical and implementation issues associated with the features are identified. Several of these features cannot be supported by EXPRESS and require extensions to the product modeling language. We consider the language extensions in the next chapter. Other features can be dealt with by proper modeling and can be covered by prescribing "good practices". Still others require specification of new features in the computing environment, such as those that might be provided by a database management system.

10.3.1 BATCH EXCHANGE OR REPOSITORY WITH UPDATES

One fundamental variation in data exchange is whether transfers consist of batch exchanges between two or more applications or alternatively if a repository exists that can be added to, incrementally read from and modified by two or more users. A related issue is whether building model data may be generated by multiple applications instead of a single one. Scenario Four consists of a batch pass-off, while all of the other scenarios assume incremental and iterated updates. It should be noted that the exchange capabilities presented in Chapters Seven, Eight and Nine all consist of batch level, serial exchanges from one to possibly several receiving applications. Those exchange capabilities provided no means for integrating incremental updates or for merging the output of multiple applications.

Batch exchanges are relatively straightforward. A dataset is prepared for exchange by the source application and user. It is received whole by one or more receiving applications. If a receiver carries earlier data describing the building, it is overwritten by the new data. The full exchange file must be managed so as not to allow corruption due to multiple users updating the file non-serially. That is, either a locking scheme must be used or "optimistic updates" must be allowed, where any time a user writes back to the file, all other users must then update their datasets with the new one. Exchanges may be in either or both directions, but they are all-or-nothing—no partial updates are allowed.

Two additional functional capabilities are required if the building model is a repository that supports selective extraction and incremental updates.

(1) *unique identification*: Each object must have a unique identification, so that the repository copy of the object can be found and updated with a new version, if necessary. The requirement for unique identification places a requirement on the application, because it must carry object identifiers that can be used in matching against those read from the repository. The Integrated Resources of Part 043 and also the IAI Resource Layer generate such identifiers, which must be carried by the application.

(2) *partial updates*: If only some of the objects in the repository are read and not others, bookkeeping must properly interpret the updates. If there are 40 walls on a floor and an application reads 25 of them, then after execution writes back 22 of them, what is implied for the 3 missing ones? A mechanism must be provided to distinguish additions and deletions, as well as modifications. If all the objects read by an application are flagged, those not written back can be assumed deleted by the application. If the application writes back new ones

without IDs, these can be assumed to be new ones. Others that are both read and then written back with matching IDs are interpreted as updates.

A central facility must assign object IDs, since they must be unique for the whole building project. This requires that they be assigned by the building model, not by an application. If the building model assigns the IDs, a means then exists to identify newly added entities. All entities with IDs written by the application to the building model must have been read earlier from the building model. Those without an ID are new and require the insertion of a new entity instance in the building model and assignment of a unique ID. During execution, applications must carry object IDs. All known mechanisms for managing incremental updates require applications to carry model-assigned IDs.

These capabilities are also needed to support merging of data from different models. If both applications carry instance IDs, then the precedence order of updates and deletions determines the contents of the shared model. However, the changes made by the first application must be propagated to the second by one of the concurrency mechanisms, such as locking, optimistic or nested transactions. Transaction methods are of great importance in supporting concurrent updates to a shared model. There is no automatic, foolproof method to merge newly created data from multiple applications; user controlled, interactive merging is required for a consistent result.

Only some of the entity instances in the building model may be read and copied by an application. These must be marked, identifying the application that has a copy of them. In order to reduce propagation of changes, only updates that actually modify the data should be treated as updates. When an update is written back to the building model, comparing the old values with the new ones identifies those that have changed. If no change was made by the application, no update should be recorded. Updates of just changed data restrict propagation to that which must be reviewed by others. Updated values are flagged. In addition, however, some entity instances may be deleted and these cannot be distinguished from those not read into the application; in neither case are they written back. In order to make this distinction, each entity instance within a building model has a flag field for each application. If it is read into an application, the flag is set. Updates involving a change are flagged one way, while no change turns off the flag. Any entity instances whose flag remains set after an update have been deleted by the application. These flags also serve to identify which users and applications must be notified when changes are made.

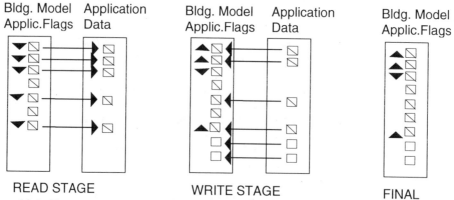

Figure 10.1: The management of updates from an application back to a model.

An example of an implementation is shown in Figure 10.1. It diagrams an example of a building model with application flags, in three steps of a revision, using one possible update protocol. The Read Stage transfers a set of instances to the external application, the Write Stage writes the updates back to the building model, with the final condition shown at the right. Within each stage, the building model and application dataset are shown as tall rectangles enclosing small boxes; the small boxes correspond to object instances. In the Read Stage, each entity instance read is flagged, shown as a triangle to the left of each entity and called the "check-out flag". A slash in a box indicates that it has an ID. Because entities are assigned IDs by the building model, all entities read from the model have them. In the example, five objects are read and two are not read. After the application runs, it makes an update back to the building model. If the entity instance written back has been modified, its check-out flag (triangle at left) is flipped. Three entities were modified. Some entities are written back without IDs and these must be new. One entity is not written back and after the Write Stage, it is left with its check-out flag as originally set by the read operation, which indicates deletion. The added and modified objects may be communicated to other users. At the beginning of the next read/write session, all flags are reset.

This bookkeeping protocol, combined with model assignment of unique object IDs, allows unambiguous updates between an application and a building model, supporting addition, modification and deletion. It covers the basic cases involving incremental updates. However, if an incremental update is to be made, the subset of the model of interest is determined at the time of reading from the building model. When data is written back, all the data read must be written, so as to allow deletions to be identified. Partial write-backs from the application to the building model are not supported by the above protocol.

These capabilities can be implemented in alternative ways. (1) They can be implemented within the building model itself, by extending the definition of Entities in the model so they carry the check-out information (a separate set must be carried for each application). (2) They can be implemented within the exchange application, on top of a building model. This involves a separate file or database facility that tracks which Entities have been accessed by which applications. (3) They can be embedded within the exchange language by providing special operators from Read and Write that do the bookkeeping. The EDM-2 product database implemented the above capabilities at the language level. This approach makes such bookkeeping simple, only requiring the user to recognize the semantics of different Read and Write operators.

10.3.2 DATA INSPECTION AND SELECTION BY THE RECEIVER

If the recipient of building model data is to be able to select the entity instances to be incorporated into a new model, then the recipient needs to be able to browse and interactively select the entities to be downloaded to his or her application. Selective acquisition of data is another needed functionality if a building model is used as a data repository. While some limited browsing facilities of product models are available, none has a user interface mature enough to support use by practicing architects or engineers. More work is needed in this area.

Querying of a building model needs to support the traditional accessing mechanisms now provided by current tools, in addition to more general extensions. Multiple accessing structures are needed. Architects and other users, for example, usually review a building based on spatial access of the data, as presented by floorplans and other sections and spatial visualizations of the building. Building specification people review a building based on the classification structure of building specifications—the CSI structure in the US—accessing entities that are part of a particular system or that have a particular function. Product suppliers and consultants today also search a set of drawings by functional systems, such as by mechanical systems, lighting, or communication systems, so as to bid on them. There needs to be multiple ways to query a building model—no one capability is likely to be appropriate for the different interest groups.

Building models are typically large, both in terms of the number of classes in the model, and also in terms of the number of instances carried within a dataset. Tracking through their structure without what database people would call query capabilities will not be satisfactory. The functional needs of such a query capability can be outlined:

1. Navigation capabilities, supporting easy user movement through the building model data structure. Three types of navigation support seem to be necessary:

   a. movement through the class-level entity and relation structure of a model, moving between Entities and their subtypes, and also from an Entity through attributes to other Entities, i.e., through relations
   b. navigation through the instance-level entity and relation structure
   c. movement about the building spatially, in terms of floor level and location within the eventual building

2. Selection of entities, singularly or in sets, after they are made visible by the navigation capabilities. Examples would include

   a. filtering the instances and relations of one or more classes, using attribute values as a selection criterion
   b. picking entity and relation instances as they are made visible during instance-level navigation
   c. selection using graphical features such as a fence drawn on the screen displaying entities spatially

3. Display of entities and instances in each of these three different modes, with movement between the nodes.

4. Importing the selected objects into the receiving application, applying a predefined external mapping.

The capabilities outlined here are not yet available in production form. However, aspects are provided by some of the STEP product vendors identified in Appendix A. The capabilities require both language extensions to support high-level queries and also application development tools that include graphical user interfaces. More work needs to be done to develop such facilities, if building modeling is to be a ready-to-use production tool.

10.3.3 POPULATING A MODEL WHERE DATA EXCHANGE IS ONLY PART OF WHAT IS NEEDED

Except in cases where an exchange is being made between similar applications, the receiving application will require additional information from sources other than the building model. The clearest expression of this condition was the scenario with the energy consultant. There, the building model was augmented by building product information, by weather data, by other external datasets and also by personal knowledge. All these sources were needed together to develop an assessment of and proposed improvements to the architect's design.

A simplified diagram of such an environment is shown in Figure 10.2. The figure illustrates that a building model is only one information source among several that are needed to support most building tasks. Most tasks require access to different data: specifications of installation procedures, building codes, engineering or design details, user data or programmatic requirements. The building information is only one part of the information support needed.

Many building-related Web browser technologies have shown the practical and effective use of general-purpose search tools capable of operating on unstructured data. The implication is that text-based search engines, "rummaging" over Web sites and digital encoding of reference manuals and other information sources, could provide effective means to access other kinds of data needed for building-related tasks. There is a growing business in developing special purpose Web sites for particular classes of users, such as architects, engineers and contractors. There are a number of efforts to use Web crawlers and other off-line search techniques to index Web sites of interest in a particular domain and provide them in a gateway site. Surely these technologies will all play a growing role in accessing data in the building industry and for populating data in domain- or application-specific building models.

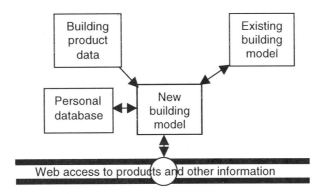

Figure 10.2: Query and selection of information for a building model to support some design task.

Search tools, developed for the Web, are available for use on one's own internal sites. For example, specifications, drawings from previous projects, images and 3D models, even video clips and other data could all be locally indexed and accessed through an Intranet web site. Building organizations are just beginning to explore these possibilities. By structuring its own materials in this way, an organization can use a single, easy-to-use browser (Netscape or Explorer) to search both corporate and external data for needed information.

The basic structure of World Wide Web data is a simple encoding language called HyperText Markup Language (HTML). HTML supports text and graphic raster formats for transmission and display, using any Web viewing program, such as Netscape Navigator or Microsoft Explorer. The HTML language and HTML viewers also support plug-ins. These are add-on applications to the base viewers that support encodings other than those native to HTML, as HTML allows other formats to be encoded within it. Some of the useful plug-ins for building product use include

Acrobat	-	book-quality formatted text and images in PDF format
Shockwave	-	dynamic, animated multimedia
!Whip	-	displays DWG files reformatted as DWF files, supports hyperlinks
Softsource	-	displays DWG and DWF files reformatted as SVF files, supports hyperlinks
Dr. DWG	-	displays and supports hyperlinks and redlining of DWG files
VRML (Virtual Reality Modeling Language)	-	3D models, with a real-time viewer
OliVR	-	real-time video with sound
Quicktime	-	real-time video with sound

A newer net format is Extensible Markup Language (XML). XML supports the development of custom formats and appears to have the capability of directly representing EXPRESS models for exchange on the Web. It will be worth watching the developments of XML as a future exchange language, especially for web-based data exchange.

This perspective suggests that building product model data, encoded in EXPRESS or other format, should be available in the same way that all the above kinds of information are available through a standard Web browser. Some CAD companies have recognized this direction and have embedded browsers and hyper-links into their CAD system environment. Some of the software companies selling EXPRESS tools include HTML- and XML-related capabilities. This promising direction is just being initiated and is likely to have a major impact in the future.

10.3.4 SINGLE VIEW MODEL OR MULTIPLE VIEWS

A quite different issue is whether or not there should be a single model of a building or multiple. In the early days of CAD, it was assumed that a variety of computer models of a product were necessary, embedded in different applications. Pairwise data exchange between this set of applications, with up to N-factorial different exchanges possible among N applications, was a motivation for developing a single "neutral model", as outlined at the beginning of Chapter Five. A neutral model was viewed as facilitating exchanges between the N different applications, requiring at most only 2*N interfaces. The early integrated CAD systems, surveyed in Chapter Two, also assumed that a single model was sufficient. We now have realized that the exchange between N applications is still difficult, even with a neutral model, and that pairwise exchanges may not be sufficient. However, the assumption of a single neutral model has been generally accepted without much critical assessment as to whether or not this is the way all parties within the building industry actually work.

In Scenario Three, the multiple construction organizations doing scheduling maintain their own models, not a single integrated one. Instead of merging all data describing a building into a single large representation, multiple representations are used, each supporting a partial description of the building. Similarly, Scenario Two, dealing with the energy consultant, involves building up a sequence of alternative energy models to be evaluated. These energy models are separate from the architect's building model. While the insights gleaned from these models will be used to advise the architect, the models themselves remain separate from those used by the architect.

Maintaining separate models is the result of many factors:
- the data is organized quite differently in the different models, making merging difficult
- the consultant may be applying procedural or analytical data that they consider proprietary, for example, the skylight values of the energy consultant in Scenario Two, or the work rates of crews used by the construction schedulers in Scenario Three
- the other people involved in the process need the results of the data, not the details

Most consultants do their own interpretation of analysis data and do not share the raw results with clients or architects. They later archive the analysis data for possible reuse and for possible issues arising later, including liability issues.

These examples illustrate a simple truth: given current building industry practices, it will not be common in the foreseeable future to have a single repository that captures all the information used in building design, in construction, or in building operation. Even within a single stage, multiple data models are likely to co-exist, given current business practices. The reasons for maintaining multiple models and repositories are not likely to disappear. This section proposes that in some areas of building, integration might better be treated as a set of interconnected aspect models, rather than a single integrated model.

The use of multiple models to represent a product is also the premise adopted by the ISO-STEP organization in their development of application protocols. Application models, as defined in the STEP architecture, are the same as what are called aspect models in Chapter Seven. If the three aspect models, CIMsteel, COMBINE and Part 225, were all used in the same project, some data carried in each aspect model would be shared and other data would be closely related. There would be a need to coordinate and possibly run evaluations on the data within pairs of models, or across all three. The coordination may be carried out in several ways:
- to use one model to initialize the other two
- to update a change made in one model so that the other two are consistent with it
- or to check that the three models are consistent with each other by applying a third model to identify and resolve conflicts in their data

In a neutral database, consistency issues are minimized by avoiding redundant data and deriving as much information as possible from a single, non-redundant source. Non-redundancy is a common criterion in both product modeling and database design. However, when multiple models are used to represent a building, consistency inevitably arises and becomes an important issue.

With multiple models, two particular sources of inconsistencies arise. One source is the computations during the mapping from one representation to another. When done manually, mapping errors are common. However, mappings can be largely automated, as we shall see. The other source of inconsistency results from incorrect propagation of changes between representations. An example is that in the mechanical design, a boiler is relocated, but its structural load and location are not updated and checked by the structural package. An automatic mechanism is needed to check that the source entities and target entities of all internal maps are consistent. If they are not, then the map must be re-executed. Of course which map to re-execute and the propagation of this update to others requires wider consideration.

At least two architectures can support multiple aspect models of a building. One of the architectures, shown in Figure 10.3, consists of a set of aspect models with mappings between them. Each aspect model supports one or a few applications.

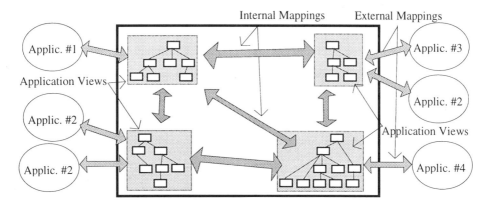

Figure 10.3: A building model schema architecture made up of multiple aspect models.

In a neutral model, all mappings are between the applications and the neutral model. When there are multiple models, there are two kinds of mappings. Those between an application and a model are called *external mappings*, while those between two models are called *internal mappings*. The SDAI, described in Chapter Five, supports external mappings. New software facilities are required to support the internal mappings. These mappings are between two or more aspect models and, as such, are completely within the building model environment. Multiple proposals for the extension of EXPRESS to support internal maps have been made and some have been implemented and are available commercially. They map the data for one Entity or a structure of Entities in a schema to another schema representing the same Entities. Language facilities for mapping between product models are reviewed in Chapter Eleven.

In Figure 10.3, a set of internal maps, one for each Entity, is depicted by darker arrows. These maps are similar to the early model of N-factorial possible internal mappings between N models. However, the task of the mappings and of the data models themselves has changed. In the new schema architecture, the aspect models are part of the data exchange environment and can be managed. The maps are between well-defined data models, each represented in EXPRESS. But in the same way that direct translation between N applications required a large number of translators, N aspect models can result in the need for a large number of internal mappings. An alternative multiple model architecture, that includes a central model to facilitate mapping between the aspect models, is shown in Figure 10.4. It reduces the number of maps required for any aspect model to two.

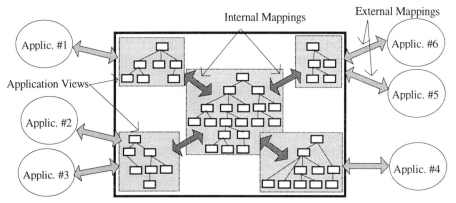

Figure 10.4: A schema architecture relying on multiple aspect models and a central building model for coordinating them.

A building model consisting of multiple aspect models, supported by a central exchange model to reduce mappings, appears to involve a complex structure. The multiple schemas mean that the complex model is large, even though in practice it is likely to be distributed over multiple machines. The same entity is represented in multiple aspect models and also in the central exchange model. In practice, the central exchange model need only carry data used by two or more aspect models. The external mappings are similar to those developed currently for models, requiring some level of tailoring to the individual application. However, with multiple models, these can be tailored to be very close to the application's own data model. The external mappings can be made so that two-way mappings between an aspect model and an application will not lose data. Specifically, these are called *isomorphic mappings*. This capability would require significant redefinition of the Aspect models reviewed in Chapter Seven.

While a building model comprised of multiple-aspect models respects aspects of current practice, it also has several other attractive features. One powerful capability of the internal maps is that they can automatically monitor consistency. Because the internal mappings are within the data model, full access and control exist over the map code and the data it reads and writes. As data is read for a map, all the variables read can be flagged. As the new data is written, these variables also can be flagged. The flags monitor the data with which they are associated. This mechanism then monitors the data associated with a map and flags the map as being invalid whenever the data is modified. In this way, the system monitors the consistency of all the maps that have been executed within it. Implementation of such a consistency monitoring technique has been demonstrated, though it hasn't been incorporated into products (yet).

The multiple view architecture also appears to hold promise in providing support for dynamic decision processes among concurrent users:

- The ability to identify inconsistencies between views allows the model environment to support collaboration strategies. As inconsistencies are identified, processes may be invoked to notify affected users or set up communication and conflict resolution procedures. The processes, for example, can notify people responsible for an inconsistency to resolve, or alternatively, to notify a supervisor that the inconsistency exists. These procedures may not be necessary when the multiple users are in the same office, but they can be very helpful if users are distributed, possibly in different time zones.

- In exploring the possible impact of a revision, it may be desirable to try changes and to review the inconsistencies created as a way to assess the impact of the change, before such a change is seriously proposed. That is, carrying multiple representations of some part of the building and propagating the impact of a change to the other representations allow assessment of the impact of the change before it is permanently executed.

The beneficial condition is that the multiple aspect models allow multiple descriptions of the design to co-exist. This redundancy means that the alternatives can be compared and possibly revised and/or merged. Another potential benefit of multiple aspect models is that they may facilitate easier integration of new applications. By packaging applications with their associated aspect model and its maps from and to a central model, many application packages may be prepackaged for easy integration into a project model. How this works in detail has not been defined, however.

Work is ongoing to develop both the system-level facilities to support mapping between aspect models and to develop good practices for using and operating within a multiple aspect product model environment. Much work remains if this is to become a production approach.

10.4 PROCESS MODELING, PROCESS PLANNING AND COORDINATION

The last issue arising out of the four scenarios at the beginning of this chapter dealt with better representation of the general flow of work. Even though an application is interfaced to a building model, it cannot be used until the instance data it accesses is generated by some previous application. (The application may be one whose sole purpose is to copy from tables or other standard sources data needed for the particular application.) As work becomes more distributed, the tacit process knowledge embedded in work will have to become more explicit. This last section reviews the different representations available for process modeling. It ends with a summary of requirements for a process representation needed to support product data modeling.

A general truth is that the process operating on some data and the product model carrying that data are interdependent. Modify the process—for example, change building fabrication from on-site to off-site factory fabrication—and the building model must change in response to different information needs.

Most of the modeling work reviewed in this book assumes that product data exchange can be realized with only high-level definition of use scenarios, without a detailed and parallel commitment to defining and modeling processes. However, is this perspective justified?

- If a new activity is undertaken, it will require data. From what other activities will the data be supplied? For example, during design, inclusion of microwave communication equipment for use within a building requires integrating new design tasks into the work flow. What information is needed to do the new activity? Where in the sequence of activities is the information available? With what other activities does it interact? These types of questions must be addressed for successful integration.

- As new computer applications are added to support various activities, each application will have assumptions built into it regarding the specific form of the data required for the application and of the context within which it assumes to operate. For example, a residential roofing program may make quite strict assumptions about the necessary format of outside walls, the definition of interior, bearing walls and the specification of any clerestory windows and skylights. Currently, the only practical way is for the same person to set up the context and to execute the application. How might such a task be distributed? The required conditions must be anticipated when maps or translations are generated to support data exchange both to and from the application.

- In general, as computer-based support for integration grows, additional process information is needed to support coordination, information distribution and change management among the members of the team.

A necessary component of these issues is how to represent for both people and computers the process being followed to realize some complex goal. The task may be driven by functional requirements, as indicated in the first bullet, requiring knowledge of general interactions and requirements of a technology. At another level, any task that becomes computer supported—that is, is supported by applications—will have more specific and stricter requirements for execution—as the second bullet indicates. Currently, all examples of large-scale integration of computer applications in building start with a fixed process model and a planned flow of information between applications.

Different processes carried out during the building lifecycle will have different requirements for process representation. Some processes, such as building maintenance, will be scheduled and planned and the process model can be defined beforehand. Construction scheduling has a well-developed set of practices and tools for process planning during the construction phase, but these change for each project. There have been some efforts to define a fixed process model for design phase activities. However, architects frequently make decisions during design that modify the future processes that are to be undertaken. Processes often change as design proceeds.

Recent work by Jeng proposes a multilevel structure for process modeling for architectural design activities. The different levels have different associated functions.

1. The top-level schedule is static, is based on interorganization agreements and is often reflected in contracts. The process plan here is a linear sequence of processes, whose completion reflects various milestones. Payments and other organizational events are associated with the completion of processes defined at this level. The AIA standard contract, discussed in Section 1.2.2, for example, outlines a five- to seven-phase process plan.

2. The second level is a task allocation level. All the top-level activities are disaggregated to this level. The specific activities typically vary from project to project but have a general structure that is often defined well in advance (but may change occasionally). This level maps the overall process to groups or individuals, who assume responsibility for the activities. The sequencing of processes may reflect some general types of dependencies, but detailed interactions are assumed to be resolved at a lower level.

3. The third level of processes is the coordination and collaboration processes. The activities are dependency driven, being initiated in response to dependencies identified at the task level. They are adaptive, changing with the structure of the design and its particular dependencies.

The activities involve communication and exchange of goals and low-level schedules as well as design data. These processes are seldom scheduled in current practice.

4. The bottom level is the atomic activities carried out by individuals. They are the low-level operations that are combined by an individual to accomplish the task laid out at the second level, augmented by the communication and collaboration tasks defined at the third level.

This four-level structure distinguishes fixed contractual coordination from scheduled task assignments that are assumed as responsibilities, separate from a dynamic coordination and communication level, with a bottom level carried out by individuals. It offers a framework for distinguishing process planning at different levels of aggregation, each level having its own functionality.

As an area of intellectual endeavor still in early development, the functionality supported by process modeling varies greatly. Some of the relevant functionality associated with process modeling are

- *Static versus dynamic definition of processes*: Some workflow models are based on repetitive workflows, requiring compilation and debugging. Others allow easy adaptation and change.
- *Explicit representation of resource allocation*: By their nature, process modelers represent processes. But when processes are allocated to machines or people, then there is a possible complementary representation of each resource that depicts its utilization over time.
- *Goal tracking*: As tasks are segmented and allocated, the goals associated with those tasks are also decomposed and allocated.
- *Change propagation*: If a process changes or a sequence of processes is reorganized, how do these changes affect other processes downstream? A change in one process implies the need to review all succeeding processes. What are the dependencies among processes?
- *Support for task execution*: Automating information used in processes can facilitate:
    - automated information distribution to enhance availability and productivity
    - review and approval processes
    - audit trail of process changes
    - resource allocation and management of bottlenecks

Different subsets of these features may be needed for representing the processes used in various building-related tasks. Process modeling is receiving renewed attention as applications attempt to embed more functionality into process models.

The functionality embedded in some computer-based process model will depend upon the representation of the processes. That is, different representations of processes have different functional capabilities. The following section reviews various process modeling representations, as they have evolved over the last century, and their associated functionality. First, we review the early efforts to structure processes, followed by some recent approaches. The recent approaches hold promise for allowing development of more powerful tools to support the integration of activities supported by a building product model.

10.4.1 EARLY FORMS OF PROCESS MODELING

The study of procedures in production and engineering began around 1900. Manufacturing procedures began to be more organized and evolved into production lines. Frederick Taylor first studied production processes and initiated a revolution in process management, planning and control. His work led to simplifying, specializing and organizing the actions and roles of production employees into assembly lines. In the 1930s, these same concepts were being applied to the operation of design offices, resulting in linear flows of work by people having specialized roles —what today we would call the workflow.

An early associate of Frederick Taylor was Henry Gantt, who first used and demonstrated the value of bonus incentive pay based on productivity. Later, in efforts to improve the productivity of munitions plants in the First World War, Gantt developed charts that mapped production over time, allowing better distribution of tasks and "load leveling" among work groups, using what is now widely referred to as a Gantt chart. An example is shown at the top of Figure 10.5. It shows four work crews. Variations of Gantt charts were explored through the 1930s and 1940s. Gantt charts showed precedence relations implicitly, with successor activities starting at the completion date of the predecessor activity. Some variations of Gantt charts explicitly connected the termination of predecessor activities and the start of successor activities.

Much later, in the 1960s, these ideas were picked up by Dupont de Nemours and Univac, who worked together to improve the planning and scheduling of construction projects using a computer. The Dupont and Univac team made explicit the dependencies between tasks. They also recognized that within a set of parallel activities some activity sequences determined the overall length of a project but that other sequences did not. It was the longest sequence among parallel subsequences that determined the overall project time. In large projects, these were not easy to identify. Computational methods to search all sequences and to identify those that determined total project time were developed and called Critical Path Method (CPM). An example is shown at the bottom of Figure 10.5. The critical path sequence of tasks is shown in dark gray. About the same time, the US Navy developed the Program Evaluation and Review Technique (PERT) for planning the Polaris submarine construction. It focused on the stochastic aspect of scheduling. CPM and PERT are the main methods for planning and managing construction processes used today. However, they have not been heavily used to plan creative work such as design or planning.

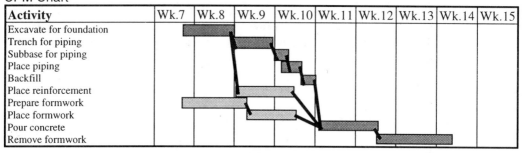

Figure 10.5: Schedule segments showing two methods of charting processes.

Given a precedence order of tasks and each task's duration, PERT and CPM scheduling programs compute expected ranges of start and finish times and identify which tasks are not on the "critical path". Resources assigned for each task, and the overall utilization rate of resources can be summarized. The process and resource allocations must be predefined; the user can change them but the computer cannot. Some scheduling programs support process decomposition.

10.4.1.1 Conceptual Models of Processes

Schedules identify activities, but do not differentiate the resources, controls or other components that comprise an activity. In the 1960s, Douglas Ross, a faculty member at MIT at the same time that Ivan Sutherland was there as a student, and who is quoted at the beginning of Chapter Two, developed the Systems Analysis and Design Technique (SADT), the first conceptual modeling method for processes. SADT was used in industry for a number of years. It was very similar to IDEFØ, which has now superceded it. IDEFØ is now the most commonly applied process diagramming method. It is used within ISO-STEP for defining the procedures to be followed in developing and adopting an ISO standard. It is widely used in the US Department of Defense and in manufacturing. It also has been used in studies of construction processes.

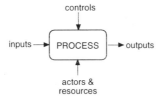

Figure 10.6: The basic structure of an activity in IDEFØ.

The basic structure of an activity in IDEFØ is shown in Figure 10.6. An activity or process is represented as a box, with inputs—material or information—coming from the left, controls from above, actors and resources from below and with outputs of the process exiting on the right.

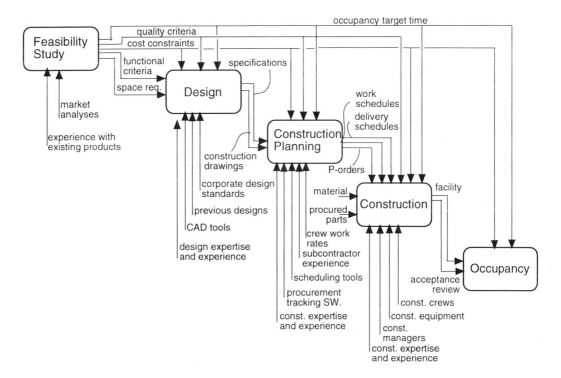

Figure 10.7: An example of a simple high-level IDEFØ model of the design and construction process.

A high-level description of the design and construction process is shown in Figure 10.7. High-level descriptions such as that shown can be decomposed, with each activity in the diagram redefined as a set of activities and process model at the next lower level of description. This decomposition can be applied any number of times, until the level of detail needed is reached.

IDEFØ is a excellent diagramming language for specifying the detail steps in large, possibly complex but fixed processes. IDEFØ does not emphasize decision-making, but rather the linear aspects of a process. It articulates coordination, control and both the information and resource flows needed by the activities and dependencies between them. IDEFØ is a descriptive language for communication between people. Though graphical computer tools facilitate producing its diagrams, IDEFØ is not executable on a computer. They embody no logic from which inferences or new knowledge can be derived. An example of an inference is the derivation of the critical path from a CPM chart. No such derivations are available from an SADT or IDEFØ chart.

10.4.1.2 State Transition Diagrams

A limitation of a schedule is that it does not show iteration, actions that are repeated a variable number of times. An early need in computer systems was to define all the different actions that a computer program must execute to solve a given problem. Each condition was defined as a path of processing operations. The result is the well-known flowchart. A simple flowchart is shown in Figure 10.8, showing an algorithm for finding the largest value in a list. A flowchart describes a single, possibly complex process and the different paths that the process may follow in different situations. The conditions are well defined.

Given n elements x[1], x[2],.....,x[j],...x[n], find m and j where m = max x[j].

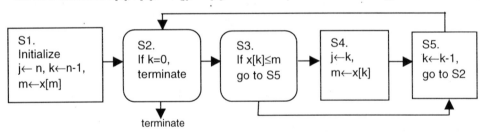

Figure 10.8: A flowchart describing an algorithm for finding the maximum value in a list.

Similar types of procedures are found in human organizations, where, for example, a design may be reviewed multiple times, or a construction step may be reviewed multiple times by a building inspector. A process representation different from a precedence chart type of schedule was needed, something like a flowchart.

Flowcharts are sometimes used in organizations, but they typically depict a single flow or "thing" going through the process. A generalization of a flow diagram, called a state diagram, was devised to represent more general conditions. A state transition diagram shows all the possible paths between different actions, does not define decision rules and allows parallel flows. An example is shown in Figure 10.9.

A state transition diagram shows a set of states and possibly a set of processes and the different sequences in which some entity may move through the states and processes. In Figure 10.9 states are boxed and processes are labeled with a C. A state is associated with each entity moving through the process and characterizes it in some way. In Figure 10.9, the states characterize a design's approval status. States are a set of mutually exclusive values that the entity may take.

They are often used to summarize complex conditions of the entity, such as test status, approvals, curing, finishing or other qualities.

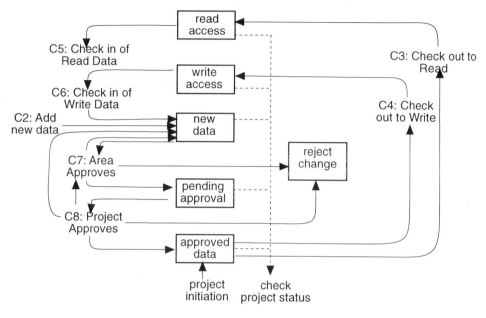

Figure 10.9: A state transition diagram of a design review and approval process.

10.4.2 RECENTLY DEVELOPED PROCESS MODELING LANGUAGES

While a major focus in construction is the coordination and allocation of resources to activities, other processes have different emphases. Sometimes we are concerned with dynamic processes that change frequently. Such changes often occur during design. We also may be concerned with the overall structure of a process, such as the conditions being matched between outputs and inputs. In other cases, we are interested in a large set of process sequences, but want to identify rules that make some sequences not meaningful or useful. Recent work in process modeling provides new capabilities not offered in traditional scheduling, IDEFØ or state representations of processes.

10.4.2.1 Petri Nets

A graph structure developed by C.A. Petri in the 1960s provides a formal representation of processes that has been usefully applied to building processes. It also supports automatic composition of processes. It also allows several other important properties that have been used in many areas of information processing.

The basic formalism consists of a directed graph with two kinds of nodes and edges. In set notation, a net N where

$$N = (S,T;F)$$

where

S is a set of states:	$S = \{s_1, s_2, \ldots s_n\}$,
T is a set of transitions:	$T = \{t_1, t_2, \ldots t_n\}$. S and T are disjoint.
Flow is:	$F \subseteq (S \times T) \cup (T \times S)$. Flows are directional and called the "flow relations" of N.

Notice that flows only connect a state to a transition or else a transition to a state.

Various logics may be applied to a Petri net. One logic useful in building information processes is to consider the net as a condition-event graph, or C/E graph. In a C/E graph, each transition has preconditions and postconditions defined by their surrounding states. Those just prior to the transition are the transition's *prestates*, while those just after are its *poststates*. More formally, these can be defined as:

$$*s = \{y \mid y \; F_n \; t\} \text{ is called the } prestate \text{ of } t,$$
$$s* = \{y \mid t \; F_n \; y\} \text{ is called the } poststate \text{ of } t.$$

(These are written in predicate logic notation. The first statement is read as "*s is equal to a variable y such that y is connected by flow of n to t and is a prestate of t".)

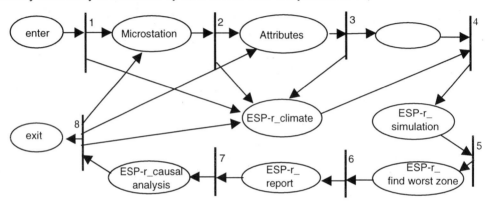

Figure 10.10: A Petri net description of processes.

Graphically, ovals or circles indicate states and a bar indicates transitions. A state is labeled by a name inside its oval; a transition is labeled by a number. An example, revised from the COMBINE project, is shown in Figure 10.10. The prestate of Transition 2 is "Micro-Station" and its poststate is "Attributes" and "ESP-r_climate". The prestates of Transition 4 are a blank state and "ESP-r_climate". Its poststate in "ESP-r_simulation". In a condition-event graph, all prestates must be fulfilled before a transition can take place. When the transition occurs, any subset of the poststates may begin. The empty node after Transition 3 is necessary to allow the optional transition from the "Attributes" state to "ESP-r_climate". The sequence consisting of Microstation and Attributes and the completion of ESP-r_climate must be realized before the transition to ESP-r_simulation can begin. After simulation, a sequence of activities must take place in order. After "ESP-r_causal analysis", iteration is facilitated by Transition 8.

This Petri net notation identifies which applications may be executed in which order. The implicit assumption is that the function of each application is already known. Thus, the execution of MicroStation is to define the geometric layout. The Attributes state has to add attributes to the geometry so as to expand the inputs required for the ESP-r simulation.

Notice that a Petri net representation differs from a state transition diagram by being more explicit about the preconditions and postconditions of some task. Figure 10.11 is a state transition diagram of the process represented in Figure 10.10. The Petri net in Figure 10.10 identifies all preconditions and postconditions that link processes; for example, the preconditions for "ESP-r_simulation" are defined—the "Microstation-Attributes" sequence and "ESP-r_climate"—while preconditions are not defined in the state transition diagram in Figure 10.11.

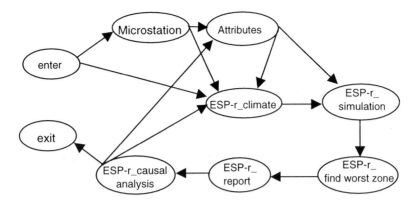

Figure 10.11: A state transition diagram of the activities shown in Figure 10.10.

Petri nets can also be used to compose processes. However, in order to do so, they need to be defined in a different way. In Figure 10.10, states are different computer applications, while the transitions are defined as the transitions between different applications. However, if we are concerned about how an application can be integrated into an existing process, we are interested in the specific conditions defining the prestates that the application requires and the conditions satisfied by the application, its poststates. That is, the application is not a state, but a transition between states. The Petri net shown in Figure 10.10 can be redefined as shown in Figure 10.12 so that applications are transitions. In this figure, transitions are defined as boxes, so that they may be labeled.

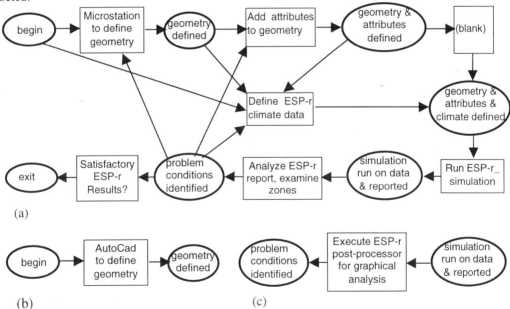

Figure 10.12: A Petri net model with applications defined as transitions between states. Two additional applications are shown at the bottom.

The blank transition is still needed, to allow an optional path to define climate conditions after geometry. It defines the same sequences of operation as shown in Figure 10.10, but makes clear the preconditions of each application. The benefit of using Petri nets in this way is that the preconditions required by an application can be represented as states. Alternative actions to achieve some state may be defined, allowing existing application sequences to be modified or new ones added. For example, two possible applications are shown at the bottom of Figure 10.12 and can be integrated with the net above it, according to their prestates and poststates. The AutoCad application (Figure 10.12(b)) has prestates and poststates that match those of the Microstation activity, allowing it to be executed as an alternative. (This assumes that the AutoCad application and MicroStation application are functionally equivalent.) Similarly, the ESP-r post-processor (Figure 10.12(c)) can be used instead of what was a manual activity to facilitate the identification of energy-related issues.

Petri nets have been used to model activities in only a few research studies in the building industry, most notably in COMBINE II. Further work is required to judge their full capabilities and limitations.

10.4.2.2 Logic-Based Activity Models

A related approach for modeling dynamic processes relies directly on the definition of preconditions and postconditions, using predicate logic. The assumption is that for any activity, one can identify the contextual information needed to undertake the activity and the conditions that the data must satisfy. For example, if a designer is to lay out electrical circuits and wiring within a space, the designer should have the following information available as input:

- general use of the space, determining the general electrical service requirements
- the loads for all electrical fixtures specified for the space, including lighting
- movable electrical equipment anticipated for the space and its location, if known
- wall, floor and ceiling construction

The layout operations would generate the following outputs:

- sizing of all wiring and connection boxes
- location of all fixtures and switches
- routing of wiring between the above

From the inputs, the actor can generate the outputs. The actor in this case may be a person working manually, a person and a design application, or an automated design system. If any of the input sources change, the electrical implications will have to be checked and possibly revised. If a change is made and its effects are not checked, it is likely that the design will be inconsistent.

The schedule of activities of a process only implicitly captures the dependencies in the process. It specifies the precedence order of activities, but does not express why a precedence relation must hold. It does not identify which values in the preceding dataset are used to make a particular decision in the succeeding one. Here we use the term dependency to identify relationships that must hold in the design. A distinction can be made between data dependencies and process dependency. *Data dependencies* refer to operation precedence where one operation generates data that is needed as input to the second operation. For example, the ESP-r simulation cannot be executed without the building's geometry being defined, the attributes assigned to the geometry, and the climatic conditions identified. All these are data dependencies. *Process dependencies* refer to precedence among operations where one operation satisfies particular logical conditions that are assumed by the second operation. The state "satisfactory ESP-r results" is a process dependency. Data and process dependencies often overlap, as when an operation generates a dataset that satisfies certain integrity constraints. Sometimes they do not overlap; for example, the requirement that no spatial conflicts occur in a piping design, as a precondition for doing a material take-off of the piping system. This constraint involves both piping and all other shape data. Process dependencies and data dependencies together impose a partial ordering on operations, restricting the sequences in which they may be meaningfully executed.

Logic-based activity models represent activities in a way allowing detailed tracking of dependencies. It also supports automatic adding to or modifying of the process flow. Most activities are expected to write some data. The written information may be new data or it may elaborate or modify existing data. In design, these may be elaboration and detailing; in contracting, these may be quantities of materials. In other cases, the activity may not write any new data, but only check existing data for certain conditions to hold. That is, it checks process dependencies. A spatial conflict check of a building model may not change the model. In such a case, the activity applies rules or conditions that here are called constraints.

In order to carry out some activity, certain information must be available. In addition, certain conditions, states or evaluations may apply to the data. These are called the Readset and Before Constraints, respectively. At the completion of the activity, some data will be added or modified. That data will possibly have some new conditions, constraints or states associated with it. We call this the activities' Writeset and After Constraint, respectively. Together, they define an activity ϕ having the following structure:

$$\phi = (\{E\}^R, \{E\}^W, \{C\}^B, \{C\}^A)$$

where

$\{E\}^R$	=	the set of entities to be read into the application
$\{E\}^W$	=	the set of entities that are written by the application
$\{C\}^B$	=	the set of constraints that must be satisfied before the application can be executed
$\{C\}^A$	=	the set of constraints that are satisfied within the application and can be relied on by later operations

The before constraints and after constraints specify a logical relationship between activities and the information and conditions the activities require. The Readsets and Writesets define data dependencies and the before constraints and after constraints identify process dependencies.

Constraints, as defined here, can have four values:

$<T> ==$ True	meaning that it has been satisfied
$<F> ==$ False	meaning that it has been evaluated and has failed
$<U> ==$ Unknown	meaning that it has not been evaluated, possibly because information is lacking to do so
$<X> ==$ Blank	meaning that changes have been made to the context, so that the state on the constraint is uncertain

Other logics have also been applied. Here, we can differentiate whether some activity has been completely successfully (its after constraints are True), has not started or is underway (its constraints are Unknown), or that it has been initiated but cannot complete because of some external condition (its after constraints are False).

The above is a class definition of an activity. The same action may be executed multiple times. Each action invocation, defined as above, is called an action instance and denoted ϕ_i, which is the ith instance of action ϕ. Each activity instance may access and modify a set of entity instances that may or may not overlap with the sets affected by other activity instances from this or other activity classes. Activity instances are denoted:

$$\phi_i \equiv (\{e\}_i^R, \{e\}_i^W, \{c\}_i^B, \{c\}_i^A)$$

where:

$\{e\}_i^R$ are instances of $\{E\}^R$

$\{e\}_i^W$ are instances of $\{E\}^W$

$\{c\}_i^B$ are instances of $\{C\}^B$

$\{c\}_i^A$ are instances of $\{C\}^A$

The properties of this activity representation in complex design and problem-solving situations have been explored. Every computer application reads some data and writes some data. These parts of the definition are obvious. The preconditions apply to both the Readset data and possibly data outside the Readset. For example, a cost estimate for a structural system will probably require that its static analysis is correct (no changes to its geometry or loading). This applies to the structural members being accessed for cost estimation, and also external entities whose loads apply to the structure. There is a significant redundancy in the $\{e\}_i^R$ and the $\{c\}_i^B$. The precondition constraints can only be evaluated if the data they access already exists, which is also tested in the Readset. Efficiency is gained if the precondition constraints are defined first, then the Readset can be defined for only those entities for which all values are acceptable. This definition of activity can apply to different levels of granularity, for example, to a whole application for a structural analysis. It can also be applied to small-scale individual entities, such as the lighting analysis for a single room, or the structural detailing of a single joint.

The power of this process representation lies in its ability to track changes. The logic of these relations is reviewed below, using predicate logic notation. (Several good references on predicate logic, used in many areas of knowledge representation, are given at the end of this chapter.) In order to evaluate constraints that may have four values, an explicit evaluation function must be introduced. We use state(c,<T>) as a function to assess constraint c to determine if its value is <T>.

Given the representation, any activity instance $\phi_i \equiv (\{e\}_i^R, \{e\}_i^W, \{c\}_i^B, \{c\}_i^A)$ is feasible when all of its before constraints have state <T> and when all of its Readset has received values:

$$\forall c_y, \forall e_x \mid (c_y \in \{c\}_i^B \wedge \text{state}(c_y, <T>)) \wedge (e_x \in \{e\}_i^R) \Leftrightarrow \text{feasible}(\phi_i)$$

Correspondingly, a successful activity instance ϕ_i is one in which a feasible activity has been successfully executed, i.e., all after constraint instances are True:

$$(\text{feasible}(\phi_i) \wedge (\forall c_y \in \{c\}_i^A : \text{state}(c_y, <T>)) \Leftrightarrow \text{successful}(\phi_i)$$

An effect of successful completion of an activity is that the Writeset $\phi_i\{E\}^W$ also has been successfully written.

An activity instance for which all before constraints are satisfied is considered feasible. However, the state of any before constraint is based on additional conditions—that the previous before constraints of earlier activities that set the current activity's before constraints have state <T>, and the before constraints of the activity that set those before constraints have state <T>, and so on recursively. This condition corresponds to the assumption that the activity instances were carried out in precedence order, beginning with activities with trivial or no before constraints, then undertaking activities whose before constraints were satisfied by the initial activities, and so on.

With this model, the integrity of the design is achieved when all activities are successful, assuming all activities are executed only when their before constraints have state <T>.

Because of the logical dependency between activities, a constraint's evaluation defines integrity only locally. Additional conditions are required to guarantee that a constraint's state is valid globally. Predecessor(ϕ_i, ϕ_j) is the relation between two activity instances such that the After constraints of ϕ_i intersect with the Before constraints of ϕ_j or the Readset of ϕ_i intersects with the Writeset of ϕ_j.

$$((\{c\}_i^A \cap \{c\}_j^B) \neq \varnothing) \wedge ((\{e\}_i^R \cap \{e\}_j^W) \neq \varnothing) \Leftrightarrow \text{predecessor}(\phi_i, \phi_j)$$

Predecessor defines a relationship between one activity instance and others that have to be successful for the one to be globally valid. The predecessor relation defines a dependency relation from any one activity back to activities that have trivial or no Before constraints. In the limit, these are the initial design activities. Additionally, we note that the predecessor relation, when its arguments are reversed, defines the successor relation between activities.

A constraint instance c_x is globally_valid only if its state is committed by an activity instance that is successful and that all its precondition (Before constraints) are also globally_valid.

$$\exists\, \phi_i, \forall c_y \mid \phi_i \equiv (\{e\}_i^R, \{e\}_i^W, \{c\}_i^B, \{c\}_i^A), \phi_j \equiv (\{e\}_j^R, \{e\}_j^W, \{c\}_j^B, \{c\}_j^A) \wedge c_x \in \{c\}_i^A \wedge$$

$$\text{predecessor}(\phi_i, \phi_j) \wedge \text{successful}(\phi_i) \wedge c_y \in \{c\}_i^B \wedge \text{globally\_valid}(c_y) \Leftrightarrow \text{globally\_valid}(c_x)$$

This proposition formally expresses the obvious fact that design actions generate valid data only if the actions are based on valid design information. The benefit of this formulation is that it can be (has been) implemented and used to track design information. Basically, globally_valid constraint instances are required for meaningful communication between activities.

The overall arrangement of design activities is a sequence of activity instances that defines entities and/or satisfies integrity conditions within its postcondition constraints that are then required as readset instances and precondition constraints of subsequent activities. The activity instances incrementally build up the design model along a frontier, behind which all constraint instances are *globally_valid* and True and in front of which the integrity states are NULL or Undefined. This frontier defines the current status of the overall design project and expands as design proceeds. Design is complete when a state of total integrity is achieved for all instantiated constraints.

The proposed method of communication suggests that any activity instance is feasible when all of its Before constraints have state <T>. However, all activities previously completed as well as all those just newly ready for execution would be flagged as feasible. This indicates that any earlier activity can be iterated as well as new activities. In order to distinguish the newly feasible activity instances, we add another predicate to eliminate activities that are in a successful state:

$$\text{feasible}(\phi_i) \wedge \neg\, \text{successful}(\phi_i) \Leftrightarrow \text{newly\_feasible}(\phi_i)$$

Newly_feasible selects activity instances whose Before constraints are <T> and whose After constraints are not <T>. These may be <F>, <U> or <X>. Any of these activities may be executed, and if successful, may record their effects according to the definition of a successful activity. Such an activity advances the border defined by the newly_feasible activities by adding their Writeset entities and setting their After constraints states to <T>. Newly_feasible activities are those that can be executed to advance the frontier.

Activities can be tracked so that if data is later changed, the actions that use that data can be automatically flagged. However, making a change not only affects those activities that directly use the changed data, but also the successor activities that use the outputs of the affected actions. There may be a logical chain of affected actions. In the abstract, as activities are undertaken, based on the results of predecessor activities, we are building up a network of dependency relations, characterized in Figure 10.13(a). If a change is made to a model that revises the input information to other actions, or changes the state of a constraint relied on by later actions, then those later actions must also be reexamined.

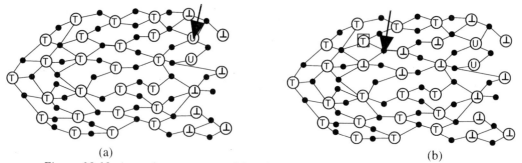

Figure 10.13: An action sequence, with actions precedent-ordered from left to right. Hollow circles are constraints, with their state inside. Black small circles are entities. Constraints have been resolved in activities with the independent entities on the left and the dependent on the right. (a) shows a design state where a constraint referenced by the arrow is infeasible, requiring iteration. (b) indicates the modified entity and the resulting propagation of invalid constraint integrity states.

Database bookkeeping operations have been devised that will track the validity of constraints and data and flag operations that have been previously executed and whose starting conditions have changed. Strategies have been developed to adaptively respond to situations where a major change is required, but full redesign would be prohibitively expensive. Strategies for patching the dependency relations have been described that allow creative correction of complex dependencies.

This representation has also been explored as a means for allowing action sequences to be dynamically composed. Here, it functions in a similar manner to Petri nets in that prestates and poststates can be matched with those that may exist in the building model. It adds conditions regarding the availability of data but distinguishes availability from the condition of the data.

The logic-based modeling of activities offers a rich representation for modeling dependencies in design situations. While in the default case, it can be used to simply track readsets and writesets of whole applications, it offers richer semantics that supports fine grain tracking at the attribute instance level. It remains to be seen if these dependencies can be reliably specified so they can be automatically managed.

10.4.3 DISCUSSION

In both Petri nets and logic-based activity modeling, the challenge is to define the contextual conditions of an action—that is, an application—in a manner that allows the conditions to be reliably matched automatically by the computer. COMBINE fixed its Combi-net approach and did not try to check state conditions to adjust the process sequence. Its process was fixed. The logic-based precondition approach has not been applied to a large test.

This review of process modeling methods is offered because of the increasingly recognized need to articulate the role of any specific data exchange according to the process it supports. Adding new applications or changing processes leads to new data exchange activity sequences, which can only be understood in the context of the larger process. More specifically, we anticipate the development of a process modeling layer that is used to coordinate data exchange activities within a product model. The process model facilitates the coordination of activities, especially those involving data exchange. The process layer is an addition to the overall data exchange architecture. It is characterized in Figure 10.14.

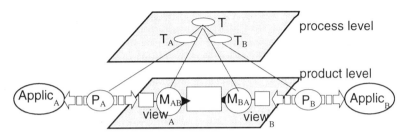

Figure 10.14: A process layer is part of a building model, used to coordinate data exchanges.

Consider a distributed design process, where different consultant teams plan their design work and collaboration over networks. In this setting, no one person may know the requirements of all programs used by the different participants. Each group needs to be able to add or subtract from their task and application sequence, adjusting their processes to be as effective for the project as possible.

Our main motivation in defining processes has been to identify a structure that can allow processes to be automatically configured and integrated, possibly within a distributed network of participants. If a user wanted to add an application to a building model with an existing suite of applications, how should the mapping of data be planned so as to allow a meaningful sequencing of the applications? Whether the building model consists of a single central model or several aspect models, a user needs information to tell when an application can be executed and where in the process it belongs. If the building model environment consists of a set of aspect models, the issue is even more important. What aspect models are to be relied upon for providing the needed data? And what successor applications will want to use the resulting newly generated data? These questions cannot be answered by current generation building models. Yet it seems that answers are necessary if effective building models are to be developed.

10.5 DISCUSSION OF EXCHANGE ARCHITECTURES

The scenarios presented in Section 10.2 point out the importance of better specifying the detailed use and user requirements for data exchange, from a human-computer interaction (HCI) perspective. While HCI is a growing area of study within the computer science field, there is a general lack of its application to the tasks within the building industry. Work is needed in this area to better understand the detailed structure and flow of information within different types of design, engineering and construction organizations.

While the EXPRESS language and associated tools have allowed significant progress to be made in the definition of data models for representing building information, this chapter suggests that the information processing environment needed to support production activities within the building industry has not been as well developed. Among the areas needing development are

new, richer forms of exchange, including incremental updates, merging the output of multiple applications and distribution from one application to several

-   exchange architectures that include multiple models which are not merged but are coordinated regarding their consistency
-   better means to track data after it has been exchanged and allowing updates if it later changes
-   coordination of data exchanges using a process model, so that critical predecessor and successor tasks are identifiable; the process model must be easily and possibly modifiable automatically

When considered closely against the requirements set forth in the scenarios, it becomes quite clear that EXPRESS was developed to explicitly support batch-type file exchanges. It was also hoped that these facilities would be sufficient for more complex forms of exchange. However, clear requirements for more complex exchanges have not yet been developed. Some of the revisions of EXPRESS proposed here are taken up in the next chapter.

Given the user scenarios at the beginning of this chapter and the eventual integration of process coordination models to support product data exchanges, we can identify the outlines of a desirable, future data exchange capability, one able to support many aspects of the presented scenarios. It allows designers or fabricators to plan data exchange processes within the scope of a project, in response to a dynamically evolving process plan. A central repository model provides support for incremental updates, change management, access and update management services. To integrate a new application in response to a change in the process plan, the participants are able to dynamically define the model subset they wish to exchange, in terms of entities, attributes and relations. The product modeling tools allow definition of the read and write mapping facilities of the agreed upon entities, attributes and relations to be defined at a high level and added to the respective application, without detailed C or C++ programming. That is, the model subset is defined dynamically. Alternatively, an existing exchange interface can be easily modified to provide additional needed data. The application interfaces should be implementable in an hour or so. With a quick test, the interface is ready to be used. Within the exchange process, users can query and browse to select the instances they need in a particular exchange. These facilities are also used to check that a new exchange interface works properly.

This scenario requires several technology developments not available today. Some are outlined in this chapter. Others are taken up in the next.

10.6 NOTES AND FURTHER READING

The general issue of concurrent access to databases, especially in the context of engineering and design, is reviewed by Barghouti and Kaiser [1991] and Elmagarmid [1992]. A discussion of methods supporting incremental and partial updates to a repository building model is presented in Eastman, Jeng and Chowdbury et al. [1997] and Jacobsen, Jeng and Eastman [1997].

A survey of classical change management techniques is presented by Katz [1990]. A more recent study directed toward design document management is Rezgui [1996]. The technology of dependency management is presented in Eastman and Shirley [1994] and more fully by Eastman, Parker and Jeng [1997].

A book on Web use in the building industry is Doherty [1997]. Two recent surveys of component software are Szyperski [1998] and Lewandowski [1998]. An introduction to XML is provided in Goldfarb and Prescod [1998] and Meltzer and Glushko [1998].

A building model architecture made up of multiple aspect models has been presented in Eastman, Jeng et al. [1997]. The collaboration uses of multiple aspect models with internal mappings are explored by Jeng and Eastman [1998] and is more fully presented in Jeng [1998]. Another effort with similar goals is Junge, Kothe et al. [1997].

The problem of maintaining the consistency of a building model that must support applications describing the same object at different levels of aggregation has been identified by several authors, including Roseman and Gero [1996] and Young [1996]. A review and proposed solution are presented in Eastman and Jeng [1999].

The history of process modeling can be traced back to its beginning in the book of Frederick Taylor [1911] that initiated the field of production management. The work of his colleague, Henry Gantt, who developed the process diagrams that bear his name and was an early proponent of statistical quality control, is traced in the retrospective notes of Goldfarb and Kaiser [1964]. Publications from Western Electric in 1932 clearly describe engineering design as a linear process Alden [1932]. The development of critical path scheduling and PERT is reported in Marks and Taylor [1966].

The potential use of Petri nets was first explored in COMBINE II [Augenbroe, 1995]. It is presented in more detail in Amor and Augenbroe [1997]. A basic source of Petri net concepts and theory is Reisig [1992]. See also Desal and Esparza [1995].

Logic-based preconditions and postconditions have been developed and explored in Eastman [1996], and Eastman, Parker and Jeng [1997]. Two useful texts on predicate logic are Genesereth and Nilsson [1987] and Kowalski [1979].

10.7 STUDY QUESTIONS

1. Select one of the information flow descriptions for the different building lifecycle phases presented in Chapter One. For some aspect of the information flow, interview one or more people or organizations that undertake that process and develop a more detailed scenario. Assess the implications of that scenario in terms of the architecture issues raised in Section 10.3.

2. Select one of the scenarios presented in Section 10.2 and define a detailed process model characterizing its information flows. Choose one of the process modeling languages, such as IDEFØ or Petri-nets. From this analysis of the scenario, develop a list of criteria that are needed to support the scenario, in terms of the communication needed to support project-related data exchanges. For example, how might dynamic changes of flow be supported?

3. The mapping capabilities, presented in Section 10.3.4 have state values similar to the logic-based activity models described in Section 10.4.2. Why do you think this is the case?

4. Review the paper by Barghouti and Kaiser [1991] in light of the requirements for data exchange outlined in this chapter. Identify the capabilities reviewed in that paper that address or resolve the issues raised in this chapter. Identify the still open issues.

5. On paper, identify three computer applications that you would like to see integrated by data exchange. Define the Petri net model of information flows between them. Which aspects of the flows are addressed with Petri nets and which are not?

6. On paper, identify three computer applications that you would like to see integrated by data exchange. Define the logic-based exchanges, including preconditions and postconditions.

Which aspects of the exchange flows are addressed by modeling the logic of the exchanges and which are not?

CHAPTER ELEVEN
Modeling Language Issues

The sciences do not try to explain, they hardly even try to interpret, they mainly make models. By a model is meant a mathematical construct which, with the addition of certain verbal interpretations, describes observed phenomena. The justification of such a mathematical construct is solely and precisely that it is expected to work.

John Von Neumann

11.1 INTRODUCTION

The EXPRESS product modeling language, reviewed in Chapter 5, is a core capability developed within STEP to define and implement product models. It provides the syntactic features used to define the semantics of STEP product models of all types—electrical, construction, manufacturing, and so forth. It also has been used by most recent efforts in building product modeling, including the IAI effort, RATAS, CIMsteel and COMBINE. Suites of commercial products have been developed to support the use of EXPRESS in defining consistent and complete models, which can be parsed for syntactic correctness and used to map instance data to or from a file or database. Today, EXPRESS is "the" language for defining the semantics of product models. It also implicitly embodies the ISO-STEP architecture, a suite of independent models defined as Application Protocols that share common properties specified as Integrated Resource Models.

Prior to the development of EXPRESS, there was little opportunity to test the software tools and the style of model development and use that were envisioned at the initiation of the ISO-10303 Committee's work. Since 1991, however, when EXPRESS was released, a large number of product models have been defined, and many people have become experienced EXPRESS users. Inevitably, weaknesses and functional limitations have been identified and articulated in STEP meetings and the research literature.

Since 1993, an EXPRESS-2 committee has been working on developing a new version of the language in response to some of these issues. At the time of this book's publication, a revision has not been distributed. At the same time, a number of informal extensions have been proposed. Some have been submitted to the ISO-10303 Committee.

The main weaknesses and limitations of EXPRESS that have been stated include
- lack of support for model evolution
- lack of support for mapping between schemas or views
- limited support for parametric modeling
- lack of a formal specification of semantics and equivalency with application models
- lack of syntactic features that would allow more complete articulation of product semantics

This chapter surveys these different issues. The motivation for them is reviewed—why they are issues—and proposals put forward to address them are articulated. These issues are important because they identify the boundary between the current functionality of EXPRESS and identified needs. They are all open research issues.

11.2 MODEL EVOLUTION

The product modeling strategy of ISO-STEP is to define a neutral model, to and from which various external applications can translate data. Because writing translators between an application and a STEP model is time consuming, it has been assumed that it would be undesirable—indeed dangerous—to modify a model after translators for it had been developed. The changes might corrupt or destroy the translators that had been implemented.

At the same time, as new applications or application domains are developed, they may not be able to operate on data that is already loaded in the existing product model, and extensions may be necessary for the data to be utilized. If the necessary extensions could be added without destroying existing interfaces, a Part model would likely have a longer life. Extensibility also presents a smooth way to transition from an existing model to the next release.

In a database environment, the need for extensibility becomes even more obvious. A building model serving as a repository may have to evolve as decisions are made regarding building technologies or construction methods. New models may need to be added or existing ones extended. A statically defined data model cannot serve well a process that adapts to project conditions.

Building product manufacturers have another motivation to extend an existing product model. They may wish to utilize a standard product model to represent their products, for example, for presentation on the World Wide Web. They may want to add other structures and attributes to the standard model, however, that describe the company's product and its features more exactly. An extended STEP Part model would allow general Part model readers to access the standard model, but additional semantics could be added, describing special features in the product not representable in the official Part model. The company would then need to provide special readers for the extended part of the product definition. Such extensions are not unlike those allowed in almost all computing languages, such as SQL; database companies provide proprietary extensions. In the longer run, it would be the merging of these extensions that would allow the official model to be enhanced in future releases of the Part model.

11.2.1 MODEL AND INSTANCE MODIFICATION

Research on object-oriented databases has identified rigorous ways to make object schemas modifiable. First, the invariant conditions that must not be violated by any schema change are identified. These identify whether data can be deleted by a schema change, which initial values assigned for schema additions are not OPTIONAL, and so forth. Then, various rules are defined addressing how the semantic transitions from one schema state to another should be handled. Next, an exhaustive list of all possible transitions is generated. This list is then selectively reduced by eliminating those that would violate the invariant conditions. Operators are then defined for the remaining transitions to support the remaining transition rules.

This method can be applied to EXPRESS. Several invariant conditions are identified in the EXPRESS language manual and are assumed to be checked by all EXPRESS parsers. The invariant conditions include

> *Existence of Referenced Types, Entities and Functions:* the model must include all components referenced by other components.
> *DERIVE Rule Scope:* all variable components in a DERIVE rule must be accessible within the scope of the rule.
> *WHERE Rule Scope:* all variable components in a WHERE rule must be accessible within the scope of the rule.

Name Uniqueness: all Type, Entity and Function names must be unique within the
model.

Acyclic Subtype Supertype Lattice: there cannot exist a subtype/supertype relation where
any Entity is its own supertype, directly or indirectly.

Other invariant conditions may also apply.

EXPRESS structures and their possible modifications:
- create a new Type*°
- delete a Type
- change the name of a Type
- change the domain of Type, i.e., NUMBER to REAL, REAL to STRING, etc.
- modify a literal
- modify the size of a STRING (increase*°, decrease)
- change an aggregation type SET to BAG, SET to LIST, etc.
- modify the bounds of an aggregation (increase the range*°, reduce the range)
- modify the set of values of an ENUMERATION type (add to the set*°, reduce the set members)
- modify the set of types allowed in a SELECT type (add to the set*°, reduce the set members)
- modify the range of an existing domain by adding*, deleting or modifying a WHERE clause
- create an Entity*°
- delete an Entity
- change an Entity's ABSTRACTness (becomes ABSTRACT, ceases being ABSTRACT*)
- add a SUPERTYPE to an Entity*
- remove a SUPERTYPE from an Entity
- change supertype constraints, i.e., ONEOF, ANDOR, AND
- modify an Entity by:
    - deleting an attribute
    - adding an attribute*°
    - changing an attribute name
    - changing an attribute's Type
    - changing an attribute's INVERSE relationship (becomes INVERSE*, removes an INVERSE)
    - changing the function of a DERIVE attribute
    - changing the OPTIONAL constraint on an attribute*
    - changing the UNIQUE constraint on an attribute or a set of attributes*
- add REFERENCE FROM or USE FROM Schema references*°
- delete REFERENCE FROM or USE FROM Schema references

Figure 11.1: All possible modifications to an existing EXPRESS Part Model.

An initial identification of all possible transitions of the language constructs in EXPRESS is
shown in Figure 11.1. The possible transitions allowed will vary with the allowed context in which
changes are executed. All the transitions defined in Figure 11.1 might be allowed if complete
generality of changes was the objective. However, only a small subset of the transitions will be
allowed if model evolution is restricted to changes that do not modify existing model semantics. A
degree symbol (°) identifies those transitions that are thought not to modify existing model
semantics. Slightly less restrictive is the requirement to maintain access paths. Asterisks (*)
identify those transitions in Figure 11.1 that are thought not to change access for any translator. In
general, those transitions that add new Entities or attributes or add to the range of an aggregation
do not change access paths or an existing instance's semantics. New inherited Rules or changed
constraints modify the semantics of existing instances but do not change access paths. Again, the
transitions to be allowed depend on whether or not existing instances in a model are to have their
semantics modified, whether or not such instances exist, and whether or not existing access paths,
embedded in translators, are to be affected.

Computer languages can be distinguished in terms of declarative and procedural components. Declarative components of a language define the existence and properties of data. Procedural components define actions to be taken on data in some sequence, organized within some control structure. Most programming languages incorporate both kinds of components. EXPRESS relies almost exclusively on declarations. Only rules as incorporated in DERIVE and WHERE clauses utilize the procedural components of the EXPRESS language.

While most languages are declarative regarding data structures, EXPRESS requires procedural manipulation of data structures in order to support model modification. Thus a revised EXPRESS language supporting model evolution will have statements of the form:

```
MODIFY TYPE connector = ADD SELECTOR (bolt, clamp);
CREATE ENTITY door
   ADD ATTRIBUTES
   . . . .
```

If a Part Model is modified using such operations and is populated with Entity instances, then those instances are assumed to accept the corresponding changes that are applied to the Types and Entities used in their definition. That is, if an Entity type has attributes or relations added to it, then all instances will receive additional fields supporting those extensions. If attributes or relations are deleted or modified, these changes will be applied to all instances. If an Entity type is deleted, all its instances are deleted. Such changes are supported in SQL databases and also in a number of object-oriented databases. Of course, some of these transitions would be difficult to execute if the physical implementation was a flat file.

11.2.2 INSTANCE MIGRATION

After a Part model has been extended, a corollary capability is to support the migration of instances from existing Entity types to new types that carry the extensions. For example, we may create a new subtype of an existing Entity that carries extra attributes for some instances. We then are likely to want to migrate some instances from the existing Entity to the new one. Of course, if we wish to transfer a model in the other direction, then instances will be stripped of their extensions and migrated to a more restricted Entity type.

This capability has been promoted by several researchers, who have identified the general need within object-oriented databases to partition a single type of Entity into multiple types, or to merge two types of Entities into a single type. These capabilities might be needed in cases where a generic element, such as a geometric shape, is partitioned into shapes made with different types of construction. A general class of building walls, for example, may be partitioned into stud walls and concrete block walls. These capabilities require Entity instance migration, allowing movement of an Entity from one type to a subtype Entity or supertype Entity.

Several methods of migration have been proposed. One approach migrates instances to specializations or generalizations up and down the type lattice. Another approach creates multiple versions of a database, with each version supporting a different set of applications. In the second approach, an unresolved issue is managing the consistency of object definitions between the different database versions.

After an Entity instance is retrieved, for example, using the SDAI operator Find_Entity_Instance_Model, the operators would allow the instances to be migrated. Alternatively, various tests could be applied to the model first to determine if it should be migrated. The instance migration operations might be

```
SPECIALIZE(«ENTITYinst»)OF ENTITY «DEname1» TO «DEname2»;
GENERALIZE (<<ENTITYinst>>) OF ENTITY «DEname1» TO «DEname2»;
```

An important issue is that type conversion of an Entity instance should not change the internal identifier of the instance, which would result in breaking previous relations of that instance with others.

11.2.3 DISCUSSION OF MODEL EVOLUTION

These two aspects of evolution—model modification and instance migration—together provide means to evolve one product model into another. New Entities may be defined, instances from old Entities migrated to the new Entity types and, if necessary, the old Entity types deleted. If we consider that an instance can be both generalized and specialized (in some order), it can be moved from any class instance in a connected inheritance lattice to any other position in the lattice.

Instance migration into Entities with new DERIVE rules allows new attributes to be derived from existing ones. This suggests that such migration might be used more generally as a model transformation mechanism, for example, converting a model at the end of construction to one supporting construction management. The potential capabilities, limitations and ease of implementing such transformations of an instance model appear not to have been explored.

These potentially powerful tools, however, come at a cost of maintaining a model so that current translators are not broken. While these tools can be defined so they are completely general and complete, they would require controlled use to define the appropriate model changes needed to achieve the extensions envisioned.

11.2.4 DEFINING THE SCOPE OF DELETIONS

An important recognition regarding model changes is that product models involve complex structures that make deletions especially problematic. These complexities are characterized by the questions:

- When one Entity instance is deleted, what other instances also are to be deleted?
- When an Entity type is deleted, what other types should also be deleted?

In production use, without additional semantic content being embedded in the modeling language, deletions from a product model are practically impossible. The deletions might not be undertaken explicitly; rather, they may be managed automatically, using "garbage collection" as implemented in Java. However, identifying candidates for deletion still requires that the modeling language has strong conventions regarding data *existence dependencies*. Of course, type, Entity and instance deletions only become relevant when significantly modifying an existing product model.

Existence dependencies have been recognized in much recent work in object-oriented systems. Such dependencies are an explicit component of the definition of proposed standard object models, such as the Unified Modeling Language (UML). The issue addressed by existence dependencies concerns relations among Entities: Which relations refer to Entities that belong uniquely to the deleted Entity instance and should be deleted and which are independent of the deleted Entity and are possibly shared? There are three possible dependency conditions:

- independent
- dependent
- conditional

Conditional relations are shared and may be accessed from multiple Entities; the conditional Entity is deleted when the last reference to it is removed.

Some examples of relations are listed below, along with the existence dependencies:

- reference to a material library for the energy transmission properties of some construction method (independent)
- reference from a part to its geometry (dependent)
- a deleted space that refers to a set of surfaces that partially define the walls, floor and ceiling that bound it (conditional)

Dependency conditions in relations are commonly distinguished by defining different types of relations. For example, UML (which uses the term "association" to refer to a relation) distinguishes between an association and an aggregation, where the aggregation is dependent. Similarly, EDM-2, a research building product database language, distinguished between complex attributes that referred to other object instances (dependent) and compositions that deal with assemblies (independent). A challenge in defining relation types, as seen in the EDM-2 example, is that several different semantic dimensions may be defined together in the relation types. The Composition relation defines a part-of relationship, carries embedded rules regarding the well-formedness of the part-of structure and also defines dependency. In other research, dependency is defined as a constraint. That is, any relation may be constrained to be a dependent one. However, an associated constraint that should be enforced is that any object instance that is the target of a dependent relation cannot be the target of relations between it and other object instances.

With well-defined dependency relations and garbage collection, a large product model can be defined and populated with instances. Later, additional Entity instances may be added or existing ones deleted without worry regarding extra data floating within the model that has been left from incomplete deletions.

11.3 MODEL TRANSLATION

In Chapter Ten, the need was defined for a data exchange environment that supported multiple models, with mapping between them. In addition to supporting data exchange between multiple models, other uses have been proposed for mapping:
- mapping between multiple models
- deriving a new model that is a view of an existing one
- defining more complex derivations, possibly across multiple Entities, within a single model

In this section, three efforts to develop mapping languages are reviewed. Two have been proposed as extensions to EXPRESS—EXPRESS-M and EXPRESS-X—and the third is a separate effort developed in the EDM-2 environment.

11.3.1 EXPRESS-M

The need to go beyond a single model and explore the integration of multiple models has been recognized by the ISO-STEP organization. A committee, WG5 N243, was formed in 1995 and developed a working draft for a mapping language called EXPRESS-M. EXPRESS-M defines constructs for specifying mappings between two STEP schemas. It was defined as an extension to EXPRESS.

EXPRESS-M's functionality is the definition of mappings between two schemas, or two sets of schemas, which have already been defined in EXPRESS. It relies heavily on the types and statements of EXPRESS. Like EXPRESS, EXPRESS-M does not define how to read/write instance Entities to/from applications. These are left to an implementation strategy. EXPRESS-M references the existing schemas, using REFERENCE and USE statements to access definitions of the source and target Entity types. Mapping functions are its main contribution. They can be viewed as possibly complex assignment statements, where the right-hand expression derives data from *source* schema(s) and the left-hand side value is assigned to the *target* schema. Correspondingly, the name space of the expression side is the source schema, while the value name space is the target schema. An important requirement for the design of EXPRESS-M is that maps must be written so that neither the source nor target schemas are modified. Source and target schemas are standard Part models. In addition, all Entity instances must be defined previously; EXPRESS-M cannot create Entities. Also, only data can be mapped, not rules.

Sometimes it is not possible to map one instance of a source Entity to a single instance of a target Entity. For example, a solid model may be mapped to a set of surfaces in a surface model. Thus a mapping may be one-to-one, one-to-many or many-to-many. Most generally, a mapping is the conversion of an instance from one type or set of types to another type or set of types. In programming languages, a change to a value's type is often called *coercion*. In the above example, the task can be treated as coercing a solid model into a surface model.

Within the general case of defining maps between Entities in different schemas, many complex conditions can arise. The source and target Entities may vary in a variety of ways. They may be
- of different types
- of different levels of aggregation
- such that the source type maps into multiple target types, depending upon the values in the source

EXPRESS-M addresses only some of these issues.

While coercion is available in most programming languages, different languages coerce one type into another in different ways. Thus coercion must be defined explicitly. In EXPRESS-M, the coercion is defined as the infix operator <-, with the conversion made from right to left. For example,

        «INTEGER_variable» <- «REAL_variable»

is a round-off operation, rather than truncation. In EXPRESS-M, the automatic coercions that are supported among primitive types are

				SOURCE				
TARGET	binary	boolean	integer	logical	number	real	string	enum_id
binary	✓	✓	✓	✓	#	#	✗	✗
boolean	✗	✓	✗	✓	✗	✗	✗	✗
integer	✓	✓	✓	✓	#	#	✗	✗
logical	✗	✓	✗	✓	✗	✗	✗	✗
number	✓	✓	✓	✓	✓	#	✗	✗
real	✓	✓	✓	✓	✓	✓	✗	✗
string	✓	✓	✓	✓	✓	✓	✓	✓
enum_id	✗	✗	✗	✗	✗	✗	✗	✓

key:
✓   - casting with no loss of accuracy
✗   - no cast possible
- casting with possible loss of accuracy

Non-primitive types may be defined within the two schemas that are unique. A MAP Entity is used to define conversions between dissimilar non-primitive types explicitly.

```
        SOURCE :                        TARGET :

                    (* EXAMPLE 11.1 *)

    type gallon = REAL;          TYPE liters = REAL;
    END_TYPE;                       END_TYPE;

    ENTITY point                 ENTITY cartesian_point;
       x,y,z : REAL;                vector : ARRAY[0:2] OF REAL;
    END_ENTITY;                    END_ENTITY;
```

The mapping can be specified as follows:

```
    MAP_TYPE liters = gallons / 3.79;
    END_MAP_TYPE;

    MAP cartesian_point <- point;
       vector[0]  := x;
       vector[1]  := y;
       vector[2]  := z;
    END_MAP;
```

Both Types—used to define domains—and Entities—used to define object instances—can be mapped. Type Maps are defined with the equal sign (=), while Entity Maps are defined with the left arrow (<-). If multiple statements are needed for assignments within the Map, these are defined within the Map block.

Notice that in the above point mapping example, the first assignment is not

```
    cartesian_point.vector[0]  := point.x;
```

The Entity references in the left-hand and right-hand expressions, as long as they are the source and target Entities, may be implicit. Later, such a predefined map may be called using the casting notation, which is the same as C:

```
        SOURCE :                        TARGET :

                   (* EXAMPLE 11.2 *)

    ENTITY line;                 ENTITY line_vector;
       start : point;               begin : cartesian_point;
       end   : point;               terminate: cartesian_point;
    END_ENTITY;                    END_ENTITY;

    MAP line_vector <- line;
       begin        := {cartesian_point}start;
       terminate    := {cartesian_point}end;
    END_MAP;
```

In the above, the point referred to by start is cast into a cartesian_point type prior to its assignment to begin.

Other types of issues arise. Multiple Entities from the source may be composed to make a single Entity in the target. Alternatively, a single source Entity may be used to define multiple target Entities. In the following example, the source consists of two types used to define arcs and circles and the target has a single type for both.

        SOURCE: **TARGET:**

                        (* EXAMPLE 11.3 *)

```
ENTITY coord_pt                     ENTITY arc2;
   px  :  REAL;                         center[0,1]  :  REAL;
   py  :  REAL;                         radius       :  REAL;
END_ENTITY;                             begin_angle  :  REAL;
                                        end_angle    :  REAL;
                                     END_ENTITY;
ENTITY arc1
   center  :  coord_pt;
   start   :  coord_pt;
   end     :  coord_pt;
END_ENTITY;
```

A possible mapping might be

```
MAP arc2 <- arc1;
   center[0]    :=    center.px;
   center[1]    :=    center.py;
   radius            := sqrt((start.py - center.py)**2)+
                        ((start.px - center.px)**2));
   begin_angle := IF (start.py - center.py) ≥ 0.0)THEN
                        arctan((start.py - center.py)
                            /( start.px - center.px))
                  ELSE  (arctan((start.py - center.py)
                            /( start.px - center.px)) + 180);
   end_angle    := IF (end.py - center.py) ≥ 0.0)THEN
                        arctan((end.py - center.py)
                            /( end.px - center.px))
                  ELSE  (arctan((end.py - center.py)
                            /( end.px - center.px)) + 180);
END_MAP;
```

In this case, the computations to define the begin_angle and end_angle are not complicated, but require the expression capabilities of the EXPRESS language.

In a source schema made up of a single, big Entity, means are provided to define a target structure made up of several Entities, with pointer references between them (each called a LINK) in EXPRESS-M.

        SOURCE: **TARGET:**

                        (* EXAMPLE 11.4 *)

```
ENTITY couple;                      ENTITY man;
    man_name : STRING;                  name : STRING;
    woman_name : STRING;                wife : woman;
END_ENTITY;                         END_ENTITY;

                                    ENTITY woman;
                                        name : STRING;
                                        husband : man;
                                    END_ENTITY;
```

The mapping could be

```
MAP man, woman <- couple;
    man.name := man_name;
    woman.name := woman_name;
    wife := LINK(woman);
    husband := LINK(man);
END_MAP;
```

In this case, the target side of the Map has multiple Entity types and some attribute names must be defined using their full pathnames, so as to be unambiguous. Using the LINK assignment, the instance of the Entity woman created by the map is assigned to the wife attribute of the man Entity instance and the instance of the Entity man is assigned to the husband attribute of the woman. Since husband and wife are unambiguous, their pathnames are not required.

Rules can be defined to specify how different instances of an Entity are to be mapped. These are also definable using the programming language capabilities for expressions and control structures in EXPRESS. For other types of complex mappings, the inheritance operators in EXPRESS or AND, ONEOF, or ANDOR are employed.

```
                    (* EXAMPLE 11.5 *)

MAP ONEOF(a_surface, b_surface) <- x_surface;
    IF x_surface.param_type = planar THEN
        MAP a_surface <- x_surface;
            .....
            .....
        END_MAP;
    ELSE
        MAP b_surface <- x_surface;
            .....
            .....
        END_MAP;
    END_IF;
END_MAP;
```

In this example, an a_surface will be created only if its parameter_type attribute is planar, else a b_surface will be created.

The structure of relations within a map may be complex. EXPRESS-M provides a PRUNE operator that helps manage such structures. In Example 11.2, if the lines being mapped are part of a polygon or graph, the points at the ends of lines would be duplicated multiple times in the target model, even though only one shared instance of each would be necessary. The access to the points through the lines gives no clue which might be shared and which might not. The PRUNE operator has a name and a list of target Entities or attributes to which the PRUNE operation is applied. It examines the Entities that have been created by the Map from the same source Entity and eliminates duplicates. An example that revises Example 11.2 to eliminate duplicate points follows:

```
SOURCE :                          TARGET :

              (* EXAMPLE 11.6 *)


ENTITY line;                  ENTITY line_vector;
   start : point;                begin : cartesian_point;
   end   : point;                terminate: cartesian_point;
END_ENTITY;                   END_ENTITY;

MAP line_vector <- line;
PRUNE line_points
   begin, terminate;
   END_PRUNE;
   begin       := {cartesian_point}start;
   terminate   := {cartesian_point}end;
END_MAP;
```

As each begin and terminate point is cast and readied for assignment, it is first matched with all existing points. If the values of the identified attributes match, a new point is not created.

Some additional features were developed to facilitate the definition of Entity instances that are subtypes of multiple supertypes.

11.3.2 DISCUSSION OF EXPRESS-M

The ready availability of the EXPRESS-M extensions would significantly add to the functional capabilities of EXPRESS models. The development of an effective language is an incredibly subtle art, and many constructs defined for EXPRESS-M are quite elegant. Use of the casting operation, the definition of all maps at the type level and the relatively clean definition of Maps are attractive features.

On the downside, EXPRESS-M also has several apparent limitations. Possibly the most serious limitation is that it is assumed that all instances for the target model have already been created, though without assigned values. That is, the mapping function cannot create instances but can only assign values to instances already created. This means that, for example, for all polygons and solids, the numbers of edges in the polygon and the numbers of faces and edges in each face of the solid, must already be defined and created outside of and prior to execution of the mapping function.

Another limitation is that the mapping capabilities are defined only for one-way mappings. No means is provided to keep track of which Entity instance on one side of the map corresponds to an Entity instance on the other side. We call this correspondence *maintaining instance identities*. The draft version reviewed states that instance identities will be dealt with in a later version. Also, EXPRESS-M lacks the ability to create new complex structures. If some of the points in the source polygon in Example 11.6 were coincident and some were not, there would be no way to determine which points should be duplicated and which ones should not. This distinction cannot be made by matching the coordinate values of vertices.

11.3.3 VIEW DERIVATION

The View capability is an important aspect of all modern database management systems. A View is a reformatted subset of a populated database, with possible new Entities, attributes and Relations that can be derived from the source data model. Thus a View is a new data model populated with instances derived from the source data model. In databases, Views allow tailoring the appearance of data for different users or applications. They can be used to restrict data access, allowing different users or applications to see only certain parts of a database.

A view can be considered as a query permanently stored within the database and is often implemented in this way. It can include derivations of new data, as long as they can be computed from the stored data. Because Views may be added to a database schema over time and can incorporate new attributes that are derived from existing ones, Views provide a limited means of model evolution.

Recently, the distinction has been made in databases between a traditional view, stored as a query and called a *virtual view*, and a view that after derivation, stores the view data for a period of time as a subschema. This subschema and its redundant data are called a *materialized view*. The advantage of a materialized view is that further queries may be applied to it and that application interfaces can be written to it.

The main conceptual limitation of Views is that they are generally read-only. Since they carry only a subset of the full data model, they do not have access to all the semantic rules embedded in the data. In the context of EXPRESS, they will generally not be able to apply all the WHERE rules associated with a Part model. Thus all updates should be made to the base model, because only the base model's assignment and update operators are sure to know all the relations that apply to an Entity or attribute. Views are an effective tool for extracting data out of a model for use in an application, but usually not for extending or developing the model data.

11.3.4 EXPRESS-X

EXPRESS is a model definition language, not a database management system language. However, the need for Views in STEP has become apparent, as experience has been gained in developing interfaces with applications. The general goal of a product model is to be complete and unambiguous. However, most applications have no need for the complete data model and only require a subset of the Entities defined. These subsets are called a conformance class or level. Different applications are likely to require different conformance classes, so one subset is unlikely to be acceptable for the range of applications desiring to share a Part model. On the other hand, a well-defined means to extract data from one model and reformat it for another model, so that it is more easily accessed and read for a particular application, would go some way to facilitate the development of the external application interfaces. For these and other reasons, the STEP organization also formed a working committee to develop mapping capabilities to support the generation of Views.

The notion of a product view is similar to that of a database view; it is a subschema that allows a user or application access to the data needed, eliminating visibility of and access to data not needed. Like EXPRESS-M, the capabilities supporting these notions were defined as extensions to EXPRESS. The initial set of extensions developed for this purpose was called EXPRESS-V and a draft of the EXPRESS-V extensions has been developed. However, because of the similar functional requirements of both EXPRESS-V and EXPRESS-M, the STEP organization merged the two committees, or at least the functional capabilities to be supported by their work. The intent was to define a single, consistent set of mapping extensions to EXPRESS. The new combined language is called EXPRESS-X, and it draws features from both EXPRESS-M and EXPRESS-V.

Because the output of EXPRESS is a conceptual model that may be implemented in various ways, it is possible to use such a specification for virtual or materialized views. However, the emphasis has been on materialized views. Views in EXPRESS-X may be defined in either of two ways:

- The view may be defined as a separate schema, essentially in the same way as needed for view mapping, as defined for EXPRESS-M.
- Alternatively, the view may be defined for one Entity at a time and the rules for populating it with values are defined at the same time.

The functional capabilities of EXPRESS-X are to support complex mappings of these two types. The mapping operations are not part of either starting or ending schemas. Rather they belong to their own Mapping Schema. EXPRESS-X calls the source the "Base schema" and the target the "View schema". Many of the capabilities provided by EXPRESS-X to support mapping were adopted from EXPRESS-M and work similarly.

11.3.4.1 View Schemas

A map can create and populate a View schema using the VIEW statement. An example of a mapping that creates a new View Entity is

```
(* EXAMPLE 11.7 *)
VIEW ViewEntity
FROM (EntityA, EntityB)
WHEN (EntityA.attr1 > EntityB.attr2) AND (EntityA.attr2 > 0));
VIEW_ASSIGN
   v_attr1 := EntityA.attr1;
   v_attr2 := EntityB.attr2;
END_VIEW;
```

The above example creates a View schema called ViewEntity, consisting of two attributes: v_attr1 and v_attr2. Instances of ViewEntity are also defined. All combinations of EntityA and EntityB instances are considered, and when the two WHEN clauses are both True, an instance of ViewEntity is created. The FROM clause tells us that the View is derived from two base Entities, BaseA and BaseB. The WHEN clause defines two predicates that identify which instances of the two base Entities are to be translated. The VIEW_ASSIGN clause assigns values to the instances.

Beside the VIEW statement, several other additions were made to EXPRESS-M. These include a GLOBAL declaration, which defines the base and view schemas referenced in the mappings and optionally assigns them unique names that are distinct from those in the base or view.

```
GLOBAL
(* schema instances *)
DECLARE bdb INSTANCE OF base_schema;
DECLARE vdb INSTANCE OF view_schema;
END_GLOBAL;
```

Bdb and vdb can be used as references to the two schemas within the scope of the mapping function.

A small example of a complete mapping schema shows the style of programming involved:

```
(* EXAMPLE 11.8 *)
SCHEMA_MAP Mapping_Schema;

GLOBAL
    DECLARE bdb INSTANCE OF Base_Schema;
    DECLARE vdb INSTANCE OF View_Schema;
END_GLOBAL;

VIEW vdb::ViewEntity ;
FROM (bdb::BaseA, bdb::BaseB, bdb::BaseC)
WHEN ((bdb::BaseA.int_a1 = bdb::BaseB.complex_b1.int_a1) AND
    (NOT bdb::BaseC.real_c1 > 10.0 ));
VIEW_ASSIGN
    int_v1 := 100;
    real_v2 := bdb::BaseC.real_c1;
    str_v3 := 'This is a view object';
END_VIEW;


END_SCHEMA;
```

The programming style derives its approach from the SQL language. The first statement creates the MAP schema Mapping_Schema. It optionally has a global declaration section. In the example, two local variables, bdb and vdb, are created which refer to local instances of the two already existing schemas. The VIEW header declares that ViewEntity is the name of the View Entities being created and associates the local class variable vdb with ViewEntity. In general, double colons (::) associate a local variable with Entities in an existing schema. The FROM clause implicitly defines an iteration over all instances of the base Entities, generating every combination. In the example, the FROM statement associates class variable bdb with the Cartesian product of all BaseA, BaseB and BaseC instances. The WHEN clause tests each combination. In the example, bdb references a new class, which is the BaseA, BaseB, BaseC combination, followed by pathname access to Entity and attributes to be accessed in making the logical test. For those combinations satisfying the WHEN clause, the VIEW_ASSIGN clause identifies how the values for the new instance are to be derived. In the example, a series of assignments are used. All variables are accessed through an explicit schema reference. If data from the view schema is needed, vdb pathnames can be used to access it.

The VIEW statement, with its FROM and WHEN clauses, can be used to define quite elaborate mappings. This type of process, using the Cartesian product (reviewed in Section 4.2.1), is familiar to readers with background in Relational databases; it is the typical form of composition using the Relational Join operator. It is less familiar and less natural to those working with object-oriented databases, which construct entities, not by examining their combinations, but by traversing the structure of references between objects.

11.3.4.2 Maps Between Two Existing Schemas
In the above examples, the View is defined as a separate schema and populated in the Mapping function. The more general use of EXPRESS-X is to map between two existing schemas. The MAP construct is used for this purpose. A small example of a complete Map follows:

```
(* EXAMPLE 11.9 *)
MAP TestMap FOR v_inst:vdb::ViewEntity ;
FROM (bdb::BaseA, bdb::BaseB, bdb::BaseC)
WHEN ((bdb::BaseA.int_a1 = bdb::BaseB.complex_b1.int_a1) AND
   (NOT bdb::BaseC.real_c1 > 10.0 ));
BEGIN_MAP
   LOCAL
         valueten : INTEGER := 10;
   END_Local;
   v_inst.int_v1 := 100/valueten;
   v_inst.real_v2 := bdb::BaseC.real_c1;
   v_inst.str_v3 := 'This is a view object';
END_MAP;
```

The MAP declaration tells how one or more types are to be mapped to one or more instances of the View model. A Map optionally has a name, allowing it to be explicitly called (in the above example, the MAP is called TestMap). The View Entity class vdb is optionally named, using (::) and an instance variable v_inst is also optionally declared, using (:). A View Entity is named where there is a single type that is the Map's target. When a WHEN clause evaluates a particular combination of Entities composed in the FROM clause to be True, a new View Entity instance is created. Here the View is assigned to v_inst. A MAP schema may define one or many Maps or Views; here, only one is declared. The Map optionally includes local variables; here, a local constant valueten is initialized. The rest is similar to the View Map, described earlier, with the target instance variable being used as part of a pathname.

There are many forms of elaboration of the above basic Map structure. An optional SUBTYPE keyword in the FROM clause creates all subtypes of the specified types. The FROM clause can reference a mix of both Base and View Entities, and the Cartesian product is generated over this mix. The creation sequence generated by the FROM clause may also be ordered.

Maps are often not one to one. In the example below, a many-to-many mapping is involved. Different subtypes of person in the Base schema are mapped to a schema with a different combination of attributes and relations.

Base EXPRESS schema: **EXPRESS View Schema:**

```
              (* EXAMPLE 11.10 *)
ENTITY person                              ENTITY male_alien;
 SUPERTYPE OF(ONEOF(male,female)           END_ENTITY;
   AND ONEOF(citizen,alien));
END_ENTITY;                                ENTITY female_alien;
                                           END_ENTITY;
ENTITY male
  SUBTYPE OF (person);                     ENTITY male_citizen;
END_ENTITY;                                END_ENTITY;

ENTITY female                              ENTITY female_citizen;
  SUBTYPE OF (person);                     END_ENTITY;
END_ENTITY;
```

```
ENTITY citizen
  SUBTYPE OF (person);
END_ENTITY;

ENTITY alien
   SUBTYPE OF (person);
END_ENTITY;
```

Possible mappings are

```
MAP male_alien <- male AND alien;
END_MAP;

MAP female_citizen <- female AND citizen;
END_MAP;

MAP male_citizen <- male AND citizen;
END_MAP;

MAP female_alien <- female AND alien;
END_MAP;
```

In this example, the attributes are all defined in the Entity `person`. Every instance of `person` is examined. All instances that are both `male AND alien` are created in the View schema. (This matching relies on the instance types, not the Entity names.) The same construction is applied to all instances that are both `male AND citizen`, `female AND alien` and `female AND citizen`. This mapping function shows the power of the `AND` supertype operation within EXPRESS. (This same mapping could have been defined in EXPRESS-M.)

When View Entities are organized in a structure that cannot be defined in a single pass, EXPRESS-X provides an explicit `COMPOSE` declaration for iterating through already created Entity instances. The `COMPOSE` statement identifies the Entity types, then the attributes to be included and finally, the `WHEN` condition required to qualify their inclusion. In the `VIEW_ASSIGN` clause, there may be added multiple `COMPOSE` clauses allowing multiple iterations over the set of instances created, for use in defining relations over the set defined. An example follows:

```
(* EXAMPLE 11.11 *)
COMPOSE example1 FOR inst_2:vdb::ViewEntity ;
WHEN (vdb::ViewEntity.real_v1 > vdb::ViewEntity.int_v1);
VIEW_COMPOSE
        inst_2:int_v1   := inst_2:real_v2;
END_COMPOSE;
```

In this statement, the `COMPOSE` statement is given the name `example1`. Like the Map statement, `COMPOSE` may include a `FROM` clause, designating the schemas to be operated on. The `COMPOSE` statement iterates over all instances of `ViewEntity`, which is referenced by the class variable `vdb` and the instance variable `inst_2`. For those instances encountered that satisfy the `WHEN` condition, the `BEGIN_COMPOSE` statements define how values for the attributes of the View Entity are to be treated.

When defining complex structures in a View schema, a group of Entity instances may sometimes be defined, for example, as the components of an assembly. A MEMBER declaration statement is provided in EXPRESS-X that creates a group construct and associates it with the View Entity type that contains the attributes. It can also be used for pruning out unwanted duplicates from the newly created assembly data.

The beginning MEMBER statement identifies a name to be used for the group of Entities represented by the MEMBER declaration and the view Entity type that contains the attributes. The MEMBER statement has two clauses. The INCLUDE clause identifies which attributes and types are to be included in the group. It also gives a unique label to each attribute. Because each attribute may refer to other Entities used as attributes, each included attribute is considered the root of a tree of attributes, and the whole tree is assumed to be included when an attribute is assigned. To deal with the trimming of this tree, an EXCLUDE clause is provided, with the same syntax as the INCLUDE clause. The VIEW_ASSIGN statement creates new instances that are derived from base instances according to selection rules for inclusion and exclusion. An example follows:

```
(* EXAMPLE 11.12 *)
MEMBER part_membership FOR wall;
INCLUDE
   framing : wood_stock  := vert_member;
   blocking : wood_stock := horiz_member;
   plates : wood_stock := bottom_member;
EXCLUDE
   WHEN ((SELF.vert_member.material IS wood) OR
     (SELF.horiz_member.material IS wood) OR
     (SELF.bottom_member.material IS wood));
   BEGIN
   attr_off1 : INTEGER := nail_count;
   attr_off2 : bonding := glue_type;
   END;
END_MEMBER;
```

In the example, the three attributes of type wood_stock—framing, blocking and plates—group the three parts of a wall carried in vert_member, horiz_member and bottom_member. For those elements whose material = wood, the attributes nail_count and glue_type and any details below them are excluded. This is a general mechanism for pruning detail or eliminating duplications in a mapping. A WHEN clause can also be embedded in the INCLUDE clause.

New Entities may be explicitly declared within the GLOBAL section, using standard EXPRESS syntax:

```
(* EXAMPLE 11.13 *)
ENTITY female;
   name : STRING;
   age  : INTEGER;
   husband : male;
   children : LIST [0:?] OF person;
END_ENTITY;
```

Instances also may be manually created, also within the GLOBAL section. These are assigned instance references using the instance naming scheme in EXPRESS

```
#bachelor = female('Mary Jones', 22, $, $ );
#w1 = female('Betty Smith', 34, #h2, (#c1,#c3));
```

The `#´ signifies a qualified identifier which is assigned a new Entity instance with the attributes shown. #h2, #c1 and #c3 are assumed to have been already defined elsewhere, corresponding to a male and two children.  The `$´ is a new EXPRESS-X symbol, denoting the NULL value. Addition of instances within a View schema is very useful. The values for an instance need not be constants; they could be values looked up on a table and read in. In this way, EXPRESS-X could be used to add new material properties to a View used by an application, or to add environmental or context conditions required by some form of analysis.

EXPRESS-X is defined as an executable language, with temporary variables, dynamic memory allocation, and other features necessary for generating executable code. It relies on the EXPRESS procedural language features for most of its detail coding. It provides means to instantiate Subtype Entities, as needed to populate Part models defined in EXPRESS. It addresses many of the complexities encountered in implementing complex maps between dissimilar data types. It is powerful, allowing multiple traversals of the base or view schemas and applying complex rules to members of the Cartesian product.

The EXPRESS-X specification identifies three conformance levels:
-   *Class 1:* processes only Map declarations. This level allows application of Map declarations to a base schema, creating instances of a View schema.
-   *Class 2:* processes Map, View, Compose and Member declarations. It supports multiple passes over the Base or View schemas both before and after Member statements have been used for grouping of Entities.
-   *Class 3:* is a superset of the current EXPRESS-X schema, for use in other Part development efforts.
Several Class 1 implementations are commercially available.

11.3.5 DISCUSSION OF EXPRESS-X
EXPRESS-X provides a production-level language that can be used to develop mappings between product models. It allows the creation of View or other schema instances. It is significantly more ambitious in its coverage than EXPRESS-M.

EXPRESS-X has eliminated the most significant shortcoming of EXPRESS-M, its inability to create new Entities and instances in the View Model. As a result, it provides an alternative way to create an instantiated EXPRESS model, independent of a more traditional Part model specification. That is, EXPRESS-X can create a new View schema and populate it with instances, with a regular EXPRESS specification. The implications of this new model definition and instantiation process have only begun to be explored. Currently, mappings are in one direction only. It has been stated that a future release will support two-way mappings.

EXPRESS-X does not support maintenance of the identity between Entities within the two schemas. Because EXPRESS-X relies on global searches in the COMPOSE clause to build relations, it seems very difficult to track the identity of Entities on both sides of the mapping. As a result, EXPRESS-X does not support updates to an already existing model. Rather, it is restricted to generating a view model from an existing base model.

The syntax for EXPRESS-X is not intuitive and writing maps in it is arduous. Its language constructs and semantics are still undergoing review and refinement. EXPRESS-X allows the product modeling community to gain experience and develop an understanding of the role and uses of mapping in a product modeling environment.

11.3.6 FUNCTIONAL REQUIREMENTS OF A MAPPING LANGUAGE

The two mapping efforts reviewed above began with EXPRESS models as their source and usually their target product models. Their goal was to develop mapping facilities within the context of EXPRESS and their main focus dealt with the issue of mapping different EXPRESS constructs, such as SELECT types, ENUMERATION types and subtype constraints. Other efforts developing mapping languages have adopted a different starting point, using models defined in languages other than EXPRESS.

The fundamental idea in maps is the "coercing" of one type into another. A mapping language allows the definition of complex coercions. Mappings may go in multiple directions, from A to B and B to A. The mappings may be approximate, as well as exact. As experience with EXPRESS-X and other mapping languages has been gained, issues regarding development of a general-purpose mapping language have become better understood. They include

(i)   Iteration of mapping operations in both directions requires instance identification of source and target, so that the proper instance can be identified for updating as well as creating and initializing. Source and target identification allows the update of just the relevant Entity and integration of the update with other data carried by the receiving Entity instance(s).

(ii)   Maps must deal with conditions where the map is not between one source and one target instance. Two different kinds of variations exist:
- One source class may be mapped to multiple target classes. This occurs when a single Entity class represents multiple Entity classes in the receiving schema. An example is a schema having a single regular polygon structure, defined by a center-point, radius and number of sides, being mapped to a schema with rectangle, triangle and polygon Entities. Each mapping creates one of several possible Entity classes. The reverse also may occur, where multiple source classes may be mapped to a single target Entity class. These two scenarios involve one-to-many and many-to-one mappings, respectively. When there is no clean way to partition out the source or target mappings, they may have to be many-to-many.
- A source instance incorporating an aggregation structure maps to a receiving Entity class with a different aggregation structure. In this case, the source Entity part-of structure must be converted to a quite different part-of structure. The example above applies to this case also: a regular polygon inscribed in a circle has a different aggregation structure from one defined as an explicit sequence of connected points. In this case, the mapping is from a single Entity to an aggregation of Entities.

In all of these mapping conditions, the identity relation between source and target must be defined and maintained.

(iii)   Maps may involve a complex network of relations that cannot be represented as a tree, requiring multiple traversals of the Entity structure. This may include merging several redundant Entities into one or one into many. For example, multiple coincident polygon vertices in one representation may be defined as a single vertex in another representation.

(iv)   Mappings must deal with updates and also additions and deletions of instances, as reviewed in Section 10.3.1. That is, one of the models connected by maps may delete or add Entities, in addition to modifying them. Deletion and creation bookkeeping techniques, such as those described earlier, must be part of the map capabilities.

No one mapping language has yet realized all these necessary conditions. EXPRESS-X has addressed the one-to-many, many-to-one and many-to-many mappings and also mappings between different types of aggregations. However, it does not maintain the identity of mapped objects, so as to support two-way mappings. With one-way updates, it does not have to address deletions from the view schema and it treats every object in the view as newly created. References to other efforts to develop mapping capabilities for product models are given in Section 10.7.

These issues suggest that there is still additional work to be done in mapping languages, if they are to become practical aspects of a data exchange environment.

11.4 PARAMETRIC MODELING

A limitation of ISO-STEP models is that they only support models defined with literal values for data. The values defining an Entity are assigned through two mechanisms: (1) direct assignment, or (2) a derivation clause. On the other hand, there are many cases where a product part is defined in a more complex manner. Parametrics is a general methodology in product modeling that allows variable parameters to be delimited by constraints which produce a well-defined, self-consistent model when values are assigned to the parameters.

There are several forms of parametric model. A *generative parametric model* records the sequence of constructions used to define a model. A new construction can be based on the context defined by previous constructions. A generative model records "how" a model is to be updated. A *variational parametric model* describes through constraints and other rules what the final part model is to be. After values are assigned to some variables, the variational approach resolves the various rules and constraints to define the other values. A variational model tells "what" the model is to be. Many modern CAD systems incorporate some mixture of generative and variational parametric modeling.

The current main application of parametric modeling is to geometry. CAD systems such as Pro-Engineer®, I-deas®, Mechanical Desktop® and Modeler® rely heavily on parametric modeling. The desire to exchange model-based information between these systems, not just a fixed geometric shape that is one instance of a model, was the impetus to form the ISO-STEP Working Group WG12 on parametric modeling. It recently produced a draft ISO Report. The scope of the report is 2D and 3D shape models, constraints between shapes, covering parametric shapes and positioning. Out of scope were conceptual design issues, form features, process-related form issues and shape associations with non-shape issues, such as behavior. Only a preliminary specification for the Parametric Framework has been proposed so far.

11.4.1 EXPRESSIONS

A basic issue that had to be addressed was how to deal in EXPRESS with the anticipated algebraic and logic expressions that are required for parametrics. Currently, the EXPRESS implementation facilities only exchange EXPRESS Entity instances with values specified as constants. The Parametrics Committee chose to temporarily use the approach of the Parts Library (PLIB) Part 20 effort. PLIB and the current parametrics effort use strings to carry expressions. In order to do this, some special types of strings must be defined, as well as the syntax within strings to represent expressions. An example of such a string is `'\e #22.radius+11.5'`. The `'\e'` flags the string as containing EXPRESS. The `'#22'` designates an instance that has an attribute `radius` to which is added `11.5`.

The syntax for these strings is

```
TYPE express_in_string = STRING;
END_TYPE;
TYPE express_in_statement = express_statement;
```

Because types cannot be inherited in EXPRESS, various other types had to be defined, such as:

```
TYPE assignment_in_string = express_statement;
TYPE logical_expression = express_expression; (*TRUE, FALSE,
                                            UNKNOWN *)
TYPE boolean_expression = logical_expression; (*TRUE,FALSE*)
TYPE entity_instance_expression = express_expression;
TYPE value_assignable_expression = express_expression;
TYPE subroutine_call_statement = express_statement;
```

Each type of expression identifies a subtype of an express_expression. The assignment_in_string is an expression that includes an assignment. Logical_expression is an expression that evaluates to TRUE, FALSE, UNKNOWN. A boolean_expression is one that is always evaluated to TRUE or FALSE. Entity_instance_expression is an EXPRESS expression that always evaluates to the full definition of an Entity instance, for example, '\e cartesian_point ("Left",[1.2,x,2*y])'. Value_assignable_expression is an EXPRESS expression whose result can serve as the left-hand side of an assignable statement. Express_expressions may apply to an Entity type and thus to all its instances. Alternatively, an expression may be defined so as to apply to a specific set of instances. Because expressions apply to some part of a geometric entity, we need a way to refer to those Entity parts. The word "element" is used here, and in the Parametric Framework documentation. Given the primitives above and other type primitives, an explicit constraint schema can be defined.

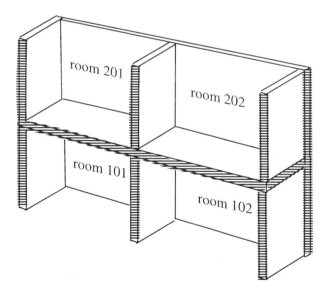

Figure 11.2: The expressions and constraints reference existing elements of a model, without changing the model.

An assumption of this early stage report was that all parametric Entities would reference non-parametric data, not the other way around; parametrics will not involve changing existing geometric entities, such as those in Part 042. This is shown diagrammatically in Figure 11.2. A parametric Entity's scope is the Entities within some `representation` instance with its associated `representation_context` and `representation_item` Entities.

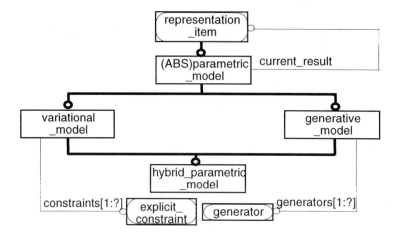

Figure 11.3: Parametric model is specialized into two subtypes.

11.4.2 FORMS OF PARAMETRIC MODEL

At a higher level, the parametric model EXPRESS schema begins with a `parametric_model`, defined as a subtype of the `representation_item` defined in Part 043. The EXPRESS-G diagram is shown in Figure 11.3. This places it parallel with `geometric_representation_item` and `topological_representation_item` in the type lattice. These are the schemas upon which the parametric model extensions build. The `parametric_model` has two major components: (1) a more detailed parametric model defined below it and (2) `current_result`, the current evaluated form of the model. Thus either the unevaluated parametric model may be accessed to read or modify parameters, or the current evaluated form of the model may be accessed to read out explicit geometry or shape properties.

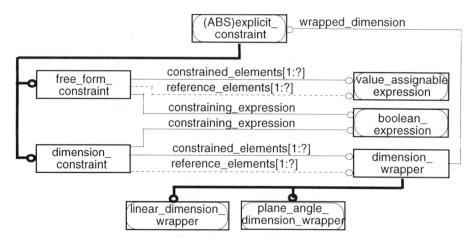

Figure 11.4: The subschema for explicit constraints.

11.4.3 VARIATIONAL PARAMETIC MODELS

The variational parametric model extensions are considered first, and shown in Figure 11.4. They are defined as a set of `explicit_constraints`. `Explicit_constraints` come in four major forms: `free_form_constraints`, `dimension_constraints`, individual constraints and structural constraints The first two types are shown in Figure 11.4. The `free_form_constraints` are defined as a set of `value_assignable_expressions` that identify the elements being constrained, and optionally a set of `reference_elements`, the independent elements that do the constraining. The `constraining_expression` references the constrained and reference elements in a Boolean expression. If there are multiple `constrained_elements`, they will be given with no implied order. An example of a free form `constraining_expression` might be

> `'\e (#46.radius) = (x + #47.radius)'`.

This is not an assignment, but a Boolean expression.

The second type of explicit constraint is the `dimension_constraint`. It also defines a set of `constrained_elements` and optionally a set of `reference_elements` as the dependent and independent elements. The constraining expression, like the `free_form_constraint`, is a `boolean_expression` that references the `constrained_elements` and the constraining_elements. In this case, however, the elements over which the `boolean_expression` applies are dimensions. The dimensions—as a presentation of a length or angle—include format information. This is abstracted away in the `dimension_wrapper`. The `wrapped_dimension` references the fully defined dimension and makes available the relevant dimension value and its control points. The `dimension_wrappers` may abstract away a `length_measure` or a `plane_angle_measure`. (The details of these entities are not presented here.)

The third major set of constraints used in variational parametric modeling consists of explicit geometric constraints. These apply to the `geometric_representation_item` in Part 042. They generally apply to simple geometric types, such as points, curves, surfaces and their specializations (instead of variables). Constraints can reference elements that are part of a geometric shape or to auxiliary or constructional geometry, such as a grid, used in defining a shape but not being part of it.

In order to reference different types of geometric Entities that play the same role in an explicit constraint, several SELECT types are defined. These are shown in Figure 11.5. They define Entities that can be used as axial elements, the general class of geometric elements, and so forth. These are then used to define a broad set of different subtypes of `explicit_geometric_constraints`, presented in Figure 11.6.

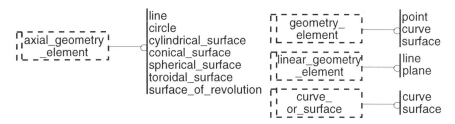

Figure 11.5: Some Select types are defined to facilitate the definition of geometric constraints.

Some of the geometric constraints will be reviewed in detail, to show the flavor of how they are meant to work. We will leave others to be examined by the reader.

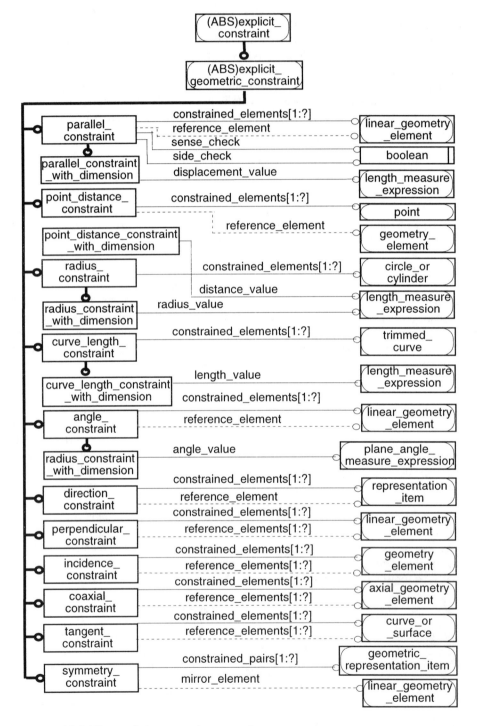

Figure 11.6: The explicit geometric constraints.

The `parallel_constraint` specifies that two or more lines or planes are parallel. It references the set of constrained elements and optionally a reference element that is the independent element to which the other elements are to be parallel. Two Boolean flags indicate whether the direction sense of the elements must also be parallel and whether the side sense of the elements are the same. The `parallel_constraint_with_dimension` subtype optionally adds to the `parallel_constraint` a length that specifies the distance between the parallel entities.

`Point_distance_constraint` constrains the distance between two points or optionally between a `reference_element` and a set of points. `Point_distance_constraint_with_dimension` specifies the distance required between points.

`Radius_constraint` requires that the radii of all members of a set of `circle_or_cylinder` entities have the same value. The specialization having a dimension defines the required length defining the circle's radius. The `curve_length_constraint` asserts that all members of a set of trimmed curves are equal. `Curve_length_with_dimension` also specifies what the length along the curve should be. There are similar constraints with subtypes for curve lengths and angles.

`Direction_constraint` delimits the direction of a `direction` or `vector`. The set of `constrained_elements` must have the same direction. An optional `representation_item` may be defined as a `reference_element`, in which case this element determines the direction of the `constrained_element` set. This is provided to define vertical, horizontal and other preset directions in 2D and 3D space.

The `perpendicular_constraint` can accept `constrained_elements` and `reference_elements` of type `linear_geometric_elements` (lines or planes, see Figure 11.5). A WHERE clause delimits the arity of the sum of both sets to 2, if the elements are two-dimensional, or 3, if they are three-dimensional. Another WHERE clause limits all the `constrained_elements` and `reference_elements` to be of the same type, i.e., all lines or all planes.

The `incidence_constraint` is powerful and general. Like the others, it references a set of `constrained_elements` and `reference_elements` of type `geometric_element` (see Figure 11.5). This can be used to constrain a set of points to lie on a plane or on a curve. Conversely, it can constrain a line to pass through a set of points, or a surface to pass through a set of points. Intersecting lines are not incident according to this constraint; all of the `reference_elements` must lie within the `constrained_element`.

Given a set of constraints applied to a geometric layout, the variational approach provides constraint solvers to find satisfactory layouts with regard to the constraints. Suppose that the initial layout is generated that satisfies the constraints and then an element is moved. The constraint solver operates to define a new arrangement that satisfies the constraints. For example, suppose a part file fragment has:

```
#2  = CARTESIAN_POINT ('P2',15.0,30.5);
#3  = CARTESIAN_POINT ('P3',43.5,13.7);
#4  = CARTESIAN_POINT ('P4',15.5,-2.6);
#5  = CARTESIAN_POINT ('P5',44.6,0.0);
#31 = EDGE ('E1', #2, #3);
#32 = EDGE ('E2', #3, #4);
#33 = EDGE ('E3', #4, #5);
#34 = EDGE ('E4', #5, #2);
```

```
#33 = PARALLEL_CONSTRAINT ('E1 PARALLELS E3', (#31, #33), (),
.F., .T.);
#34 = PARALLEL_CONSTRAINT ('E2 PARALLELS E4', (#32, #34), (),
.F., .T.);
```

Instance #33 is an example encoding of a parallel constraint. As defined here, any of the four points may receive new values to satisfy the constraints. In addition, the relations between edge and point may be changed; that is, the reference identifiers may be modified.

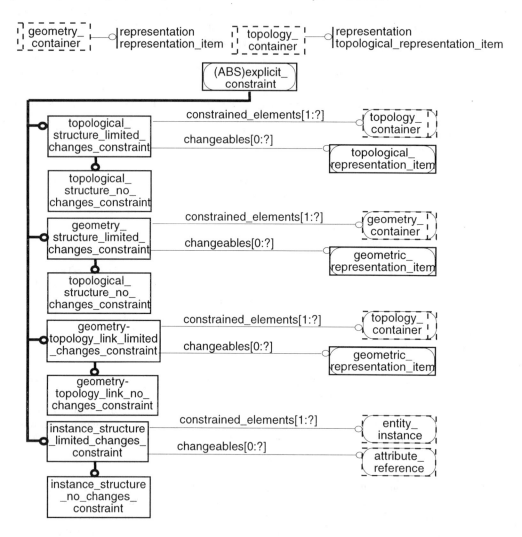

Figure 11.7: Explicit structural constraints for variational parametric models.

In order for the variational models to be effective, structural constraints are needed to define which elements are fixed and which can be changed. The structural constraints are shown in Figure 11.7. They are interpreted by the constraint solver and used as a guide in determining what entities may not be changed (in different circumstances). Three sets of structural constraints are defined: *topological*, *geometric* and *geometric_topology_link*. Each comes in two forms: limited_changes and no_changes. Limited_changes allow for exceptions, while no_changes do not. These would allow adding further constraints, such as

```
#34 = geometry_topology_link_constraint ('line to points
                    fixed', (#31,#32,#33,#34), (#2,#3,#4,#5))
#35 = instance_structure_no_changes_constraint ('fixed point',
                                                 (#2), () );
```

These fix the topology and the location of the first point.

In a variational model, all the constraints are applied together. These four sets of constraints are thought to capture the major aspects of variational parametric modeling and allow this type of model to be stored in a neutral format for use in other applications.

11.4.4 GENERATIVE PARAMETRIC MODELS

In contrast to the variational model, a generative parametric model addresses the procedural operations used to define a shape. Each generative parametric model will define an ordered set of operations that specifies a construction function for each defined Entity. The procedural operations will rely on a suite of low-level geometric construction operations and a means to define and use local variables. They also may utilize any of the high-level shape definition operators, to define a curve, surface or primitive solid.

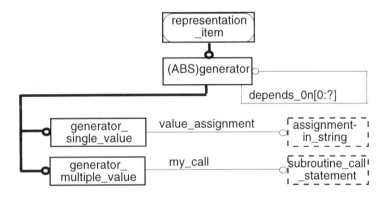

Figure 11.8: The parametric framework provides two types of generators, a single-valued one and a multi-valued one.

A generator executes an express_statement (encoded in a string) and calculates a value for one or more items, such as the value of an attribute or an entity instance. The generator instances are organized in a dependency graph.

The construction operations used in a generator rely on variables that are assigned values to create instances of the parametric model. The structure for carrying variables is shown in Figure 11.9. Variable_context identifies the representation in which it applies. If variable_context is limited access, then it can only be accessed from the identified representation_contexts. Variable_semantics carries the essential variable information. This includes a label, description and type_name. It also has a default_value and a flag defining whether it is used as input or output.

The generative model provides a set of maps for generating representation_items, that is, instances of shapes. These are defined as subtypes of representation_map, referenced from Part 043 (see Figure 6.2). The parametric_representation_map evaluates a set of assignments_in_string, using the mapped_item's subtype entity, which provides the input arguments to the evaluation. (Mapped_item is also referenced from Part 043.)

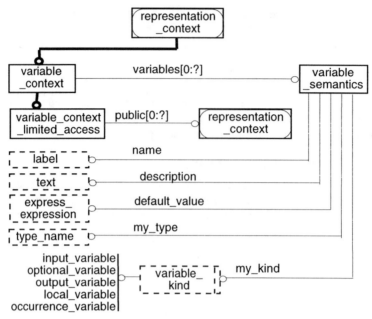

Figure 11.9: The structures for carrying variables for generators.

The `representation_map_3d_from_2d` maps a 2D Entity into a 3D coordinate space. The position of the mapped item is determined by the `mapped_item's axis2_placement_3d` (see Figure 6.7). The `typed_mapped_item` is dynamically typed as an indirect reference mapped item. The `indirect_reference` was created to allow dynamic typing.

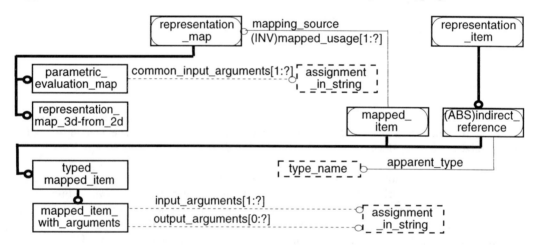

Figure 11.10: The main mapping Entities used in the generative parametric model.

With this basic structure, generative subroutines may be defined with embedded assignments. These rely on low-level construction functions that include those listed at the bottom of Figure 11.11.

The generative parametric model includes several means to project shapes from 3D to 2D, a basic operation in many construction operations. The low-level construction operations are those typically found in any geometric construction library, for line and circle construction, for computing distances, areas and volumes, and for projecting shapes from 3D to 3D and evaluating a CSG solid to derive its BRep evaluated shape. They are assumed to be core functions in a generative parametric model.

The parametric framework also includes a history-based model for allowing storing and editing of history logs, another mechanism that captures some aspects of a parametric model.

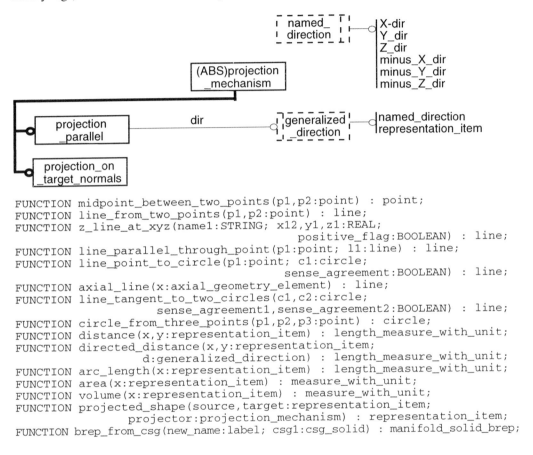

```
FUNCTION midpoint_between_two_points(p1,p2:point) : point;
FUNCTION line_from_two_points(p1,p2:point) : line;
FUNCTION z_line_at_xyz(name1:STRING; x12,y1,z1:REAL;
                                     positive_flag:BOOLEAN) : line;
FUNCTION line_parallel_through_point(p1:point; l1:line) : line;
FUNCTION line_point_to_circle(p1:point; c1:circle;
                              sense_agreement:BOOLEAN) : line;
FUNCTION axial_line(x:axial_geometry_element) : line;
FUNCTION line_tangent_to_two_circles(c1,c2:circle;
                sense_agreement1,sense_agreement2:BOOLEAN) : line;
FUNCTION circle_from_three_points(p1,p2,p3:point) : circle;
FUNCTION distance(x,y:representation_item) : length_measure_with_unit;
FUNCTION directed_distance(x,y:representation_item;
                d:generalized_direction) : length_measure_with_unit;
FUNCTION arc_length(x:representation_item) : length_measure_with_unit;
FUNCTION area(x:representation_item) : measure_with_unit;
FUNCTION volume(x:representation_item) : measure_with_unit;
FUNCTION projected_shape(source,target:representation_item;
                projector:projection_mechanism) : representation_item;
FUNCTION brep_from_csg(new_name:label; csg1:csg_solid) : manifold_solid_brep;
```

Figure 11.11: Projection mechanisms and generation functions.

11.4.5 DISCUSSION OF THE PARAMETRIC FRAMEWORK

The types, Entity and functions presented here are from a draft release of the Parametric Framework model. While aspects of the framework are certain to change, the style and approach will remain.

The major concept here is the "patching" of embedded expressions onto the current EXPRESS architecture facilities as specially denoted text strings. This certainly is an inelegant "hack" until steps are taken to develop a truly executable form of EXPRESS.

The development of an executable form of EXPRESS raises significant issues:
- EXPRESS is meant to be a neutral representation that supports multiple physical implementations, such as files and databases. How is this possible if it includes language-specific code? More specifically, how would it map to C, C++ and Java?

- In practice, parametric models that will be defined using these constructs may be implemented in different languages. How could an executable implementation of EXPRESS facilitate translation between different parametric models, possibly written in different languages?

These issues go to the heart of current research in data exchange, as distinct from data definition and modeling. They emphasize that the issue is one of code translation, not just model definition and construction.

At a lower level, the two forms of parametric model outlined in the parametric framework are different from those defined in the attribute-driven geometry developed in the IFC building model. The IFC model has a strong generative flavor; it does not support local variables, but instead relies on attributes, especially for input variables. This appears to be a very common-sense approach for the input variables. While a straight parametric modeler will rely on input variables defined specifically for the model, as defined above, any application that uses these shapes will associate them with user-defined parameters, allowing them to be easily changed. On the other hand, temporary variables and even output variables are likely to be needed by the IFC. These features could be adopted from the Parametric Framework.

Comparing the Parametric Framework and IFC parametric models more abstractly makes clearer the philosophy of the IFC. It is more than a data exchange model. By defining specific constructs for defining geometry, the IFC has specified a core application development library rather than an exchange mechanism.

11.5 SEMANTICS OF DATA MODELS

A fourth area of proposals for the revision of EXPRESS involves the semantics that can be communicated by the language. Originally, EXPRESS was developed for a particular role: as the language for implementing an application interpreted model (AIM). Separate modeling languages were used for application reference models (ARMs)—IDEF1x and NIAM. However, two changes came about:

(1) EXPRESS-G was developed and became a popular ARM modeling tool. It had automatic and exact translation to EXPRESS.
(2) EXPRESS itself began to be used as an AIM, to define the data structures needed for multiple applications to be integrated. This allowed models to be implemented and tested before becoming final. It facilitated the interpretation process of integrating the new model with the Integrated Resources and with other models.

Since that time, a large literature on data modeling has been published. Other data models have been developed with richer semantic foundations than EXPRESS. There has been growing interest in process models, as well as data models, as reviewed in Chapter Ten. Process models are not dealt with well in EXPRESS. In particular, there have been efforts to standardize object-oriented data and design modeling using UML (Unified Modeling Language).

As discussed in Chapter Four, there are many different "flavors" of data modeling languages. Some languages were developed to capture the requirements of a situation and the domain knowledge of experts. These are meant to be read by humans. Other languages have been conceived as graphical programming languages that use back-end processors to generate code. These are machine-oriented languages with a clean textual format and precise machine execution semantics.

EXPRESS is a machine-oriented data modeling language and its primary role is to generate object libraries *via* code generators, or to generate data dictionaries that can be used at run-time to look up the structure of some object. UML, on the other hand, is more of a requirements-oriented language. There are associated implementation recommendations and some environments now automatically generate C++ object definitions from UML.

One line of discussion is *should UML should be considered as an ARM language?* This issue is being discussed within the ISO-10303 committees and also within the IAI. UML is a richer language, in terms of constructs, than EXPRESS-G. UML deals with dependencies and automatic creation and deletion, as discussed in Chapter Ten. It supports more kinds of relations (called associations in UML) than does EXPRESS-G. It supports constraints across attributes as well as across inherited subtypes. These data modeling features could lead to richer ARMs than can be defined in EXPRESS-G or the other currently supported graphical data modeling languages. Standard implementation of some of its semantic constructs is still being worked out and a similar effort would have to be made if it became an official ARM. This would involve definition of the mapping between ARM constructs defined in UML and an AIM in EXPRESS. By implication, each of the AIM constructs defined would eventually have a C++ or Java implementation also.

Implicit in this discussion is the fundamental issue of where the requirements for a building product model come from. Does a building model get defined based on the knowledge of what users in the domain express as being needed, even though the data is not yet used in applications? Or does the building model get defined by a consortium of application vendors that incorporates the data their applications rely on, including some technical data that is not of high importance to end users? We called these the idealistic and the pragmatic approaches, respectively. We have reviewed examples of both. The IAI has developed an idealized model, independent of applications. CIMsteel and COMBINE, on the other hand, both developed models whose direct users were existing applications. Of course, this issue does not demand a binary choice but offers a spectrum between two extremes.

11.5.1 DATA MODEL CONSTRUCTS

The developers of EXPRESS went to some effort to address how issues of arity, inheritance and well-formedness could be expressed within it. These were among the major semantic issues of the 1980s. As experience with various responses to these issues has been developed and tried out, practices for dealing with them have become recognized. EXPRESS is a good example of a careful response to these issues.

In the 1990s, other semantic issues were identified as being of significance and are being responded to. These are newer semantic issues that were not addressed in EXPRESS. These include

- *stronger semantics of relations:* Relations are implicit in EXPRESS. There is no structural difference between an attribute value and a relation. One solution, used in all models when relations became complex, was to objectify the relations. That is, the relations are represented by Entities. This allows a relation to include various issues that it could not otherwise represent: attributes and constraints (WHERE clauses). However, when relations are objectified, the model becomes more complex and the issue of automatic creation and deletion grows. A new construct specifically defined as a relation object, with an explicit dependency definition, would allow more complex models to be developed and managed more easily.

- *locating constraints:* An intrinsic part of a data model is the set of constraints that tell if the instance data loaded into some Entity has a meaningful set of attribute values. These are defined in EXPRESS in WHERE clauses. This text has not dwelt on the WHERE clauses and the functions used to define them that are integral to each product model. Many of these are quite complex. Constraints may apply to the well-formedness of a single Entity, in which case the constraint quite naturally is carried by the Entity. In other cases, the constraint is across a set of different Entities and the structure that should carry the constraint is not apparent. In almost all cases, constraints across different Entities can be defined by and carried in a relation.

- *constraint definitions:* Constraints, as defined in WHERE clauses, are often complex. They may require several pages of code and rely on other functions. Such constraints are defined in a rather *ad hoc* way, without a lot of care in the reuse of functions. A first step in simplifying constraint definitions would be to formalize the function library available in writing WHERE clauses. They should be defined as Integrated Resources to be used by a family of Part Models. After stabilization of such a library, a further step in simplification would be to incorporate the functions as primitives in the data modeling language. For example, if the construction functions defined in the Parametric Framework (see Figure 11.10) were included as primitives in a modeling language, more subtle and complex constraints could be defined.

- *modeling of behavior:* A significant problem in mechanical equipment simulation and design is the modeling of the behavior of different pieces of mechanical equipment. The behavior may be described at different time scales, such as average performance over time, behavior under a sample set of environmental contexts, or as a continuous mathematical function over its full behavior range. Regular ways to model such behavior are needed so that one application could specify or select equipment with defined behavior and another application could execute a simulation combining the behavior of several pieces of equipment, and define their aggregate system-level behavior. An approach is needed to cover the different time slice cases and to characterize behavior for each kind of case.

11.6 SUMMARY

EXPRESS was conceived with a well-defined role as an AIM language in ISO-10303. It was defined to represent complex data structures that can be physically implemented in various ways. The issues raised in this chapter all address slightly expanded roles of the EXPRESS language. One set of changes addresses the role of a building model as a repository, for which model modifications may be required. A repository may need both schema changes and the migrating of instances up and/or down the type lattice. It may require mappings from one part of a schema to another or between different schemas. Another kind of extension deals with the different kinds of knowledge used in a model. EXPRESS allows the representation of knowledge defined as data. However, some kinds of knowledge can only be represented procedurally. Parametric models are based on procedural knowledge of geometry. Rules and constraints are also represented procedurally. The language functionality issues reviewed in this chapter can only be resolved by both respecting the role of EXPRESS and thinking strategically regarding the evolution of its role in the overall data exchange process.

We can expect that someday there will be an EXPRESS-2 language that responds to many or all of the issues described here. It would then make the job of product modeling in general, and building modeling in particular, much easier.

11.7 NOTES AND FURTHER READING

The original work and model for object-oriented database schema evolution was presented by Banergee, Kim et al. [1987]. An extensible building product model has been demonstrated by Eastman, Assal and Jeng [1995]. Object-oriented databases supporting schema evolution include ORION, VERSANT and UniSQL. A critique of EXPRESS has been presented in Eastman and

Fereshatian [1994]. A more general critique of current building modeling efforts is given in Augenbroe and Eastman [1998].

Instance migration up or down the specialization lattice has been proposed by Osborn [1989]. Separate versions of a database have been proposed by Skarra and Zdonik [1987] and Narayanaswamy and Rao [1988].

Proper deletions in object-oriented databases, with well-defined dependencies, have been studied by several researchers. The Unified Modeling Language provides different types of relations that distinguish those relations that are dependent from those that are not [Muller, 1997]. This approach is also advocated by Kotz-Dittrich and Dittrich [1995] and Eastman and Jeng [1999]. Kemper and Moerkotte develop the General Object Model (GOM) that uses explicit constraints to define dependencies [1994, Chapter 9]. Mapping between levels of aggregation has been identified as a problem by Rosenman and Gero [1996].

The documentation on EXPRESS-M is Bailey [1996]. The earlier work on EXPRESS-V is available in Hardwick, Spooner et al. [1994]. The specification for EXPRESS-X, which integrates EXPRESS-M and EXPRESS-V, is [LIII,1996]. Two popular articles on EXPRESS-X are Hardwick, Spooner et al., [1996] and Spooner and Hartwick [1997]. Other documentation on EXPRESS-X, written very clearly, is available from EPM Technology [1997]. There has been much recent interest in mapping languages. Reviews of mapping languages have been made by Verhoef, Liebich, Amor [1995] and Khedro, Eastman, Junge, Liebich [1996]. Another interesting mapping language effort is Robert Amor's [Amor, Hosking and Mugridge,1995]. A different approach to communicating between views is presented in Cutkosky, Englemore et al. [1993].

Views are a well-developed aspect of relational databases [Ullman, 1988, Chapter 4], [Groff and Weinberg, 1990, Chapter 1]. Views have been treated pragmatically in object-oriented systems and have been conceived with rigorous foundations [Kim, 1992, pg. 214-5], [Kim and Kelley, 1995].

The Parametric Framework is presented in ISO TC184/SC4/WG12 N022 [1998].

The EXPRESS-2 Website is http://www.nist.gov/sc4/parts/part011e2/current/

11.8 STUDY QUESTIONS

1.   Review the EXPRESS-X language definition or one of its implementations. Study how the identity between base and view objects might be treated, so that maps could be written and executed in both directions, with partial updates. What mechanisms can be used for different types of object mappings:
    a.   when the mapping is one-to-one?
    b.   when the mapping is one-to-many or many-to-one?

2.   Review the data modeling constructs available in UML, for example, as defined in Muller [1997]. Considering only the data modeling part of UML, make a list of the semantic constructs that it supports for which equivalent constructs exist or do not exist in EXPRESS-G. Discuss which of these seem to be important for product models and which, if any, do not.

3.   If implemented, WHERE clauses in EXPRESS return a logical value (TRUE, FALSE, UNKNOWN). These constraint tests are supposed to be useful in real data exchange situations. However, there are not many proposals regarding how they should be handled. For example, should a data exchange be rejected if any WHERE clause is FALSE? Consider the

WHERE clause in two different contexts and propose a set of rules regarding how they should be used:

    (a) In a flat file exchange situation, where the exchanges are defined by an application programming interface. What might an API function look like that would allow checking the WHERE clauses and responding appropriately?

    (b) In a database context, where there may be interactive querying of a data model. Would a general WHERE clause in a database system, that queries the WHERE clause values of a model be sufficient, or are other facilities needed in addition? What might those facilities be?

4.   Examine the Part 041 model that supports the unique ID capability in STEP. Review the policies for naming. What is needed to maintain the identify of objects when carried in different Views? Does it serve the purpose for keeping track of different Views of the same object?

5.   The IFC Version 1.5 includes a limited form of parametric shapes—the Attribute-Driven Solids described in Section 9.4.2. Compare these with the parametric solids proposal reviewed in this chapter. Which aspects are similar? Which of the Attribute-Driven Solids could not be implemented in the parametric modeling example?

APPENDIX A
STEP Tool Resources

Listed below are most of the companies that currently offer software tools supporting STEP technology, especially the EXPRESS modeling language.

EPM Technology

EPM offers several STEP tools, including EDMvisual Express, a complete graphical editor for EXPRESS, using an extended EXPRESS-G graphical model. It allows generation and editing of EXPRESS schemas. They also provide tools for checking datasets for conformance to an EXPRESS model and libraries of SDAI calls, in C and C++. These tools run on most UNIX and all MS Windows workstations. Both client and server versions are available.

Jotne EPM Technology
Grenseveien 107
N-0663 Oslo Norway
http//:www.epmtech.jotne.com
Email: info@epmtech.jotne.com

ICEBREAKER Communications, Inc.

Icebreaker proves a means to link graphical objects with EXPRESS entities, allowing definition of products for distribution on Internet or email. The graphical objects may be video clips, raster images, or solid models developed in a CAD system. It runs on MS Windows workstations.

ICEBREAKER Communications, Inc.
Ed Wagner
Email: Ewags@tiac.net

ICEBREAKER Communications, Inc.
Ulf Lindquist
Email: Ulfi@cascade.se

Product Data Integration Technologies, Inc. (PDIT)

PDIT offers graphical modeling tools running in Microsoft Windows® environments. They support direct graphical generation and editing of EXPRESS-G models and the automatic generation of EXPRESS models from them.

Product Data Integration Technologies, Inc. (PDIT)
100 W. Broadway
Suite 540
Long Beach CA 90802 USA
Email: info@pdit.com

ProSTEP Production Technology Gmbh

ProSTEP offers an EXPRESS-G graphical editor for all major Unix and MS Windows platforms. It also offers its EXPRESS Workbench with parser, compiler, linker and library, with SDAI early and late binding libraries.

ProSTEP Production Technology Gmbh
Julius Reiber Strasse 15
D-64293 Darmstadt Germany
http//:www.prostep.de

Email: info@prostep.de

ProSTEP Production Technology Gmbh
Rudower Chaussee 5
D-12489 Berlin Germany

InterData Access Inc. (IDA)
1127 Mannheim Road
Suite 305
Westchester IL 60154 USA

STEP Tools Inc.

STEPTools provides a suite of products. They include an EXPRESS parser and a graphical visualizer of EXPRESS code. They offer libraries for moving EXPRESS data into and out of various databases, including Oracle, ObjectStore and Versant object database. They also provide SDAI toolkits for C and C++ early and late bindings. They offer programming tools to transfer ACIS solid models to and from STEP model AP203 and to visualize STEP models in OpenGL and VRML. They also offer tools to publish STEP models as applets. They also provide EXPRESS-X language facilities for mapping data from one EXPRESS schema to another.

STEP Tools Inc.
Rensselear Technology Park
Troy New York 12180
http//:www.steptools.com
Email: info@steptools.com

REFERENCES

Aalami A, Fischer M, 1998, "Joint Product and Process Model Elaboration Based on Construction Method Models," in *The Life-Cycle of Construction Innovations: CIB Working Conference* Eds. B C Björk and A Jägbeck, June 3-5, Stockholm Sweden pp 1-11

Adrian J J, 1973, *Quantitative Methods in Construction Management* (American Elsevier, New York)

Akin O, Sen R, Donia M, Zhang Y, 1995, "SEED-Pro: Computer-Assisted Architectural Programming in SEED," *Journal of Architectural Engineering*, **1** (4) pp 153-161

Alden J L, 1932, "Machine Design Management," *Proceedings Society of Mechanical Engineers*, SMP-54-9, pp. 105-118

Alexander C, Ishikawa S, Silverstein M, 1977, *A Pattern Language* (Oxford University Press, New York)

Alexander J, Coble R, Drogemuller R, Newton P, 1998, "Information and Communication in Construction: Closing the Loop," in *The Life-Cycle of Construction Innovations: CIB Working Conference* Eds. B C Björk and A Jägbeck, June 3-5, Stockholm Sweden pp 35-43

Alferes M, Seireg A, 1996, "An integrated system for computer-aided design and construction of reinforced concrete buildings using modular forms," *Automation in Construction* **5** pp 323-341

Alshawi M, Underwood J, 1996, "Applying object-oriented analysis to the integration of design and construction," *Automation in Construction* **5** pp 105-121

Alting L, 1994, *Manufacting Engineering Processes* (Marcel Dekker, New York)

America P, 1990, "Designing an Object-oriented Programming Language with Behavioral Subtyping," in *Lecture Notes in Computer Science 489* Eds. J W De Bakker, W P De Roever, G Rozenberg, (Springer-Verlag) pp 60-90

American Institute of Architects (AIA), 1994, *The Architect's Handbook of Professional Practice* Volume 2, AIA Document B162 (American Institute of Architects, Washington, DC)

Ames A, Nadeau D, Moreland J, 1997 *VRML 2.0 Sourcebook* (John Wiley, New York)

Amor R, Augenbroe G, 1997, "Project Control in Integrated Building Design Systems," *Proceedings CIB W-78 Conference: Information Technology Support for Construction Process Re-engineering*, (Cairns, Australia), pp 53-66

Amor R, Hosking J, Mugridge W, 1995, "A Declarative Approach to Inter-Schema Mappings," *Proceedings CIB Workshop on Computers and Information in Construction*, Eds, M Fisher, K Law, B Luiten, (Stanford University) pp 223-232

Augenbroe G (Ed.) 1993, COMBINE 1 Final Report, EU DG XII JOULE Report

Augenbroe G (Ed.) 1995, COMBINE 2 Final Report, EU DG XII JOULE Report

388

Augenbroe G, Eastman C M, 1998, "Product Modeling Strategies for Today and the Future," in *The Life-Cycle of Construction Innovations: CIB Working Conference* Eds. B C Björk and A Jägbeck, June 3-5, Stockholm Sweden pp 191-208

Augenbroe G, de Wit S, 1997, "Communicating performance assessments of intermediate building states," *Proceedings Building Simulation'97,* Vol. II (Prague, Czech) pp 429-436

Augenbroe G, Rombouts W, 1994, "Extended Topology in Building Design Systems," in *ASCE First Congess on Computing in Civil Engineering* (Washington, DC) pp 1196-1203.

AutoCAD Customization Manual, 1992, AutoCAD Release 12, Publication 100891-01

Baer A, Eastman C M, Henrion M, 1979, "Geometrical modeling: A survey," *Computer-Aided Design* **11** (5) pp 253-272

Bailey I, 1996, *EXPRESS-M Reference Manual* ISO TC184/SC4/WG5 N243 (CIMIO Ltd, Brunel Science Park, Surry, England)

Banerjee J, Kim W, Kim H, Korth H, 1987, "Semantics and implementation of schema evolution in object-oriented databases," *Proceedings ACM SIGMOD 1987 Annual Conference* **16**(3) pp 311-322

Barghouti N S, Kaiser, G E, 1991, "Concurrency Control in Advanced Database Applications," *ACM Computing Surveys*, **23** (September), pp 269-317

Barrett G V, Blair J, 1982, *How to Conduct and Analyze Real Estate Market and Feasibility Studies* (Van Nostrand Reinhold, New York)

Baumgart B G, 1974, *Geometric Representation for Computer Vision* Stanford University Computer Science Dept Technical Report AIM-249, STAN-CS-74-463

Baumgart B G, 1975, "A polyhedron representation for computer vision," in *Proceedings AFIPS 1975 National Computer Conference* pp 589-596

Bertino E, Martino L, 1991, "Object-Oriented Database Management Systems: Concepts and Issues," *IEEE Computer* **24** (April) pp 33-47

Bezier P E, 1993, "The First Years of CAD/CAM and the UNISURF CAD System," in *Fundemantal Development of Computer-Aided Geometric Modeling* Ed. L Piegl (Academic Press, New York) pp 13-26

Bijl A, 1979, "Computer aided housing and site layout: experience of research software in a production environment," in *Proceedings, PARC'79, International Conference on the Applications of Computers in Architecture, Building Design and Urban Planning* Berlin pp 283-304

Bijl A, 1989, *Computer Discipline and Design Practice: Shaping our Future* (Edinburgh University Press, Edinburgh)

Bijl A., Renshaw T, Barnard D, Wyatt S, Burney D, 1971, "Application of computer graphics to architectural practice," ARU CAAD Studies Report, project A25/SSHA-DOE, Architectural Research Unit, Edinburgh U., December

Bijl A., Shawcross G, McLeish G, Weir G, Stone D, O'Regan S, 1974, "SSHA-DOE Site Layout Project: Phase I," Final Report, CAAD Studies, Edinburgh University, January

Bijl A, Shawcross G, 1975, "Housing site layout system," *Computer-Aided Design* **7** (1) pp 2-10

Björk B-C, 1989, "Basic structure of a proposed building product model" *Computer-Aided Design* **21** (2) pp 71-78

Björk B-C, 1992a, "A Unified Approach for Modelling Construction Information" *Building and Environment* **27** (2) pp 173-194

Björk B-C, 1992b, "A conceptual model of spaces, space boundaries and enclosing structures," *Automation in Construction* **1** (3) pp 193-214

Björk B-C, 1995, *Requirements and information structures for building product models,* Technical Report, Publication 245, VTT Technical Research Centre of Finland

Björk B-C, 1997, "ISO DIS 13567 - the proposed international standard for structuring layers in computer aided building design" *Electronic Journal of Information Technology in Construction* **2** (electronic journal http://itcon.org)

Björk B-C, Penttila H, 1991, "Building Product Modelling Using Relational Databases, Hypermedia Software and CAD Systems" *Microcomputers in Civil Engineering* **6** (4) pp 267-279

Bloor M S, Owen J, 1995, *Product Data Exchange* (UCL Press, London)

Bobrow D, Collins A (Eds.), 1975, *Representation and Understanding* (Academic Press, New York)

Booth W, 1992, *Design Build Marketing* (Van Nostrand Reinhold, New York)

Borkin H, McIntosh J, McIntosh P, 1981, "ARCH:Model: A Geometric Modeling Relational Database System," *ARCH:Model User's Manual* (Architectural Research Laboratory, University of Michigan, Ann Arbor, Michigan)

Brachman R, 1983, "What is-a is and isn't?," *IEEE Computer* **16** (10) pp 30-36

Braid I C, 1973, *Designing with Volumes* (Cantab Press, Cambridge, UK)

Braid I C, 1975, "The synthesis of solids bounded by many faces," *Communications of the ACM* **18** (4) pp 209-216

British Standards Institution, 1992, *BS1192 Part 5: Construction Drawing Practice: Guide for the structuring of computer graphic information* (British Standards Institution, Milton Keynes, UK)

Brodie M L, 1981, "Association: a database abstraction for semantic modeling," in *Entity-Relationship Approach to Information Modeling and Analysis* Ed. P P Chen (ER Institute)

Burkett W C, Yang Y, 1995, "The STEP Integrated Information Architecture," *Engineering with Computers* **11** pp 136-144

Burnett M M, Goldberg A, Lewis T, 1995, *Visual Object-Oriented Programming* (Manning, Greenwich, CT)

Caldwell B, Stein T, 1998, "Beyond ERP: New IT Agenda" *Information Week* (30 Nov 1998) pp 30-38

Carrera G, Kalay Y, Novembri G, 1994, "Knowledge-based computational support for architectural design" *Automation in Construction*, **3** (2-3) pp 157-166

Chandhry P K, 1992, "NIAMEX: a NIAM Compiler and EXPRESS Pre-processor," in *Proceedings EUG'92 - Second Int. EXPRESS User Group Conference*, Ed P Wilson, Dallas, TX

Chen P, 1976, "The entity-relationship model: towards a unified view of data" *ACM Transactions on Database Systems* **1** (1) pp 9-36

CIMsteel: The Logical Product Model, Version 3.3, 1993, University of Leeds, Release 2.0

Cinar U, 1975, "Facilities planning: a systems, analysis and space allocation approach," in *Spatial Synthesis in Computer-Aided Building Design* Ed C M Eastman (Elsevier Applied Science Publishers, London) pp 19-40

Clarke J, MacRandel D, 1993, "Implementation of Simulation Based Design Tools in Practice," in *Proceedings Building Simulation'93* Adelaide, Australia

Coad P, North D, Mayfield M, 1995, *Object Models: Strategies, Patterns and Applications* (Prentice Hall, Englewood Cliffs, NJ)

Cook W R, Hill W, Canning P, 1990, "Inheritance is not subtyping," in *Proceedings 17th Annual ACM Symposium on Principles of Programming Languages* (San Francisco, CA), pp 125-135.

Coons S A, 1967, *Surfaces for Computer-Aided Design of Space Forms* Technical Report TR-41, Project MAC, MIT, Cambridge, MA

Coplien J O, Schmidt D (Eds.), 1995, *Pattern Languages of Program Design* (Addison-Wesley, Reading MA)

Coyne R, Rosenman M, Radford A, Balachandran M, Gero J, 1989, *Knowledge-Based Design Systems*, (Addison Wesley, Reading, MA)

Crowley A, Watson A, 1997, "Representing Engineering Information for Constructional Steelwork" *Microcomputers in Civil Engineering,* **12** pp 69-81

Crowley A, Watson A, Christodoulakis D, 1997, "From Data Exchange to Information Sharing," in *Concurrent Engineering in Construction: Proceedings 1st International Conference, Institution of Structural Engineers* Eds. C Anumba, N Evbuomwan, London, pp 131-140

Cuff D, 1991, *Architecture: The Story of Practice* (MIT Press, Cambridge, MA)

Cutkosky M, Englemore R, Fikes R, Genesereth M, Gruber T, Mark W, Tanenbaum J, Weber J, 1993, "PACT: a experiment in integrating concurrent engineering systems," *IEEE Computer*, **26**(1) pp 28-38.

Dahl O-J, Dijkstra E, Hoare C A R, 1972, *Structured Programming* (Academic Press, New York)

Dahl O-J, Nygaard K, 1966, "SIMULA - An ALGOL-Based Simulation Language," *Communications of the ACM* **9** pp 671-678

Danforth S, Tomlinson C, 1988, "Type Theories and Object-Oriented Programming," *ACM Computing Surveys* **20** (1) pp 29-72

Danner W F, Yang Y, 1992a, *STEP Development Methods: Specification of Semantics for Information Sharing*, NISTIR 4915, NIST, National Engineering Laboratory, Gaithersburg, MD

Danner W F, Yang Y, 1992b, *STEP Development Methods: Resource Integration and Application Interpretation*, ISO TC1184/SC4/WG5 N31

Dayal U, Bernstein P, 1982, "View definition and generalization for database integration in a multidatabase system," *IEEE Transactions on Software Engineering* **10** pp 628-645

De Chiara J, Callender J H (Eds.), 1990, *Time-saver Standards for Building Types* (McGraw-Hill, New York)

De Troyer O, 1989, "RIDL: A tool for the computer-assisted engineering of large databases in the presence of integrity constraints," in *Proceedings 1989 ACM SIGMOD Conference, SIGMOD Record* pp 418-429

De Waard M, 1992, *Computer Aided Conformance Checking* (doctoral thesis) Civil Engineering Department, Delft University.

Desel J, Esparza J, 1995, *Free Choice Petri Nets*, (Cambridge University Press, UK)

Doherty P, 1997, *Cyberplaces: The Internet Guide for Architects, Engineers and Contractors* (R.S. Means, Kingston, MA)

Duerk D P, 1993, *Architectural programming: information management for design* (Van Nostrand Reinhold, New York)

Dym C L, Leavitt R, 1991, *Knowledge-Based Systems in Engineering* (McGraw Hill, New York)

Eastman C M (Ed.), 1975a, *Spatial Synthesis in Computer-Aided Design* (Elsevier Applied Science Publishers, London)

Eastman C M, 1975b, "The use of computers instead of drawings in building design," *Journal of the American Institute of Architects* (March) pp 46-50

Eastman C M, 1978, "Representation of design problems and maintenance of their structure," in *Artificial Intelligence and Pattern Recognition in Computer Aided Design,* Ed. J C Latombe (North-Holland Press, New York) pp 335-366.

Eastman C M, 1980a, "Systems facilities for CAD databases," in *Proceedings 17th Design Automation Conference* pp 50-56

Eastman C M, 1980b, "Prototype integrated building model," *Computer Aided Design*, **12**(3), pp 115-119

Eastman C M, 1981, "The design of assemblies," *SAE International Conference* Technical Paper no. 81097

Eastman C M, 1990, "Vector versus raster: a functional comparison of drawing technologies," *IEEE Computer Graphics and Applications* **10** (Sept.) pp 68-80

Eastman C M, 1993a, "Editorial: Uses of Data Modeling in Engineering," *Journal of Computing in Civil Engineering*, **7** (1) pp 1-5

Eastman C M, 1993b, "Conceptual Modeling in Design," in *Fundamental Developments of Computer-Aided Geometric Modeling*, Ed. L Piegl, (Academic Press, London) pp 185-202

Eastman C M, 1996, "Managing Integrity in Design Information Flows," *Computer Aided Design* (May), **28** pp 551-565.

Eastman C M, Assal H, Jeng T S, 1995, "Structure Of A Product Database Supporting Model Evolution," in *Proceedings 1995 CIB W-78 Symposium*, Stanford, CA pp 327-338.

Eastman C M, Baer A, 1975, "Database Features for a Design Information System," *ACM Workshop on Databases for Interactive Design,* Eds. W van Cleemput and J Linders, University of Waterloo, Canada pp 45-53

Eastman C M, Bond A, Chase S, 1991, "A data model for engineering design databases," *1st International Conference on A.I. in Design*, Ed. J Gero, Butterworths-Heinemann, (June), pp 339-365

Eastman C M, Chase S, Assal H, 1993, "System architecture for computer integration of design and construction knowledge," *Automation in Construction* **2** (2) pp 95-108

Eastman C, Fereshatian N, 1994, "Information Models for Product Design: a Comparison," *Computer-Aided Design*, **26**(7) pp 551-572

Eastman C M, Henrion M, 1977, "GLIDE: a language for design information systems," *SIGGRAPH'77 Proceedings Computer Graphics* pp 24-33

Eastman, C M, Jeng T S, Chowdbury A R, Jacobsen K, 1997, "Integration Of Design Applications With Building Models," *Proceedings CAADfutures'97*, Munich Germany, pp 45-59

Eastman C M, Jeng T S, 1999, "A database supporting evolutionary product model development for design," *Automation and Construction* **8**(3), pp 305-324

Eastman C M, Lividini J, Stoker D, 1975, "A database for designing large physical systems," in *Proceedings AFIPS Press, National Computer Conference* pp 603-611

Eastman C M, Parker D S, Jeng T S, 1997, "Managing the Integrity of Design Data Generated by Multiple Applications: The Principle of Patching," *Research in Engineering Design* **9** pp 125-145.

Eastman C M, Shirley G, 1994, "The management of design information flows," in *Management of Design: Engineering and Management Processes,* Eds. S Dasu, C M Eastman (Kluwer Press, New York)

Eastman C M, Siabiris A, 1995, "A Generic Building Product Model Incorporating Building Type Information," *Automation in Construction* **3** (1) pp 283-304

Eastman C M, Weiler K, 1979, "Geometric modeling using the Euler operators," in *Proceedings Conference on CAD/CAM*, Ed. D Gossard, Cambridge, MA, pp 248-262

Eckholm A, Fridquist S, 1996, "Modeling of user organizations, buildings and spaces for the design process," in *Construction on the Information Highway: Proceedings of the CIB W78 Workshop*, Ed. Z Turk Bled, Slovenia

EDIF, 1985, *Electronic Design Interchange Format* Version 1.1, Dallas TX, EDIF Users' Group

Elmagarmid, A K (Ed.), 1992, *Database Transaction Models for Advanced Applications* (Morgan-Kaufmann, San Mateo, California)

Enkovaara E, Salmi M, Sarja A, 1988, *RATAS Project - Computer Aided Design for Construction* (Building Books Ltd., Helsinki)

EPM Technology, 1997, *EPMxpx - a Structural Data Mapping Language for structured data sets defined in EXPRESS* (EPM Technology, Oslo Norway) see Web site: http://www.epmtech.jotne.com

Faux I, Pratt M, 1979, *Computational Geometry for Design and Manufacture* (John Wiley, New York)

Fenves S, Flemming U, Hendrickson C, Maher M L, Quadrel R, Terk M, Woodbury R, 1994, *Concurrent Computer Integrated Building Design* (Prentice Hall, Englewood Cliffs, NJ)

Fischer M, Luiten G, Aalami F, 1995, "Representing project information and construction method knowledge for computer-aided construction management," in *Modeling of Buildings Through Their Lifecycle: Proceedings, CIB Publication 180* Eds. M Fischer, K Law, B Luiten, Palo Alto, CA, pp 404-414

Flegg H G, 1974, *From Geometry to Topology* (English Universities Press, London)

Flemming U, Woodbury R, Akin O, Fenves S, Garrett J, et al.1995, *Special Issue on Computers in Building Design, Journal of Architectural Engineering* **1** (4)

Foley J, Van Dam A, Feiner S, Hughes J, 1991, *Computer Graphics: Principles and Practice* (Addison Wesley, Reading, MA)

Fowler M, 1997, *UML Distilled: Applying the Standard Object Modeling Language* (Addison Wesley Longman, Reading, MA)

Froese T, 1995, "Core Process Models and Application Areas in Construction," in *Modeling of Buildings Through Their Lifecycle: Proceedings, CIB Publication 180*, Eds. M Fischer, K Law, B Luiten, Palo Alto, CA, pp 427-436

Froese T, Grobler F, Yu K, 1998, "Development of Data Standards for Construction—an IAI Perspective," in *The Life-Cycle of Construction Innovations: CIB Working Conference* Eds. B C Björk and A Jägbeck, June 3-5, Stockholm Sweden, pp 233-244

Froese T, Rankin J, Yu K, 1997, "Project Management Application Models and Computer-Assisted Construction Planning in Total Project Systems," *International Journal of Construction Information Technology* **5** (1) pp 39-62

Fulton R E, Yeh C P, 1988, "Managing Engineering Design Information," in *Proceedings AIAA'88 Aircraft Design, Systems and Operations Conference* Atlanta, GA

Galle P, 1994, "Product Modeling for Building Design: Annotated Bibliography," Computer Aided Building Design Unit, Institute of Building Design, Technical University of Denmark, Building 118, DK-2800, Lyngby, Denmark.

Galle P, 1995, "Towards integrated 'intelligent' and compliant computer modeling of buildings," *Automation and Construction* **4** (3) pp 189-211

Genesereth M, Nilsson N, 1987, *Logical Foundations of Artificial Intelligence* (Morgan Kaufmann Publishers, Palo Alto, CA)

Gero J, Lee H, Tham K, 1991, "Behavior: A Link Between Function and Structure in Design," in *Proceedings IFIP WG 5.2 Working Conference on Intelligent CAD* Columbus, OH pp 201-230

Giblin P J, 1977, *Graphs, Surfaces and Homology* (Chapman and Hall, London)

Gielingh W, 1988a, *General AEC Reference Model (GARM)* ISO TC184/SC4 Document 3.2.2.1 TNO Report B1-88-150, Delft, The Netherlands

Gielingh W, 1988b, "General AEC Reference Model (GARM)," in *Proceedings CIB Conference on The Conceptual Modeling of Buildings*, Ed. P Christensson, Lund Sweden, pp 165-178

Goel A, Gomez A, Grue N, Murdock N, Recker M, Govindaraj T, 1996, "Design Explanations in Interactive Design Environments," in *Proceedings Fourth International Conference on AI in Design* Palo Alto, CA

Goldberg A, Robson D, 1983, *Smalltalk 80: The Language and Its Implementation* (Addison Wesley, Reading MA)

Goldfarb N, Kaiser W, 1964, Chapter 1, in *Gantt Charts and Statistical Quality Control,* Eds. N Goldfarb, W Kaiser, Hofstra University Yearbook of Business 1:1 (March, 1964)

Goldfarb C, Prescod P, 1998, *XML Handbook,* (Prentice Hall, NJ)

Groff J, Weinberg P, 1990, *Using SQL,* (Mcgraw Hill, New York)

Gutman R, 1988, *Architectural practice: a critical view* (Princeton Architectural Press, Princeton, NJ)

Guttag J V, 1977, "Abstract data types for the development of data structures," *Communications of the ACM* **20** pp 307-315

Hales H L, Ed, 1985, *Computerized Facility Planning* (Industrial Engineering Management Press, Norcross GA)

Halpin T, 1995, *Conceptual Schema and Relational Database Design*, 2nd *ed.* (Prentice Hall, Sydney, Australia)

Halpin T A, Nijssen G M, 1989, *Conceptual Schema and Relational Database Design* (Prentice Hall, New York)

Hardwick M, Spooner D, 1987, "Comparison of some data models for engineering objects," *IEEE Computer Graphics and Applications* **7** (2) pp 56-65

Hardwick M, Spooner D, Kilty M, Jiang Z, 1994, *Mapping EXPRESS AIMs to ARMs Using Database Views* Design and Manufacturing Institute Report 94041, Rensselear Polytechnic Institute, NY

Hardwick M, Spooner D, Rando T, Morris K, 1996, "Sharing Design and Prototyping Information in Virtual Enterprises," *Communications of the ACM*, **39** (2) pp 46-54

Harris F, McCaffer R, 1995, *Modern Construction Management* (Blackwell Science, Oxford)

Harvey R C, Ashworth A, 1993, *The Construction Industry of Great Britain* (Butterworth Heinemann Ltd., Oxford)

Hill F S, 1990, *Computer Graphics* (Macmillan, New York)

Hoffman C, 1989, *Geometric and Solid Modeling: an introduction* (Morgan Kaufmann, San Mateo, CA)

Hofstadter D R, 1979, *Godel, Escher, Bach: an eternal golden braid* (Basic Books, New York)

Hoskins E M, 1973, "Computer aids in system building," in *Computer-Aided Design*, Eds. J Vlietstra, R Weilinga (North-Holland Press, New York) pp 127-140

Hoskins E M, 1977, "The OXSYS System," in *Computer Applications in Architecture*, Ed. J. Gero (Applied Science, London) pp 343-391

Hoskins E M, 1979a, "Design Development and Description Using 3-D Box Assemblies," in *Proceedings PARC Conference* Berlin, pp 688-724

Hoskins E M, 1979b, "Descriptive Databases in Some Design Manufacturing Environments," *Computer-Aided Design*, **11** (3) pp 151-157

Hull R, King R, 1987, "Semantic database modeling: survey, applications and research issues," *ACM Computing Surveys*, **19** (3) pp 201-260

IATC, 1993, *Interoperability Acceptance Testing Methodology IGES Guidelines, Version 1.01*, Interoperability Acceptance Testing Committee, National Institute of Standards and Technology, Gaithersburg, MD

IDEFØ, 1981, "ICAM Architecture Part II, Vol. 5," *Information Modeling Manual (IDEFØ)*, Report number AFWAL-TR-81-4023, Vol. 5 Mantech Technology Transfer, Center WL/MTX.

IDEF1x, 1985, *Integrated information support system, Volume V: Common data model subsystem*, Mantech Technology Transfer Center, WL/MTX, Wright-Patterson AFB, OH; also D. Appleton Company Inc., 1986, IDEF1X manual, USAF Integrated Computer-Aided Manufacturing Program, El Segundo, CA

IGES, 1991, *Initial Graphics Exchange Standard (IGES), Version 5.1*, NISTIR 4412, U.S. National Bureau of Standards, Gaithersburg, MD

International Alliance for Interoperability, 1997, *Industry Foundation Classes, Release 1.5* IAI, 2980 Chain Bridge Road, Suite 143, Oakton, VA

ISO DIS10303 Part 11, 1991 *EXPRESS Language Reference Manual: External Representation of Product Definition Data*, Ed. D Schenck, ISO TC184/SC4/WG5, Document N14

ISO DIS10303 Part 225, *Building Elements Using Explicit Shape Representation* ICS 25.040.40

ISO TC184/SC/WG1, 1989, *AEC Building Systems Model*, Working Paper, Ed. J Turner, ISO TC184/SC/WG1

ISO TC184/SC4/WG4 N34, 1992, *Guidelines for the Development and Approval of STEP Application Protocols, Version 1.0*, NIST, Gaithersburg, MD

ISO TC184/SC4/WG12 N101, 1995, *ISO 10303 Part 042*, Draft ISO TC184/SC4/WG12

ISO TC184/SC/WG1 1996, *Part 106 Draft T100* Building Core Construction Model

ISO TC184/SC4/WG12 N101, 1997, *ISO 10303 Part 043*, Supersedes ISO TC184/SC4/WG12 N092, ISO/CD 10303-43

ISO TC184/SC4/WG12 N107, 1997, *ISO10303 Part 041*, Supersedes ISO TC 184/SC4/WG 12 N094, ISO/CD 10303-41

ISO TC184/SC4/WG12 N022, 1998, *Parametrics Framework*, Eds. M Pratt, N Christensen, NIST, Gaithersburg, MD

396

ISO TC10/SC8/WG13, 1996, *Implementation of ISO13567 CAD Layer Standard*, (DRAFT) World Wide Web http://www.ce.kth.se/fba/bit/cadlayer.htm

ISO WD 10303-Part 22, 1993, *Standard Data Access Interface*, SOLIS Repository, NIST, Gaithersburg, MD

Jägbeck A, 1998, *IT Support for Construction Planning—a system based on integrated information* (doctoral thesis), Royal Institute of Technology, Construction management and economics, Stockholm.

Jacobsen K, Jeng T S, Eastman C, 1997, "Information management in Creative Engineering Design and Capabilities of Database Transaction," *Automation and Construction*, **7**(1) pp 55-69

Jeng T S, 1998, *Design transactional flow management: structuring design process for CAD frameworks*, Ph.D. Thesis Georgia Institute of Technology, Atlanta GA

Jeng T S, Eastman C, 1998, "Database Architecture for Design Collaboration," *Automation and Construction*, **7**(6) pp. 475-484

Junge R, Kothe M, Schulz K, Zarli A, Bakkewren W, 1997, "The VEGA Platform," in *CAADfutures1997 Proceedings*, Ed. R Junge, Munich, Germany, pp 591-616

Junge R, Liebich T, 1997, "Product Data Model for Interoperability in a Distributed Environment," in *CAADfutures1997 Proceedings*, Ed. R Junge, Munich Germany, pp 571-589

Karhu V, 1997, "Product Model based Design of Precast Facades," *Electronic Journal of Information Technology in Construction*, **2**, an Internet journal at http://itcon/org/

Katz R, 1990, "Toward a unified framework for version modeling in engineering databases," *ACM Computing Surveys*, **22**(4) pp. 375-408

Kemper A, Moerkotte G, 1994, *Object-Oriented Database Management: Applications in Engineering and Computer Science* (Prentice Hall, NJ)

Khedro T, Eastman C, Junge R, Liebich T, 1996, "Translation Methods for Integrated Building Engineering," *ASCE Conference on Computing*, (Anaheim, CA)

Kim W, 1992, *Introduction to Object-Oriented Databases* (MIT Press, Cambridge, MA)

Kim W, Banerjee J, Chou H, Garza T, Woelk D, 1987, "Composite object support in an object-oriented database system," in *OOPSLA '87 Proceedings* Orlando, FL pp 118-125

Kim W, Bertino E, Garza J, 1989, "Composite objects revisited," in *Proceedings 1989 ACM SIGMOD Conference* pp 337-347

Kim W, Kelley W, 1995, "View support on Object-Oriented Database Systems," in *Modern Database Systems: The Object Model, Interoperability and Beyond*, Ed. W Kim (Addison Wesley, New York)

Kotz-Dittrich A, Dittrich K, 1995, "Where object-oriented DBMSs should do better: a critique based on early experiences," in *Modern Database Systems*, Ed. W Kim (Addison-Wesley, New York) pp 238-254

Kowalski R, 1979, *Logic for Problem Solving* (North Holland, NY)

Krishnamurthy K, Law K H, 1997, "A Data Management Model For Collaborative Design in a CAD Environment," *Engineering With Computers,* **13** (2) pp 65-86

Kutay A, Eastman C M, 1983, "Transaction management in engineering databases," in *SIGMOD Conference on Engineering Databases Proceedings* (ACM New York) pp 73-80

LII (Laboratory for Industrial Information Infrastructure), 1996, *EXPRESS-X Reference Manual (Draft),* ISOTC184/SC4/WG5

Lakatos I, 1977, *Proofs and Refutations* (Cambridge University Press, Cambridge, UK)

Lange J, Mills D, 1979, *The Construction Industry: Balance Wheel Of The Economy* (Lexington Books, Lexington, MA)

Levy S M, 1993, *Japan's Big Six: Inside Japan's Construction Industry* (McGraw Hill, New York)

Levy S M, 1994, *Project Management in Construction, 2nd Ed.* (McGraw Hill, New York)

Lewandowski S, 1998, "Frameworks for Component-Based Client/Server Computing," *ACM Computing Surveys,* **30** (1), pp 3-27

Lewis R, Sequin C, 1998, "Generation of 3D Building Models from 2Dl Architectural Plans," *Computer-Aided Design,* **30** (10) pp 765-780

Liggett R, 1989, "Facility Layouts Via the Computer," *IFMA Journal* (June) pp 8-19

Lockley S, Sun M, 1995, *COMBINE II- Task 3: The Development of the Data Exchange System* Department of Architecture, University of Newcastle upon Tyne, UK

Luiten G T, 1994, *Computer Aided Design for Construction in the Building Industry* (doctoral thesis) Civil Engineering Department, Delft University.

MacKeller B K, Peckham J, 1992, "Representing design objects in SORAC: a data model with semantic objects, relationships, and constraints," in *Proceedings 2nd International Conference on Artificial Intelligence and Design,* Ed. J Gero pp 201-220

Mantyla M, 1988, *Introduction to Solid Modeling* (Computer Science Press, Rockville, MD)

Marks N, Taylor H L, 1966, *CPM/PERT: A Diagrammatic Scheduling Procedure* Bureau of Business Research, Graduate School of Business, University of Texas, Austin

Marti M, 1981, *Space Operation Analysis* (PDA Publishers Corp., West Lafayette IN)

Mayer R, 1987, "IGES: One answer to the problems of CAD database exchange," *Byte* (June) pp 209-214

McIntosh P, 1982, "The Geometric Set Operations in Computer-aided Building Design," Doctor of Architecture thesis, College of Architecture, University of Michigan

Meager M A, 1973, "The application of computer aids to hospital building," *Computer-Aided Design,* Eds. J Vlietstra, R Weilinga (North-Holland Press, New York) pp 423-452

Meltzer B, Glushko R, 1998, "XML and Electronic Commerce: Enabling the Network Economy," *SIGMOD Record 27* (December) ACM Press, pp 21-24

Menges A, 1998, *Frank O. Gehry: Guggenheim Bilbao Museo,* (Edition Axel Menges, Stuttgart)

Meyer B, 1988, *Object-Oriented Software Construction* (Prentice-Hall, Englewood Cliffs, NJ)

Meyer B, 1990, *Introduction to the Theory of Programming Languages* (Prentice-Hall, New York)

Minsky M, (Ed.), 1969, *Semantic Data Models* (MIT Press, Cambridge, MA)

Mitchell W, 1977, *Computer-Aided Architectural Design* (Petrocelli Charter, New York)

Mitchell W, McCullough M, 1991, *Digital Design Media* (Van Nostrand Reinhold, New York)

Miyagawa T, 1997, "Construction manageability planning: a system for manageability analysis in construction planning" *Automation in Construction* **6** (3) pp 175-19.

Muller P A, 1997, *Instant UML* (Wrox Press, Birmingham, UK)

Myer T, 1970, "An information system for component building," in *EDRA TWO, Proceedings of the 2nd Annual Environmental Design Research Association Conference,* Eds. J Archea, C M Eastman, Pittsburgh, PA, pp 299-308

Myllymaki R, 1998, "Cost Estimating as the Integrator Between Design and Production," *Life-Cycle of Construction IT Innovations, CIB-W78 Conference,* Eds. B C Björk and A Jägbeck, June 3-5, Stockholm, Sweden, pp 307-318

Narayanaswamy K, Bapa Rao K V, 1988, "An incremental mechanism for schema evolution in engineering domains," in *Proceedings 4th International Conference on Data Engineering* Los Angeles pp 294-301

Navathe S B, Cornelio A, 1990, in "Modeling Engineering Data by Complex Structural Objects and Complex Functional Objects," in *Extending Database Technology* Eds. F T Bancilhon, C Tsichritzis, C Dionysios (Springer-Verlag, Berlin)

NEDO, 1989, *Information Transfer in Buildings*, NEDC Publications, London

Norrie M C, Wunderli M, 1995, "Modelling in Coordination Systems," *International Journal of Cooperative Information Systems* **4** (2-3) pp 189-211

Ojwaka P, 1995, "Process Modelling Approach to Facilities Management," in *Modeling of Buildings Through Their Lifecycle: Proceedings, CIB Publication 180* Eds. M Fischer, K Law, B Luiten, Palo Alto, CA, pp 293-304

Osborn S L, 1989, "The role of polymorphism in schema evolution in an object-oriented database," *IEEE Transactions on Knowledge and Data Engineering* **1** (3) pp 310-317

Palmer M, Reed K, 1992, *3D Piping IGES Application Protocol Version 1.01*, NISTR 4797, National Institute of Standards and Technology, Gaithersburg, MD. 20899

Panero J, Zelnik M, 1979, *Human Dimensions and Interior Space,* (Whitney Library of Design, New York)

Parnas D, 1973, "On the criteria to be used in decomposing systems into modules," *Communications of the ACM* **15** (12) pp 1053-1058

PDDI, 1984, *Product Definition Data Interface* (5 vols.) obtainable from CAM-I Inc. Arlington TX

Peckham J, Maryanski F, 1988, "Semantic data models," *ACM Computing Surveys* **20** (3) pp 153-190

Penrose R, 1989, *The Emperor's New Mind: Concerning Computers, Minds and the Laws of Physics* (Penguin Books, New York)

Pohl J, Myers L, 1994, "A distributed cooperative model for architectural design," *Automation in Construction* **3** (2-3) pp 177-185

Poyet, P, 1993, *XPDI Manual*, CSTB (Sophia Antipolis, France)

Pratt T, 1975, *Programming Languages: Design and Implementation* (Prentice-Hall, Englewood Cliffs, NJ)

Preiser W (Ed.), 1985, *Programming the Built Environment* (Van Nostrand Reinhold, New York)

Rasmussen J, Pejtersen A M, Goodstein L P, 1994, *Cognitive Systems Engineering* (John Wiley and Sons, New York)

Reisig W, 1992, *A Primer in Petri Net Design* (Springer Verlag, New York)

Requicha A, 1980, "Representations of rigid solids: theory, methods and systems," *ACM Computing Surveys* **12** (4) pp 437-466

Rezgui Y, 1996, "An Integrated Appraoch for a Model Based Document Production and Management," *Electronic Journal of Information Technology in Construction*, **1**, an Internet journal at http://itcon/org/

Rivard H, Fenves S, Gomez N, 1995, "An Information Model for Multiple Views of Buildings," in *Modeling of Buildings Through Their Lifecycle: Proceedings, CIB Publication 180* Eds. M Fischer, K Law, B Luiten, Palo Alto, CA, pp 248-259

Roos D, 1966, *ICES System* (MIT Press, Cambridge, MA)

Rosen D, 1993, "Feature Based Design: Four Hypotheses for Future CAD Systems," *Research in Engineering Design* **5** (3-4) pp 125-139

Rosenman M, Gero J, 1996, "Modelling multiple views of design objects in a collaborative CAD environment," *CAD Journal* **28** (3) pp 193-192

Ross D, 1977, "Structured Analysis and Requirements Definition," *IEEE Transactions of Software Engineering* **3** (1) pp 6-34

Rowe P, 1987, *Design Thinking* (MIT Press, Cambridge, MA)

Rutherford J, Maver T, 1994, "Knowledge-based design support" *Automation in Construction* **3** (2-3) pp 187-202

Sanvido V, 1984, *A framework for designing and analyzing management and control systems to improve the productivity of construction projects*, Dept. of Civil Engineering, Center for Integrated Facilities Engineering, Report No. 282

Sanvido V, 1990, *An Integrated Building Process Model*, Technical Report No. 1, Computer Integrated Construction Program, Pennsylvania State University

Schenk D A, Wilson P R, 1994, *Information Modeling the EXPRESS Way* (Oxford University Press, New York)

Schlechtendahl E G (Ed.), 1988, *Specification of a CAD*I Neutral File for CAD Geometry, Wireframes, Surfaces, Solids, Version 3.3*, Espirit Project 322, Vol. 1 (Springer, Heidelberg)

Schley M K (Ed.), 1990, *CAD Layer Guidelines* (The American Institute of Architects Press, Washington DC)

Sedgewick P, 1974, "Computer Graphics for Drafting," *Computer Graphics and Image Processing* **3** pp 91-124

SET: z-68-300, 1989, *Industrial automation—external representation of product definition data—data exchange and transfer standard specification* version 89-06, L'association francaise de normalisation (AFNOR) Tour Europe Cedex 7, 92080 Paris-la-Defense, France

Shaw N K, Bloor M S, de Pennington A, 1989, "Product data models," *Research in Engineering Design* **1** (1) pp 43-50

Sherwood R, 1978, *Modern Housing Prototypes* (Harvard University Press. Cambridge MA)

Shriver B, Wegner P (Eds.), 1987, *Research Directions in Object-Oriented Programming* (MIT Press, Cambridge, MA)

Skarra A, Zdonik S, 1987, "Type Evolution in an Object-Oriented Database," in *Research Directions in Object-Oriented Programming*, Eds. B Shriver, P Wegner (MIT Press, Cambridge MA) pp 393-415

Smith B, 1983, "IGES: A Key to CAD/CAM Systems Integration," *IEEE Computer Graphics and Applications* **3** (November) pp 78-83

Smith B, Rinaudot G, Wright T, Reed K, 1988, *Initial Graphics Exchange Specification (IGES) Version 4.0*, Center for Manufacturing Engineering, National Bureau of Standards, ANSI Y14.26M

Smith J M, Smith D C, 1977a, "Database abstractions: aggregation and generalization," *ACM Transactions on Database Systems* **2** (2) pp 105-133

Smith J M, Smith D C, 1977b, "Database abstractions: aggregation," *Communications of the ACM* **20** (6) pp 405-413

Sowa J F, 1984, *Conceptual Structures: Information Processing in Mind and Machine* (Addison Wesley, Reading MA)

Spitler J D, Hensen J L M (Eds.), 1997, *Building Simulation '97*, Sponsored by the International Building Performance Simulation Association, Prague (available on CD-ROM)

Spooner D L, Hardwick M, 1997, "Using Views for Product Data Exchange," *IEEE Computer Graphics and Applications* **17** (September/October) pp 58-65

Staub P, 1995, "Integrated Product and Process Modeling for Facility Management," in *Modeling of Buildings Through Their Lifecycle: Proceedings, CIB Publication 180* Eds. M Fischer, K Law, B Luiten, Palo Alto, CA, pp 280-292

Stein T, Sweat J, 1998, "Killer Supply Chains," *Information Week* (9 Nov 1998) pp 36-46

Stone H, 1973, *Discrete Mathematical Structures and Their Applications* (Science Research Associates, Palo Alto, CA)

Sutherland I, 1963, "SKETCHPAD: A Man-Machine Graphical Communication System" in *Proceedings 1963 Spring Joint Computer Conference* pp 329-346.

Svensson K, 1998, *Integrating Facilities Management Information—A process and product model approach* (doctoral thesis), Royal Institute of Technology, Construction management and economics, Stockholm.

Szyperski C, 1998, *Component Software: Beyond Object-oriented Programming* (Addison Wesley, Reading MA)

Taivalsaari T, 1996, "On the Notion of Inheritance," *ACM Computing Surveys* **28** (3) pp 438-479

Tanijiri H, Ishiguro B, Arai T, Yoshitake, R, Kato, M, Morishima, Y, Takasaki, N, 1997, "Development of an automated weather-unaffected building construction system," *Automation in Construction* **6** (3) 215-227

Tarandi V, 1998, *Neutral Intelligent CAD Communication—Information exchange in construction based upon a minimal schema* (doctoral thesis), Royal Institute of Technology, Construction management and economics, Stockholm.

Taylor F, 1911, *The Principles of Scientific Management* (Harper Bros., New York)

Teague L, 1968, "Research in Computer Applications to Architecture," in *Computer Application in Architecture and Engineering* Ed. C N Harper (McGraw-Hill, New York)

Teorey T J, 1990, *Database Modeling and Design: The Entity-Relationship Approach* (Morgan Kaufmann, Menlo Park)

Thompson B, Webster G, 1978, "Progress on CEDAR3: a computer-aided building design system for the sketch plan stage," *CAD78: Third International Conference on Computers in Engineering and Building Design* (ICL Press, London)

Thompson B, Lera S, Beeston D, Coldwell R, 1979, "Application of CEDAR3: case studies from the pre-production trials of a large interactive computer-aided building design system in United Kingdom Central Government organisations," *ParC79, International Conference on the Applications of Computers in Architecture, Building Design and Urban Planning*, (AMK Berlin, Germany) pp 321-333

Tombasi P, 1997, "Travels From Flatland: The Walt Disney Concert Hall and the Specialization of Design Knowledge in the Building Sector," Ph.D. Thesis, School of Arts and Architecture, UCLA, Los Angeles

Turner J A, 1988, "A systems approach to the conceptual modeling of buildings," in *CIB Conference on the Conceptual Modeling of Buildings, TC 5.2 and 7.8* Eds. P Christiansson, H Karlsson, Lund, Sweden, pp 179-187

Turner J A, 1997, "Some Thoughts on the Existence of a Generic Building Object," in *CAADfutures 1997* Ed. R Junge, Munich, Germany, pp 533-552

Tzichritzis D, Klug A (Eds.), 1978, "The ANSI/X3/SPARC DBMS Framework: report of the study group on database management systems," *Information Systems* **3** pp 173-191

Ullman J D, 1988, *Principles of Database and Knowledge-Base Systems, Vol.1* (Computer Science Press, Rockville, MD)

VDA-FS, 1987, "VDA surface interface version 2.0," Verband der Automobilindustrie e. V (VDA) W-6000 Frankfurt am Main Westendstrasse 61, Germany

Verheijen G M A, Van Bekkum J, 1982, "NIAM: An Information Analysis Method," in *Information Systems Design Methodologies: A Comparative Review,* Eds. T W Olle, H G Sol, A A Verrijin-Stuart (North-Holland Press, New York)

Verhoef M, Liebich T, Amor R, 1995, "A Multi-Paradigm Mapping Method Survey," in *Modeling of Buildings Through Their Lifecycle: Proceedings, CIB Publication 180,* Eds. M Fischer, K Law, B Luiten, Palo Alto, CA, pp 233-247

Warszawski A, Sacks R, 1995, "The Project Model of an Automated Building System," in *Modeling of Buildings Through Their Lifecycle: Proceedings, CIB Publication 180,* Eds. M Fischer, K Law, B Luiten, Palo Alto, CA, pp 77-89

Watanbe S, 1994, "Knowledge integration for architectural design," *Automation in Construction,* **3** (2-3) pp 149-156

Watson, A, 1995, "Product Models and Beyond," in *Integrated Construction Information,* Eds. P Brandon, M Betts (E & F Spon, London) pp 159-172

Watson A, Crowley A, Boyle A, Knowles G, 1993, "Practical Problems in the Development of Product Models," in *Management of Information Technology in Construction,* Eds. K Mathus, M Betts (World Scientific Publishing, Singapore) pp 571-583

Watson A, Crowley A, 1997, "Deploying the CIMsteel Integration Standards," in *Competitive Deployment of Product Data Technology, Proceedings of the 3rd European Conference on Product Data Technology* pp 39-46.

Wegner P, 1987, "The Object-Oriented Classification Paradigm," in *Research Directions in Object-Oriented Programming,* Eds. B Shriver, P Wegner (MIT Press, Cambridge, MA) pp 479-560

Wegner P, Zdonik Y, 1988, "Inheritance as an incremental modification mechanism or what like is and isn't like," in *ECOOP'88: European Conference on Object Oriented Programming* pp 55-77

Weiler K, Eastman C M, 1979, "Geometric Modeling Using the Euler Operators," *Proceedings Conference on Computer Graphics in CAD/CAM Systems,* Ed. D Gossard, MIT pp 248-259

Williams S, 1989, *Hongkong Bank: The Building of Norman Foster's Masterpiece* (Little, Brown, Boston)

Wilson P R, 1987, "A short history of CAD data transfer standards," *IEEE Computer Graphics and Applications* **7** (6) pp 64-67

Wilson P (Ed.), 1992, *EUG'92, Second International EXPRESS User Group Conference,* Dallas, TX

Yasky Y, 1980, "Transforming a set of building drawings into a consistent database," in *Proceedings CAD80 Conference* pp 101-108

Yasky Y, 1981, "A Consistent Database for an Integrated CAAD System: Fundamentals for an Automated Design Assistant" Ph.D. Thesis, Department of Architecture, Carnegie-Mellon University

Young R, 1996, "Supporting multiple views in design for manufacture" in *Knowledge Intensive CAD Vol. II,* Eds. M Mantyla, S Finger, T Tomiyama (Chapman and Hall, London) pp 259-268

Zeckendorf W, 1970, *The Autobiography of William Zeckendorf*, with Edward McCreary (Holt, Rinehart and Winston, New York)

Zimring C, Ataman O, 1994, "Incorporating Guidelines into A Case-Based Architectural Design Tool" *Proceedings of the Annual Meeting of the Association of Computer Aided Design in Architecture*, St. Louis, MO pp 87-101

Index